North Carolina Wesleyan College
WISDOM AND COURAGE THROUGH CHRISTIAN EDUCATION
1956
ROCKY MOUNT, N.C.

ADVANCED CARDIOVASCULAR EXERCISE PHYSIOLOGY

ADVANCED EXERCISE PHYSIOLOGY SERIES

Denise L. Smith, PhD
Skidmore College

Bo Fernhall, PhD
University of Illinois at Urbana-Champaign

Human Kinetics

Library of Congress Cataloging-in-Publication Data

Smith, Denise L.
Advanced cardiovascular exercise physiology / Denise L. Smith and Bo Fernhall.
p. ; cm. -- (Advanced exercise physiology series)
Includes bibliographical references and index.
ISBN-13: 978-0-7360-7392-9 (hard cover)
ISBN-10: 0-7360-7392-2 (hard cover)
1. Cardiovascular system--Physiology 2. Exercise--Physiological aspects. I. Fernhall, Bo. II. Title. III. Series: Advanced exercise physiology series.
[DNLM: 1. Exercise--physiology. 2. Cardiovascular Physiological Phenomena. 3. Cardiovascular System--anatomy & histology. 4. Resistance Training. WE 103]
QP114.E9S65 2011
612.1--dc22

2010032221

ISBN-10: 0-7360-7392-2 (print)
ISBN-13: 978-0-7360-7392-9 (print)

The Web addresses cited in this text were current as of May 2010, unless otherwise noted.

Acquisitions Editors: Michael S. Bahrke, PhD, and Amy N. Tocco; **Developmental Editor:** Kevin Matz; **Assistant Editors:** Melissa J. Zavala and Brendan Shea; **Copyeditor:** Joyce Sexton; **Indexer:** Betty Frizzell; **Permission Manager:** Dalene Reeder; **Graphic Designer:** Joe Buck; **Graphic Artist:** Denise Lowry; **Cover Designer:** Keith Blomberg; **Photographs (interior):** © Denise L. Smith and Bo Fernhall, unless otherwise noted.; **Photo Production Manager:** Jason Allen; **Art Manager:** Kelly Hendren; **Associate Art Manager:** Alan L. Wilborn; **Printer:** Thomson-Shore, Inc.

Printed in the United States of America 10 9 8 7 6 5 4 3 2

The paper in this book is certified under a sustainable forestry program.

Human Kinetics
Web site: www.HumanKinetics.com

United States: Human Kinetics, P.O. Box 5076, Champaign, IL 61825-5076
800-747-4457
e-mail: humank@hkusa.com

Canada: Human Kinetics, 475 Devonshire Road Unit 100, Windsor, ON N8Y 2L5
800-465-7301 (in Canada only)
e-mail: info@hkcanada.com

Europe: Human Kinetics, 107 Bradford Road, Stanningley, Leeds LS28 6AT, United Kingdom
+44 (0) 113 255 5665
e-mail: hk@hkeurope.com

Australia: Human Kinetics, 57A Price Avenue, Lower Mitcham, South Australia 5062
08 8372 0999
e-mail: info@hkaustralia.com

New Zealand: Human Kinetics, P.O. Box 80, Torrens Park, South Australia 5062
0800 222 062
e-mail: info@hknewzealand.com

E4414

Contents

Preface

There have been remarkable advances in our scientific understanding of the function of the cardiovascular system in the last couple of decades, many of which have occurred as a result of scientific research aimed at understanding cellular and molecular aspects of the cardiovascular system. In addition to better understanding of the mechanisms of cardiovascular function, in recent years there has been extensive research and a concomitant increase in our understanding of how exercise affects the cardiovascular system. These research efforts have led to far better appreciation of the mechanisms by which exercise provides cardioprotection. Thus, the purpose of this text is to provide a single resource that (1) offers a clear and concise explanation of each component of the cardiovascular system—the heart, the vasculature, and the blood; and (2) systematically details the effect of acute exercise (aerobic and resistance) and chronic exercise training (aerobic and resistance) on each of the components of the system. An additional goal is to highlight the complex interaction of the components of the cardiovascular system, both at rest and during exercise.

This text relies heavily on the latest scientific and medical research to describe physiological functioning, exercise responses, and adaptations. The text is richly illustrated with figures to elucidate physiological mechanisms. Graphic presentations are extensively used to convey scientific data and to depict exercise responses and training adaptations. Although this text is intended primarily for graduate students (or advanced undergraduate students) who are studying the effects of exercise on the cardiovascular system, health care professionals and clinicians will also benefit from this compilation of research that documents the myriad effects of exercise on this system. In the text, specific attention is paid to the beneficial effects of exercise on the various components of the cardiovascular system and the mechanisms by which regular exercise provides cardioprotection. It is presumed that readers of this text will have had courses in basic anatomy, physiology, and exercise physiology.

This text is divided into two sections, one dedicated primarily to describing the structure and function of the cardiovascular system, and one devoted to detailing the effects of exercise on the system. The first section focuses on cardiovascular physiology and provides a concise description of the structure and function of each component of the cardiovascular system—namely, the heart, the vasculature, and the blood. The first chapter is an integrative chapter on the normal function of the cardiovascular system that provides a theoretical foundation for the detailed discussions that follow and emphasizes how the various components of the cardiovascular system function together as an intact and highly interdependent organ system, both at rest and during exercise. Chapter 2 presents the heart as a pump and emphasizes the role of the heart in delivering oxygen-rich blood to the body, as well as the need to regulate cardiac output to match the metabolic demands of the body. Chapter 3 details the structure and function of the myocardial cells, the myocytes, that are ultimately responsible for the contractile force of the heart. Chapter 4 addresses the electrical activity of the heart, both within the specialized conduction system of the heart and within the cardiac myocytes. Chapter 5 describes the standard electrocardiogram (ECG) and details the clinical relationship between electrical activity in the heart and the waveforms visible on the ECG. Chapter 6 provides an organ-level description of the function of the vasculature, discussing the important topics of hemodynamics in general and the regulation of blood flow and blood pressure in particular. Chapter 7 delves into the relatively new science of vascular biology and details the structure and function of the endothelium and the vascular smooth muscle. This chapter relies heavily on relatively recent discoveries to describe how substances released by the endothelium can control vessel diameter and, ultimately, blood flow. Chapter 8 details the hemostatic function of blood, describing platelet function, coagulation, and fibrinolysis. This chapter emphasizes the delicate balance that must be maintained between coagulation and fibrinolysis in order to prevent unnecessary blood clotting while simultaneously being able to prevent blood loss when a vessel is damaged.

The second section of the book systematically details the effect of exercise on the cardiovascular system—including acute response and chronic adaptations to aerobic and resistance exercise. Chapter 9 describes the effect of acute aerobic exercise on cardiac function, vascular function, and hemostatic variables. Chapter 10 presents the chronic effects of a systematic program of aerobic exercise training on cardiac structure and function, vascular structure and function, and hemostatic variables. Following the same pattern, chapter 11 describes the effect of an acute bout of resistance exercise on cardiac function, vascular structure and function, and hemostatic variables. Finally, chapter 12 documents the chronic effects of a systematic program of resistance exercise training on cardiac structure and function, vascular structure and function, and hemostatic variables.

While it is clear that a single textbook cannot comprehensively cover all that is known about the cardiovascular system, it is our hope that the information presented in this text will provide readers with a framework for understanding how all the components of the cardiovascular system function together to support exercise, and how those components adapt to a systematic program of exercise training. Students who wish to conduct research related to the effects of exercise on the cardiovascular system may find a direction for their research by noting gaps in our current knowledge as identified in the text.

Series Preface

Having a detailed knowledge of the effects of exercise on specific physiological systems and under various conditions is essential for advanced-level exercise physiology students. For example, students should be able to answer questions such as these: What are the chronic effects of a systematic program of resistance training on cardiac structure and function, vascular structure and function, and hemostatic variables? How do different environments influence the ability to exercise, and what can pushing the body to its environmental limits tell us about how the body functions during exercise? When muscles are inactive, what happens to their sensitivity to insulin, and what role do inactive muscles play in the development of hyperinsulinemia and type 2 diabetes? These questions and many others are answered in the books in Human Kinetics' Advanced Exercise Physiology Series.

Beginning where most introductory exercise physiology textbooks end their discussions, each book in this series describes in detail the effects of exercise on a specific physiological system or the effects of external conditions on exercise. Armed with this information, students will be better prepared both to conduct the high-quality research required for advancing scientific knowledge and to make decisions in real-life scenarios such as the assement of health and fitness or the formulation of effective exercise guidelines and prescriptions.

Although many graduate programs and some undergraduate programs in exercise science and kinesiology offer specific courses on advanced topics in exercise physiology, there are few good options for textbooks to support those classes. Some instructors adopt general advanced physiology textbooks, but such books focus almost entirely on physiology without emphasizing *exercise* physiology.

Each book in the Advanced Exercise Physiology Series addresses the effects of exercise on a certain physiological system (e.g., cardiovascular or neuromuscular) or in certain contexts (e.g., in various types of environments). These textbooks are intended primarily for students, but researchers and practitioners will also benefit from the detailed presentation of the most recent research regarding topics in exercise physiology.

Acknowledgments

There are many people who helped to make this book possible and we are indebted to all of them, but a few warrant special mention. Our acquisition editor, Mike Bahrke, provided the encouragement and insight expected from an editor, and also offered the kindness and patience of a friend throughout the preparation of the manuscript. Our developmental editor, Kevin Matz, provided good advice, offered consistant support, and effectively used his gentle nature to keep us organized throughout the last stages of preparing this text. Discussions with Tom Rowland helped frame our thinking in several places, and he kindly served as a reviewer of several chapters. Hannah Segrave has been indispensible in preparing the text; her skills strengthened the manuscript, and her good humor made the project much more fun.

We would like to acknowledge the role of our students in pushing us to frame a logical approach to understanding the complex and intricately interrelated components of the cardiovascular system and the effects of exercise upon those various components. Furthermore, we gratefully recognize the central role that our students have played in providing us with the inspiration and the motivation to pursue the daunting task of writing this text.

Finally, we would like to thank our families for their unwavering support and encouragement through the long process of preparing this text, a process that kept us from them for more than we would have liked.

SECTION I

Cardiovascular Physiology

The cardiovascular system is composed of the heart, the vasculature, and the blood. The cardiovascular system responds to exercise in a complex and integrated way that allows it to meet the metabolic needs of the working muscles, preserve needed levels of homeostasis for bodily function, and respond to potential bodily threats.

Section I provides a concise explanation of the structure and function of each component of the cardiovascular system (the heart, the vessels, and the blood), placing considerable emphasis on how the cells of organs function and how their functions are controlled. The second section of the book describes how all the components of the cardiovascular system respond in an integrated fashion to aerobic and resistance exercise and to training programs. It is useful to keep the integrated response to the stress of exercise in mind throughout each chapter in Section I.

CHAPTER 1

Essentials of the Cardiovascular System

The human cardiovascular system is a fascinating system that has inspired awe and provoked serious investigation among clinicians and researchers for hundreds of years. In ancient times, the heart was seen as the seat of our emotions, and even today the image of the heart is tied to the notion of sentimental feelings. In 1628, William Harvey proposed that the heart propelled blood through a closed vascular circuit (Fye, 2006). Today, every high school student has a rudimentary understanding of the role of the cardiovascular system in sustaining life. Nonetheless, researchers continue to make exciting new discoveries about the cardiovascular system every day, with recent discoveries focused largely on cellular and molecular aspects of cardiovascular function.

The cardiovascular system is a complex organ system that functions with multiple other physiologic systems in an integrated way. The cardiovascular system is composed of three overlapping and interrelated components: the heart, the vasculature, and the blood. Together, these components provide the basic function of the cardiovascular system: the delivery of oxygen and nutrients to the cells of the body and the elimination of waste products from the cells. The cardiovascular system serves multiple functions, which may be categorized into several major and sometimes overlapping categories as follows:

1. Transport and delivery
    - The transport and exchange of respiratory gases (oxygen and carbon dioxide)
    - The transport and exchange of nutrients and waste products
    - The transport of hormones and other chemical messengers
2. Hemostatic regulation
    - Fluid balance among various fluid compartments

- The maintenance of pH balance
- The maintenance of thermal balance
- The regulation of blood pressure

3. Protection
   - Prevention of blood loss through hemostatic mechanisms
   - Prevention of infection through white blood cells and lymphatic tissue

These essential functions are achieved because of the close functional relationships between the cardiovascular system and other major systems of the body, notably the neural, respiratory, endocrine, digestive, urinary, skeletal, and integumentary systems. As seen in the schematic presented in figure 1.1, the cardiovascular system provides blood flow to the pulmonary circulation and the systemic circulation. The pulmonary circulation delivers partially deoxygenated blood from the right ventricle to the pulmonary capillaries, where it becomes oxygenated and is returned to the left atrium. The pulmonary circulation highlights the important interrelationship between the cardiovascular system and the respiratory system. In short, the respiratory system is responsible for bringing oxygen into the alveoli, whereas the cardiovascular system is responsible for distributing oxygen to the cells of the body. Likewise, the cardiovascular system delivers carbon dioxide that is produced at the cellular level to the pulmonary capillaries, where it diffuses into the lungs to be exhaled. The systemic circulation

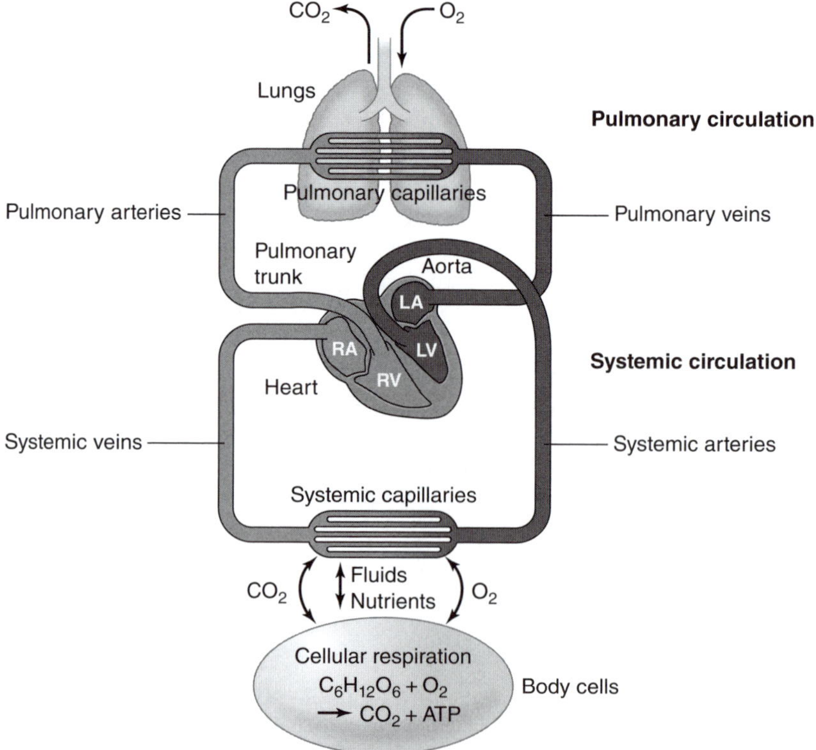

Figure 1.1 Overview of pulmonary and systemic circulations. The pulmonary circulation delivers blood to the lungs in order to eliminate carbon dioxide and oxygenate hemoglobin.

distributes blood to all the major systems and tissues of the body. The systemic circulation interacts extensively with other bodily systems, notably the digestive, urinary, and integumentary systems, to accomplish the major functions of the cardiovascular system.

Many of these functions play an important role in homeostatic balance, including maintenance of adequate blood pressure to perfuse body tissues and thus supply optimal levels of oxygen; maintenance of pH balance within tight limits; thermal regulation, through both the formation of sweat (derived from plasma) and increase in cutaneous blood flow; and metabolic regulation, particularly in terms of blood glucose levels.

COMPONENTS OF THE CARDIOVASCULAR SYSTEM

This chapter briefly reviews the structure and function of the components of the cardiovascular system to provide the reader with an appreciation of the extent to which all the components of the system must operate together to achieve the functions just described. Subsequent chapters will discuss the structure and function of each component in greater detail and explain how each component responds to the stress of exercise.

Heart

The cardiovascular system is composed of the heart, the vasculature, and the blood. The heart serves as the pump for the system and provides the contractile force necessary to distribute blood to the various organs (figure 1.2 provides a schematic of the heart structure). The atria serve as the receiving chambers, receiving blood from the superior and inferior vena cava. The right ventricle pumps blood to the lungs (pulmonary system), whereas the left ventricle pumps to the entire body (systemic circulation). The muscular wall of the heart is termed the myocardium, meaning "heart muscle." Properly functioning valves ensure the one-way flow of blood through the heart. Although the heart is a relatively small organ, weighing approximately 300 to 350 g in healthy adults, it receives about 4% of resting blood flow and accounts for approximately 10% of resting oxygen consumption.

MAXIMAL OXYGEN CONSUMPTION ($\dot{V}O_2$MAX)

Maximal oxygen consumption ($\dot{V}O_2$max) is one of the most common measures of an individual's overall fitness. It is also a functional measure of the entire cardiovascular system. Maximal oxygen consumption reflects the capacity of the cardiovascular system to deliver blood (and the oxygen it contains) to working muscle and the ability of that muscle to use the oxygen delivered. This requires an increase in the total amount of blood pumped by the heart (increased cardiac output) and a redistribution of the blood so that a greater percentage is directed to the exercising muscle and a lesser percentage to the non-exercising muscle and organs that do not require as much oxygen at that time (e.g., kidney and gastrointestinal tract). Cardiac output at rest is approximately 5 L/min; during maximal exercise it may increase to values over 30 L/min. Equally impressive is how the cardiac output is distributed in each case. In particular, at rest, approximately 20% of the cardiac output (or 1 L/min) is distributed to skeletal muscle. During maximal exercise, approximately 90% of cardiac output (or 27 L/min) is directed to the skeletal muscle.

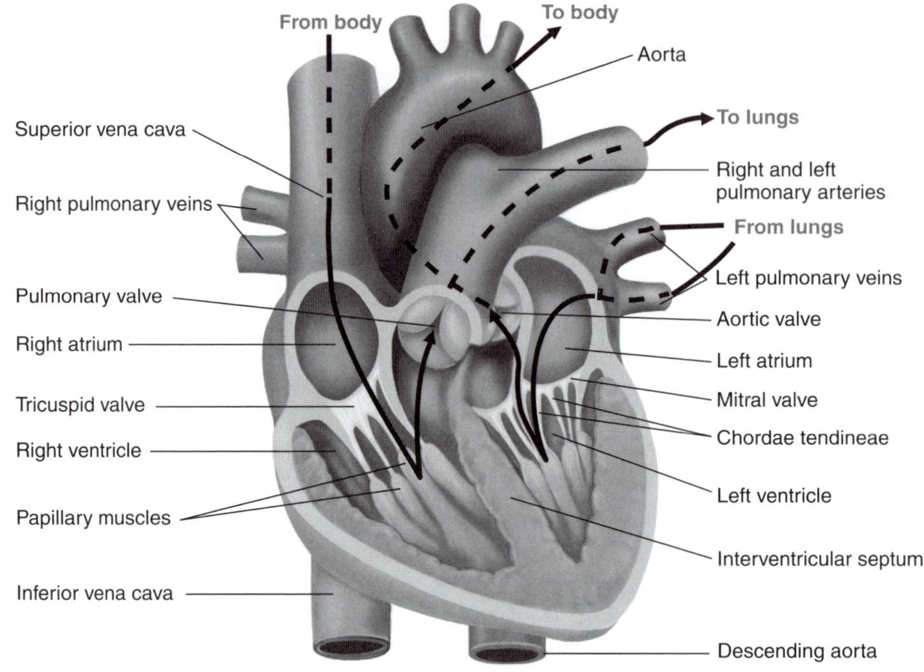

Figure 1.2 Heart structure. Valves play a major role in ensuring one-way flow of blood through the heart.

Reprinted, by permission, from J.W. Wilmore, D.L. Costill, and W.L. Kenney, 2008, *Physiology of Exercise and Sport,* 4th ed. (Champaign, IL: Human Kinetics), 125.

Cardiac output is the amount of blood ejected from the ventricles in one minute. It is a measure of the heart's ability to pump blood to support the needs of the body on a per minute basis. Cardiac output is determined by the product of heart rate (number of beats per minute) and stroke volume (amount ejected per beat). Under normal, resting conditions, cardiac output is approximately 5 L/min, depending largely on body size, but this value can change quickly to meet the changing needs of the body. For instance, during strenuous exercise, cardiac output may increase four to five times to meet the metabolic demands of working muscles.

Vessels

The vessels are responsible for distributing oxygen, nutrients, and myriad other substances throughout the body. Figure 1.3 provides a schematic view of the circuitry of the cardiovascular system. Although circulation to most of the organs is in parallel, the liver and renal tubules are in series. The relative distribution of blood that is delivered to each of the circulations is intricately controlled by the degree of vasoconstriction or vasodilation in the arterioles that supply the organs. The degree of smooth muscle contraction is, in turn, determined by extrinsic (neurohormonal) and local control (i.e., the metabolic needs of the tissue).

Far from being simple conduits, the vessels are dynamic organs that constantly alter their diameter to change blood flow to meet their requirement for blood flow. The vessel wall also releases a number of chemical mediators that participate in blood clotting and the inflammatory response.

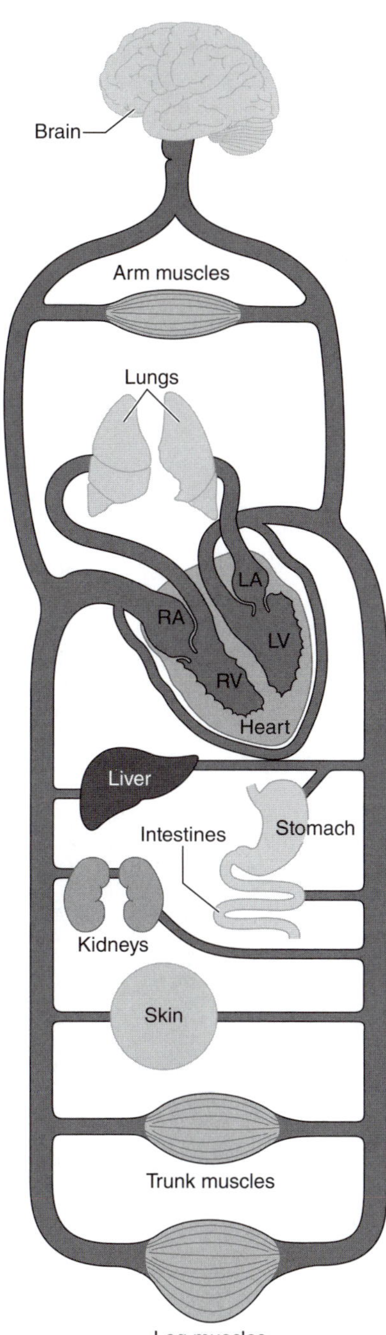

Figure 1.3 Schematic of the major circuitry of the cardiovascular system. Circulation to most systemic organs is in parallel, but the liver and kidney tubules are in series.

Reprinted, by permission, from J.W. Wilmore, D.L. Costill, and W.L. Kenney, 2008, *Physiology of Exercise and Sport,* 4th ed. (Champaign, IL: Human Kinetics), 175.

Blood Velocity and Pressure Through the Vascular System

Figure 1.4 presents a schematic of velocity, blood pressure, and resistance of blood flow through the systemic circulation and relates these variables to the cross-sectional area of the vessels. On an individual basis, the aorta (the largest artery in the body) is larger than regular arteries, arteries are larger than arterioles, and arterioles are

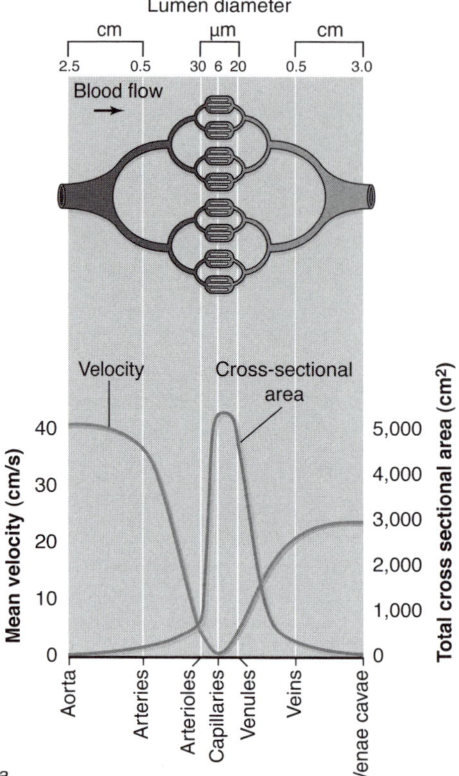

Figure 1.4 Cross-sectional area *(a)*, velocity *(a)*, blood pressure *(b)*, and resistance *(c)* throughout the vascular system. The entire cardiac output passes each dashed line per minute. Blood pressure is pulsatile in the arteries but becomes continuous as blood moves through the arterioles (resistance vessels).

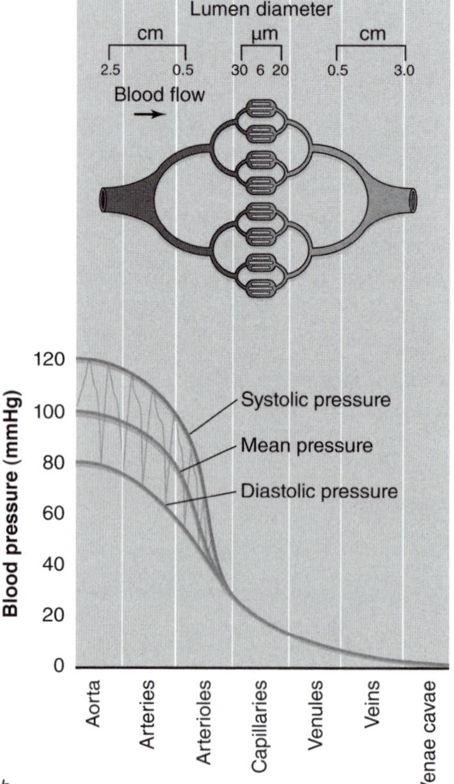

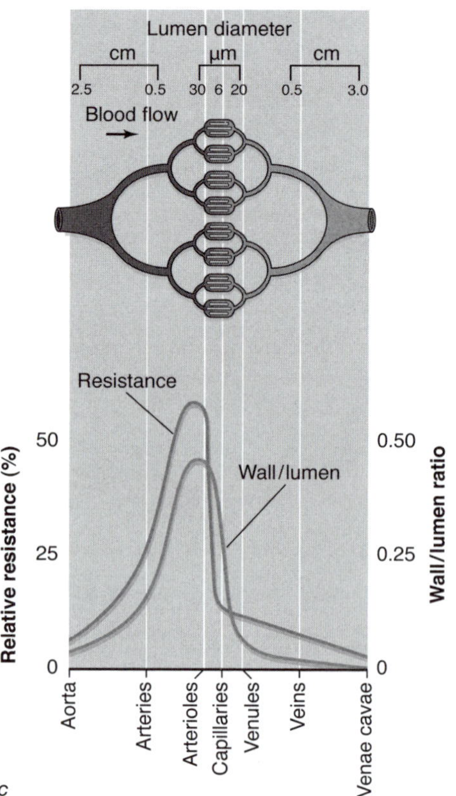

larger than capillaries. Given the extensive branching of the vascular system, however, the capillaries have by far the greatest total cross-sectional area. Blood pressure and blood velocity are pulsatile in the large arteries owing to the alternating periods of myocardial contraction (systole) and relaxation (diastole). Although pressure in the ventricles falls to nearly 0 mmHg during diastole, pressure in the arteries does not generally fall much below 70 mmHg because of the pressure created as the vessel walls recoil during diastole (Windkessel effect). Blood pressure and velocity drop drastically as blood enters the arterioles (resistance vessels). Blood flow through the capillaries is slowest because of their large cross-sectional area—a situation that makes the capillaries well suited to their role as the site of gas exchange. Blood pressure is also low in the capillaries, which is beneficial as high blood pressure can easily damage the vessel wall of the tiny capillaries. Blood velocity increases on the venous side as the total cross-sectional area of the vessels decreases. However, pressure on the venous side of the system is very low because the blood is so far away from the contractile force of the heart. One-way valves in the veins and the muscular and respiratory muscle pump are necessary to ensure the adequate return of venous blood to the heart given the low pressure gradient driving blood back to the heart.

Functional Classification of Vessels

Blood vessels can be identified by functional classes, which are determined largely by the structural characteristics of the vessel wall. Table 1.1 summarizes the diameter, wall thickness, and composition of the various types of vessels.

Elastic (central) arteries are the largest arteries (such as the aorta and iliac arteries) and are very distensible owing to large amounts of elastin in the tunica media. **Muscular,** or **conduit, arteries** include many of the named arteries that students routinely study in anatomy, such as the radial, ulnar, and popliteal arteries and the cerebral and coronary arteries.

Muscular arteries are so named because they contain a large amount of smooth muscle in the tunica media layer. Relaxation of the smooth muscle causes vasodilation and increased blood flow (assuming pressure remains constant). This is seen during exercise when smooth muscle surrounding muscular arteries and arterioles supplying skeletal muscle relax, and increased blood flow to the active skeletal muscle.

Table 1.1 Structural Comparison of Blood Vessels

	Elastic arteries	Muscular arteries	Arterioles	Capillaries	Venules	Veins
Internal diameter	1.5 cm	0.4-0.6 cm	30 μm	9 μm	20 μm	0.5 cm
Wall thickness	1 mm	1 mm	6 μm	0.5 μm	1 μm	0.5 mm
Approximate composition of vessel wall (%)						
Endothelial tissue	5	5	10	95	20	10
Elastic tissue	60	10-15	10			5
Smooth muscle	20-30	65	60		20	30
Fibrous tissue	15-25	20	20	5	60	60

Terminal arteries and arterioles are termed resistance vessels because they are the sites of greatest resistance within the vascular tree. **Resistance vessels,** primarily arterioles, are responsible for determining local blood flow; they vasodilate and vasoconstrict in order to ensure that local blood flow matches local metabolic demand.

Exchange vessels, composed primarily of capillaries but also including the smallest vessels on either side of the capillaries, perform the ultimate function of the cardiovascular system—gas and metabolite exchange. Gas and nutrient exchange is facilitated across the capillary wall because of

- the large total cross-sectional area of the capillaries;
- the thin vessel wall, consisting essentially of a single layer of endothelium, which offers little impedance to diffusion; and
- the decreased velocity of blood as it flows through the capillaries, allowing adequate time for exchange.

In addition to the exchange of gases and nutrients across the capillary wall, these tiny vessels also permit the exchange of fluids, and this helps maintain fluid balance.

Capacitance vessels include the venules and veins and are so named because they contain approximately two-thirds of the blood volume at any given time. The ability to accommodate large quantities of blood is directly related to characteristics of the vessel wall; venules and veins have a thin wall that contains a thin tunica media with a small amount of smooth muscle and collagen. Because of their thin wall, veins can easily become distended or collapse and thus can act as a reservoir for blood in the venous system. Because the veins are innervated by vasoconstrictor nerve fibers, the volume of blood in the venous system can be actively controlled.

Blood

Blood is composed of formed elements and plasma. As shown in figure 1.5, approximately 40% to 45% of blood volume is due to formed elements, including erythrocytes (red blood cells, RBC), leukocytes (white blood cells, WBC) and thrombocytes (platelets). The remaining portion of blood, approximately 55% to 60%, is composed of plasma.

Formed Elements

Erythrocytes are small biconcave discs (approximately 2.4 μm thick and 8 μm in diameter) that are responsible for transporting oxygen from the alveoli of the lungs to the systemic capillaries where it diffuses into cells. Red blood cells are composed almost exclusively of hemoglobin, the protein that accounts for their remarkable ability to bind to and transport oxygen.

Leukocytes constitute a small portion (approximately 1%) of the formed elements. Leukocytes play a critical role in bodily defenses, immune function, and inflammation. There are five types of leukocytes, each with specific and sometimes overlapping functions.

Thrombocytes (platelets) are very small (2-3 μm) cell fragments that play a critical role in initiating coagulation. Platelets also interact with the endothelial lining of blood vessels and are involved in the pathological process of atherosclerotic plaque formation. Platelets are derived from very large cells, megakaryocytes, that break up into small fragments and circulate in the blood as platelets.

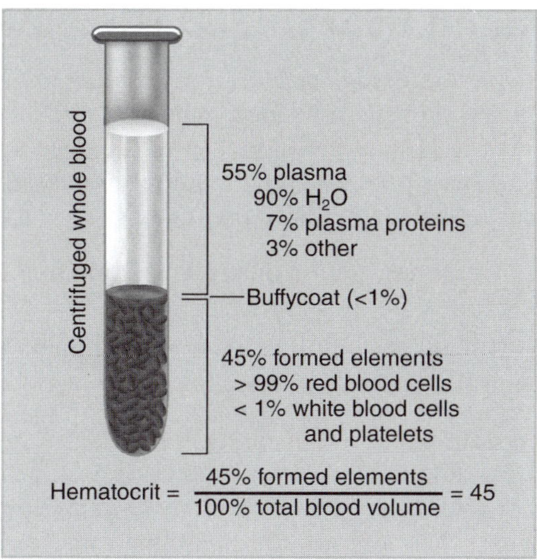

Figure 1.5 Components of blood. Blood is composed of plasma and formed elements. Formed elements include the erythrocytes, leukocytes, and platelets; but erythrocytes account for the vast majority of the formed elements.

Reprinted, by permission, from J.W. Wilmore, D.L. Costill, and W.L. Kenney, 2008, *Physiology of Exercise and Sport,* 4th ed. (Champaign, IL: Human Kinetics), 140.

Plasma

Plasma is the fluid portion of the blood that is responsible for transporting and distributing the formed elements, as well as nutrients, hormones, and other substances. Plasma is composed mostly of water but also contains many critical proteins, nutrients, electrolytes, respiratory gases, and hormones (table 1.2).

Table 1.2 Composition of Plasma

Constituent	Description and importance
Water	90% of plasma volume; dissolving and suspending medium; absorbs heat
	Solutes
Plasma proteins	8% (by weight) of plasma volume; all contribute to osmotic pressure and help maintain fluid balance; specific functions for each protein
• Albumin	60% of plasma proteins; main contributor to osmotic pressure
• Globulins	36% of plasma proteins
Alpha, beta	Transport proteins that bind to lipids, metal ions, and fat-soluble vitamins
Gamma	Antibodies released by immune cells during immune response
• Fibrinogen	4% of plasma protein; forms fibrin threads of blood clot
Nonprotein nitrogenous substances	By-products of cellular metabolism, such as urea, uric acid, creatinine, and ammonium salts
Nutrients (organic)	Materials absorbed from digestive tract
Electrolytes	Cations and anions: help to maintain plasma osmotic pressure and blood pH
Respiratory gases	Oxygen and carbon dioxide: dissolved in plasma as bicarbonate ion or CO_2, or bound to hemoglobin in red blood cells
Hormones	Chemical messengers transported in the blood

CARDIOVASCULAR RESPONSES TO EXERCISE

Briefly, exercise creates an increased need to supply oxygen to the working muscles in order to increase the generation of adenosine triphosphate (ATP) to support continued muscle contraction. At the same time, there is also the need to expel the increased carbon dioxide that has been produced as a result of increased cellular respiration. Two adjustments occur in order to achieve these goals:

- Heart rate and force of contraction increase, leading to a large increase in cardiac output.
- Blood vessels in different regions of the body adjust their diameter so that more blood is supplied to the working muscle and less to inactive areas.

These adjustments occur as a result of changes in myocardial cells and the smooth muscle cells surrounding blood vessels. In turn, these muscle cells may be responding to changes in neural stimuli (sympathetic nervous system), hormonal stimuli (such as catecholamines), local chemical stimuli (including nitric oxide), and mechanical stimuli (such as the degree of stretch).

In addition to adjustments that foster increased oxygen delivery and waste removal, the cardiovascular system adjusts to exercise in ways that maintain homeostasis and enhance the body's ability to deal with a threat (such as infection or bleeding). In order to meet these needs,

- heat dissipation is increased (due to increased sweating and increased cutaneous blood flow);
- circulating levels of leukocytes are increased; and
- coagulatory and fibrinolytic potential is enhanced.

SUMMARY

The cardiovascular system is composed of the heart, the vasculature, and the blood. The cardiovascular system responds to exercise in a complex and integrated way that allows it to meet the metabolic needs of the working muscles, preserve needed levels of homeostasis for bodily function, and respond to potential bodily threats. The integrated response to exercise is detailed in the later portion of this text.

The remaining chapters in this section provide details on the structure and function of each of the components of the cardiovascular system (the heart, the vessels, and the blood), placing considerable emphasis on how the cells of organs function and how their functions are controlled. The second section of the book describes how all the components of the cardiovascular system respond in an integrated fashion to aerobic and resistance exercise and to training programs. It is useful to keep the integrated response to the stress of exercise in mind throughout each chapter.

CHAPTER 2

The Heart as a Pump

Despite all the sentimental connotations associated with the heart, it is essentially a double-sided muscular pump that is responsible for circulating blood throughout the body. The upper chambers, the atria, receive blood from the venous system. The ventricles provide the contractile force to pump blood throughout the circulatory system, with the right ventricle pumping blood through the pulmonary circulation and the left ventricle pumping blood through the systemic circulation. The amount of blood ejected from each ventricle per unit of time (usually expressed as L/min) represents the total blood flow through the circulatory system for the same time period.

GROSS ANATOMY OF THE HEART

Figure 2.1 depicts the orientation of the heart and the heart valves in the thoracic cavity. The base of the heart consists largely of a fibrotendinous ring of connective tissue, called the annulus fibrosus, which provides an important structural framework for the heart: anchoring the atria and ventricles together and providing the fibrous structure around the heart valves. The annulus fibrosus also separates the atria from the ventricles and prevents the electrical signals from the atria from passing directly to the ventricles. The apex of the heart is formed by the inferior tip of the ventricles.

The atria are the upper chambers that receive blood from the venous system (as shown in figure 1.2, p. 6). The thick-walled ventricles serve as the pumps, generating the force needed to eject blood from the heart and distribute it through the vascular system. The left ventricular wall is typically about three times thicker than the right ventricular wall owing to the greater pressure the left ventricle must generate in order to pump blood through the entire systemic circulation.

When a cross section of the heart is viewed schematically, it becomes clear that the ventricles have different shapes. The right ventricle is smaller, and the force of ejection is produced when the anterior wall of the ventricle pushes against the ventricular septum. The anatomical structure of the right ventricle allows the thin walls of the right ventricle to eject a large volume of blood with minimal shortening against a low outflow pressure. On the other hand, the larger left ventricle has a more cylindrical arrangement, and contraction of the thicker ventricular walls ejects blood against the high outflow pressure in the systemic arterial system. The specific arrangement of the

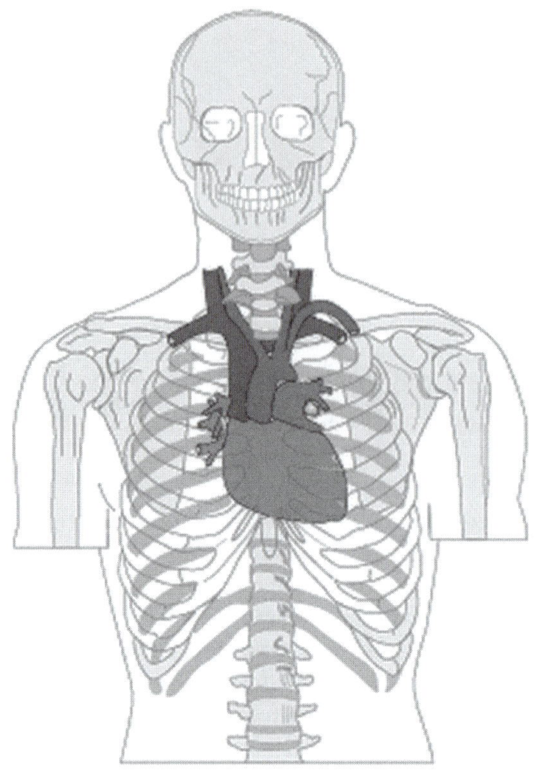

Figure 2.1 Orientation of the heart and heart valves in the thoracic cavity. The heart lies obliquely in the thoracic cavity. The four valves are grouped closely in an area behind the sternum.

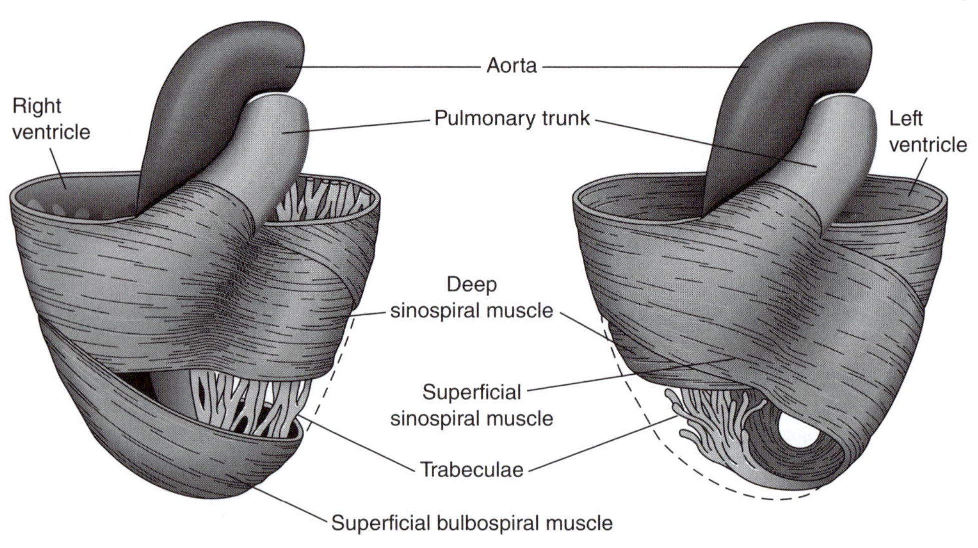

Figure 2.2 Layers of the heart. The inner circumferential layer of fibers causes a constriction of the ventricles, particularly the left ventricle. The outer layer pulls the apex of the heart upward toward the base.

myocardial muscle cells into layers of overlapping and interdigitating fibers allows the muscles to shorten and pull the apex anteriorly toward the base, and to constrict circumferentially (figure 2.2).

CARDIAC CYCLE

Figure 2.3 presents the cardiac cycle: the alternating periods of contraction and relaxation of the heart (and the associated electrical and mechanical events) that occur with each beat. Assuming a normal resting heart rate of approximately 75 beats/min, a complete cardiac cycle is completed once approximately every .8 s. Systole is the period of contraction, whereas diastole is the period of relaxation. Ventricular systole can be divided into two periods: the isovolumetric contraction period (a brief period of time during which the ventricles contract but do not eject blood) and the ejection period (the portion of systole during which blood is ejected from the heart). Similarly, ventricular diastole can be divided into two periods: the isovolumetric relaxation period (a brief period during which the ventricles are relaxing but no filling is occurring) and the ventricular filling period (the portion of diastole during which the ventricles fill with blood). The position of the heart valves is critical to understanding the flow of blood during the cardiac cycle, which is summarized in table 2.1, along with the time associated with each period of the cardiac cycle in resting humans.

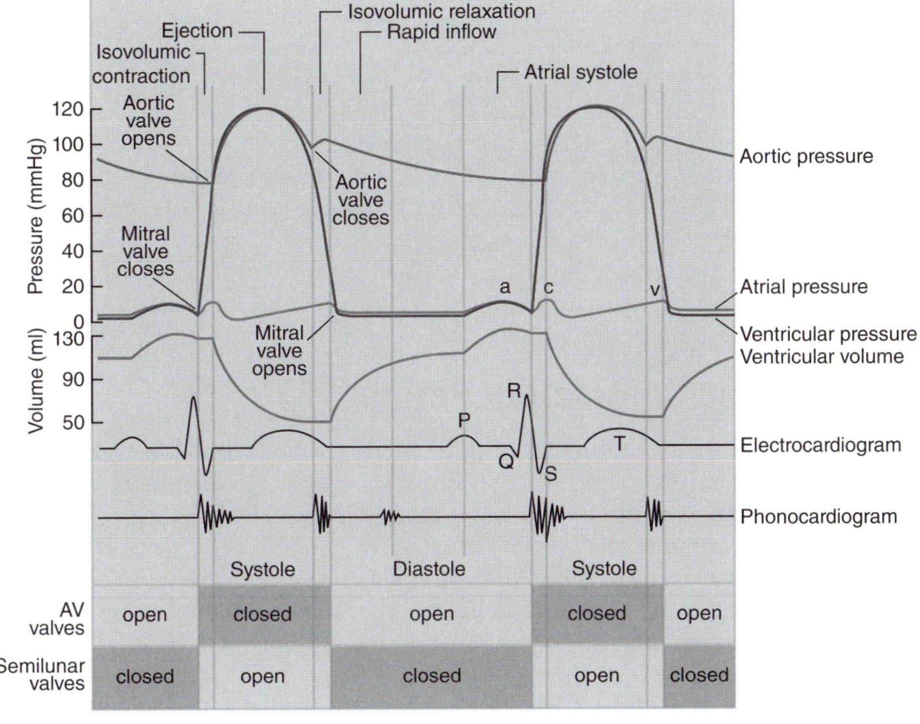

Figure 2.3 Cardiac cycle.

Adapted from U. Unglaub silvertorn, 2001, *Human physiology,* 2nd ed. (Upper Saddle River, NJ: Pearson Education), 433.

Table 2.1 Position of the Heart Valves and Duration of Each Period of the Cardiac Cycle

Period of cardiac cycle	Diastole or systole	Duration (sec)	Atrioventricular valves (tricuspid, mitral)	Semilunar valves (aortic, pulmonary)
Ventricular filling	Diastole	0.5	Open	Closed
Isovolumetric contraction period	Systole	0.05	Closed	Closed
Ejection period	Systole	0.3	Closed	Open
Isovolumetric relaxation period	Diastole	0.08	Closed	Closed

Ventricular Filling Period

Figure 2.3 begins during the **ventricular filling period** of diastole. Diastolic function is the general term that refers to the ability of the myocardium to relax and allow for adequate filling of the ventricles. **Myocardial relaxation** is an active, energy-dependent process that causes pressure in the left ventricle to decrease rapidly after the contraction and during early diastole. During active myocardial relaxation, the left ventricular pressure becomes less than the left atrial pressure, causing the mitral valve to open and diastolic filling to begin. Once the atrioventricular (AV) valves are open, blood flows freely from the atria into the ventricle. Blood flow into the ventricle is enhanced by what is considered "ventricular suction": As the ventricle is recoiling and relaxing after systole, the expansion of the ventricle enhances the pressure differential between the ventricle and the atrium, thus creating a form of "suction" to enhance the filling. This type of "passive" filling accounts for approximately 70% to 80% of ventricular filling. Near the end of the ventricular filling period, the contraction of the atria (following the P wave) completes ventricular filling and accounts for the remaining 20% to 30% of the total end-diastolic ventricular blood volume. Pressure in the ventricles is relatively low during ventricular filling because the myocardium is relaxed. Pressure increases slightly as the ventricles fill, but this pressure increase is very small relative to the increase in pressure that accompanies ventricular contraction.

Isovolumetric Contraction Period

Following the electrical stimulation of the ventricles (reflected by the QRS complex on the ECG, or electrocardiogram), the ventricles contract. Ventricular contraction immediately causes pressure in the ventricles to increase dramatically. Ventricular volume, however, remains unchanged for a very brief period of time (0.05 s)—hence the name, **isovolumetric contraction period.** Once pressure in the ventricles exceeds pressure in the aorta, the pressure forces the semilunar valves open, beginning the ventricular ejection period.

Ventricular Ejection Period

Blood is pumped from the ventricles during the **ventricular ejection period** of systole. Once the semilunar valves have been forced open by the increasing pressure generated by ventricular contraction, the volume in the ventricles immediately begins to decrease as blood in the ventricles is ejected from the heart. **Systolic function** is the general term given to the ability of the heart to adequately produce the force needed to eject blood from the ventricle. We can calculate the amount of blood ejected from each ventricle, called stroke volume (SV), by subtracting the amount of blood that is in the ventricles after contraction (end-systolic volume, ESV) from the volume of blood that filled the ventricles at the end of ventricular filling (end-diastolic volume, EDV). That is,

$$\text{SV (ml)} = \text{EDV (ml)} - \text{ESV (ml)}.$$

The percentage of blood ejected from the ventricle is called the ejection fraction (EF) and is calculated as

$$\text{EF (\%)} = [\text{SV (ml)} / \text{EDV (ml)}] \times 100.$$

Isovolumetric Relaxation Period

As the ventricles relax (following the T wave on the ECG), ventricular pressure quickly decreases. Once pressure in the ventricles is less than pressure in the aorta, the semilunar valves close. At this point, the AV valves are also closed. Because the ventricles are relaxing and there is no change in ventricular volume, this period is aptly named the **isovolumetric relaxation period.** This, however, is a brief period; as soon as the pressure in the atria exceeds the pressure in the ventricles, the AV valves are forced open, and blood begins to fill the ventricles. This starts the ventricular filling period and the cycle repeats—approximately 75 times per minute at rest!

THE VENTRICULAR PRESSURE–VOLUME LOOP

The relationship between ventricular pressure and volume throughout the cardiac cycle can be represented in the **pressure–volume loop** (figure 2.4). This graph presents volume of blood in the ventricles on the x-axis and left ventricular pressure on the y-axis. The four periods of the cardiac cycle each represent one side of the closed loop

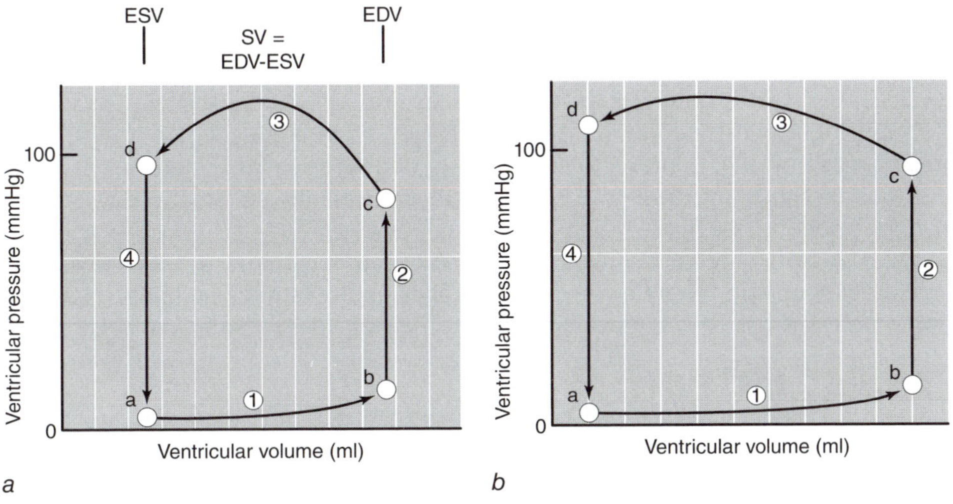

Figure 2.4 Pressure–volume loop of left ventricle *(a)* at rest and *(b)* during aerobic exercise.

that results. Ventricular filling is seen along the bottom of the loop. As blood fills the left ventricle, volume increases significantly, while pressure changes little. During the isovolumetric contraction period, depicted on the right-hand side of the loop, ventricular pressure increases dramatically, but volume remains unchanged—again, volume is unchanged because the valves are closed. During the ejection period, volume decreases because of the opening of the aortic valve and the outward flow of blood. During the isovolumetric relaxation period, pressure decreases as the ventricles relax.

The ventricular pressure–volume loop is a convenient way to summarize changes in volume and pressure during the cardiac cycle. It is also a powerful way to illustrate changes in ventricular pressure and volume with exercise (or other perturbations) and to visualize the work of the heart. The energy expended by the myocardium during contraction results in heat and mechanical work. The mechanical work generated by the myocardium is evidenced by increased pressure and volume in the arterial system. The amount of work performed by the ventricles during ejection, called stroke work, is equal to the change in pressure and volume and can be quantified as the area inside the pressure–volume loop. An increase in either stroke volume or left ventricular pressure results in a larger area inside the loop and reflects greater stroke work.

Figure 2.4 depicts expected changes in the pressure–volume loop during dynamic exercise. In this case, end-diastolic volume is greater (due to increased venous return of blood); the change of pressure is greater (reflecting increased blood pressure that follows increased sympathetic stimulation with exercise); and end-systolic volume is less (because of enhanced contractility of the myocardium). The increase in end-diastolic volume and the decrease in end-systolic volume both contribute to an increased stroke volume with exercise. The increased stroke volume and increased pressure both contribute to the increased stroke work associated with dynamic exercise.

In figure 2.4*a,* ventricular volume is plotted against ventricular pressure. During the ventricular filling period (1), the mitral valve is open (point a) and end-diastolic volume is increasing. During the isovolumetric contraction period (2), the heart valves are closed, so volume remains the same despite large increases in ventricular pressure. During the ejection period (3), blood is ejected from the heart because the aortic semilunar valve opened at point c, and ventricular volume decreases. During the isovolumetric relaxation period (4), the heart valves are closed, so volume remains the same as pressure in the relaxing ventricle decreases dramatically. End-diastolic volume is reached at point b (also where the mitral valve closes), and end-systolic volume is reached at point d (also where the semilunar valve closes). Stroke volume is equal to end-diastolic volume minus end-systolic volume. The area within the pressure–volume loop represents the stroke work of the heart.

During aerobic exercise (figure 2.4*b*), end-diastolic volume is enhanced because of increased venous return (owing to the skeletal muscle pump and venoconstriction), and end-systolic volume is reduced because of enhanced contractility (owing to sympathetic nervous stimulation). The enhanced end-diastolic volume and reduced end-systolic volume both contribute to an increase in stroke volume during exercise.

CARDIAC OUTPUT

Cardiac output is the amount of blood pumped by each ventricle per minute (L/min). It represents blood flow for the entire cardiovascular system and reflects the ability of the heart to meet the body's need for blood flow. Cardiac output is continually adjusted in response to changes in the cardiovascular system and in response to metabolic needs of the body. For instance, during strenuous physical activity, cardiac output may increase more than fivefold to meet the needs of the skeletal muscle for increased blood flow. Cardiac output ($\dot{Q}$) is the product of stroke volume (the amount of blood pumped by each ventricle with each beat) and heart rate (the number of times the heart beats per minute), that is,

$$\dot{Q}\text{ (L/min)} = \text{HR (beats/min)} \times \text{SV (ml/beat)}.$$

In order for cardiac output to adjust to the metabolic demands of the body, there must be a change in one or both of these variables.

Heart Rate

Factors affecting heart rate do so by changing the depolarization and repolarization characteristics of the pacemaker cells of the conduction system (see chapter 4 for full discussion). Heart rate is affected primarily by autonomic nervous stimulation: Sympathetic nervous stimulation increases heart rate (positive chronotropic effect), whereas parasympathetic stimulation decreases heart rate (negative chronotropic effect). Neural influences have immediate effects on heart rate and therefore can cause very rapid adjustments in cardiac output.

Stroke Volume

Stroke volume is the amount of blood ejected from each ventricle with each beat of the heart. It is determined by three primary factors:

1. Preload
2. Afterload
3. Contractility

Preload

Preload is the amount of blood returned to the heart during diastole. As more blood is returned to the heart, more blood is ejected from the heart—up to a point. The relationship between ventricular volume and stroke volume is described by the **Frank-Starling law** of the heart, which states that an increased stretch of the myocardium (reflected in increased end-diastolic volume or pressure) enhances the contractile force of the myocardium and, therefore, causes more blood to be ejected. In short, the more the ventricle is stretched in diastole, the greater the force of contraction in systole. Thus, greater venous return leads to greater cardiac output—reinforcing the dynamic interplay between the circulation and heart function—and vice versa. Figure 2.5 presents a ventricular function curve with end-diastolic volume (an index of muscle fiber length) on the x-axis and stroke volume (a measure of contractile energy) on the y-axis (other ventricular function curves use ventricular filling pressure on the x-axis and stroke work (SV × MAP [mean arterial pressure]) on the y-axis).

Preload is determined by the amount of blood in the ventricle; this is a function of venous return and filling time (heart rate, HR). Preload is dependent on venous return of blood and can be affected by any factor that alters venous return. For example, during rhythmical exercise, the skeletal muscle pump causes increased venous return, which leads to increased stroke volume. Conversely, immediately upon standing, there is a temporary decrease in venous return that results in a transient decrease in stroke volume.

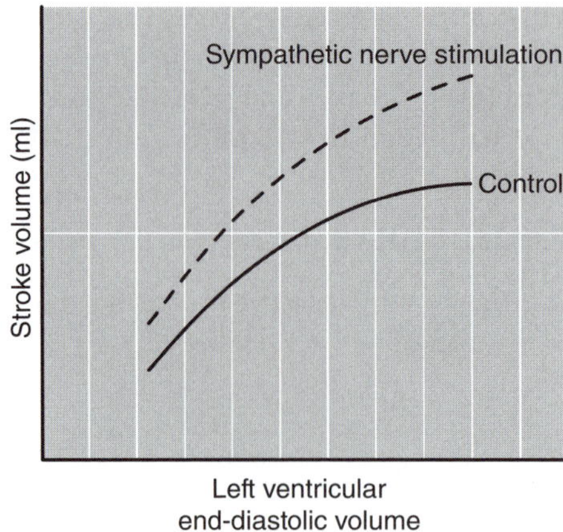

Figure 2.5 Frank-Starling law of the heart. As end-diastolic volume (and pressure) increases, the myocardium is stretched and contracts more forcefully, leading to a greater stroke volume. Also shown is the effect of enhanced contractility (increased sympathetic nerve activity). At any given end-diastolic volume, enhanced contractility will shift the curve to the left, leading to a greater stroke volume.

The Frank-Starling law of the heart is an important mechanism to equalize cardiac output from the right and left ventricles. It is essential that the output of the right and left ventricles be equal. Because the circulatory system is a closed system, even a slight imbalance in outflow from the ventricles would lead to serious (or fatal) pulmonary congestion (if RV output was greater than LV output) or insufficient pulmonary circulation (if LV output was greater than RV output). The Frank-Starling mechanism prevents such dangerous outcomes by equalizing cardiac output between the ventricles over very few beats. If left ventricular output begins to exceed right ventricular output, the increase in systemic venous return to the right atrium will increase right ventricular filling and lead to an increase in right ventricular output, thereby restoring the balance in ventricular output. Thus, the Frank-Starling mechanism provides an intrinsic mechanism to balance cardiac output of the right and left ventricles over time.

Afterload

Afterload is the pressure that opposes the ejection of blood and is usually taken to be equal to systemic arterial pressure, as reflected by mean arterial pressure. When all other factors are equal, stroke volume decreases as mean arterial pressure increases. Under normal conditions, afterload is relatively constant because blood pressure is a closely controlled variable. In disease conditions, however, such as hypertension or aortic stenosis, afterload (blood pressure) can increase significantly, causing a decrease in stroke volume, or requiring greater cardiac work to eject the same amount of blood.

Contractility

While an increased stretch of the myocardium does enhance contractile strength, **contractility** refers specifically to the force of myocardial contraction independent of preload or afterload. Factors that increase cardiac contractility are referred to as having a positive inotropic effect on the heart. Contractility, for any given level of preload, is enhanced by sympathetic nervous stimulation and circulating catecholamines. Norepinephrine is the most important physiological factor that enhances cardiac contractility. Figure 2.5 depicts the relationship between end-diastolic volume and stroke volume under normal conditions and enhanced sympathetic stimulation. For any given end-diastolic volume, stroke volume is greater when contractility is enhanced by sympathetic nervous stimulation. Of course, in the intact human organism, myocardial stretch and increased contractility may occur simultaneously to enhance stroke volume. For example, during dynamic exercise, preload is increased due to the skeletal muscle pump, and contractility is increased due to sympathetic nervous system activation.

Figure 2.6 summarizes the primary factors that affect cardiac output, with special emphasis on how the sympathetic nervous system influences both stroke volume and heart rate.

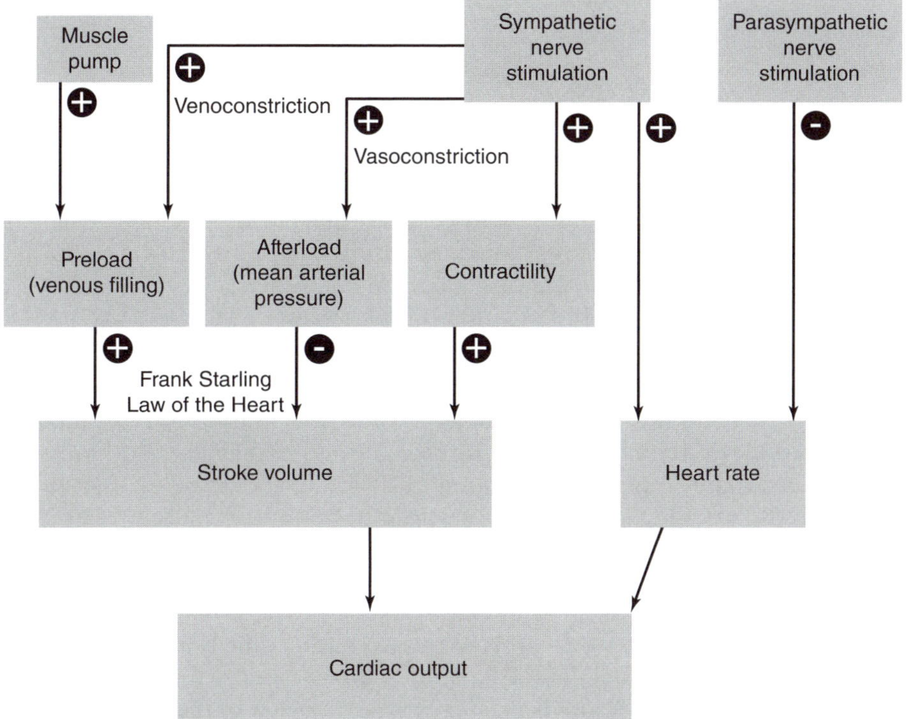

Figure 2.6 Factors affecting cardiac output. The sympathetic nervous system has a major effect on heart rate and contractility. Sympathetic nervous stimulation also affects afterload and preload. Cardiac output also has a profound effect on stroke volume through its effect on preload. Because the cardiovascular system is a closed system, heart outflow (cardiac output) must equal heart inflow (venous return) over time.

DISTRIBUTION OF CARDIAC OUTPUT

The amount of blood pumped by the heart is determined largely by the metabolic needs of the body. The metabolic needs of the body, however, change constantly in response to myriad stimuli, including environmental changes, position, muscular work, neurohormonal factors, and emotional state. For example, following a meal, it is advantageous to increase blood flow to the gastrointestinal (GI) tract, which may cause a slight increase in overall cardiac output; but it is not necessary or advantageous to increase blood supply to every organ at that time. During strenuous physical activity, cardiac output increases dramatically to provide the necessary blood and oxygen to working muscles. In these examples, the total amount of blood in the circulatory system does not change substantially (excepting changes in plasma volume secondary to activity), but the amount of blood pumped each minute (cardiac output) can change dramatically (up to three- to sixfold during maximal exercise depending on training status). Equally important, the distribution of the cardiac output varies tremendously depending on the needs of the body. Change in the diameter of blood vessels is the primary mechanism for altering the distribution of cardiac output in response to internal or external stimuli. Cardiac output represents the total flow of blood through the vascular system and can be expressed in relationship to other variables using the equation

$$\dot{Q} = \Delta P / R$$

where $\dot{Q}$ = cardiac output, ΔP = the difference in pressure (between the left ventricle and the right atrium), and R = resistance.

Thus, the total flow of blood through the cardiovascular system is determined by the difference in pressure in the left ventricle (where the driving force of blood originates) and the right atrium (where blood is returned to) divided by the resistance the blood encounters as it travels through the vascular system. Given that pressure in the right atrium is so low (0-4 mmHg), the equation is typically expressed as

$$\dot{Q} = MAP / R.$$

Figure 2.7 presents the typical distribution of cardiac output at rest and during strenuous aerobic exercise. Notice that cardiac output has increased substantially (a fourfold increase) but that the distribution of the cardiac output has also changed dramatically. In fact, in this example, blood flow to the working muscle has increased from 1,200 ml/min to 22,000 ml/min. Blood flow to critical areas, such as the cerebral circulation, is maintained, while areas that require additional blood flow are enhanced, such as the coronary and skeletal muscle beds. In order to maintain or enhance blood flow to some vascular beds, other regions, such as the renal and splanchnic circulations, receive less blood flow. Changing the distribution of the cardiac output is accomplished primarily through a change in the diameter of the resistance vessels (arterioles) supplying these vascular beds.

Similar to the principles of the equation that describes blood flow in the entire cardiovascular system, the blood flow in any region is determined by the difference in pressure entering and leaving that vascular bed divided by the resistance to flow.

$$F = \Delta P / R$$

While this equation reveals that flow can be increased by altering either pressure or resistance, in reality, it is the change in resistance that has the most dramatic effect on blood flow in most cases. This is due to the large impact that a small change in diameter has on resistance and hence flow (this concept is described fully in chapter 6).

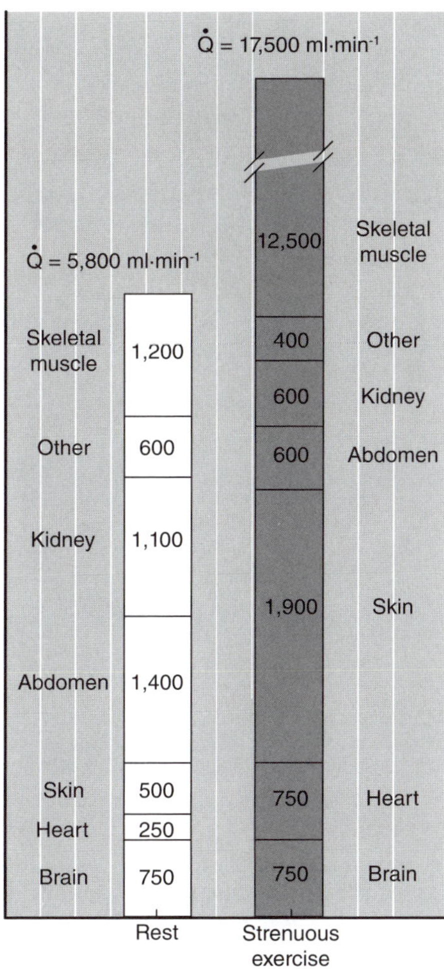

Figure 2.7 Distribution of cardiac output at rest and during exercise.

CORONARY BLOOD SUPPLY

The myocardium needs a rich supply of blood to provide the oxygen necessary to perform the work of the heart. The amount of oxygen that is used by the myocardium is affected by coronary blood flow and oxygen extraction.

Coronary Circulation

The primary work of the heart is pumping blood against resistance. This work is done continually and requires the constant production of chemical energy in the form of adenosine triphosphate (ATP). The coronary circulation supplies the heart muscle (myocardium) with the oxygen needed to produce this supply of ATP.

The right and left coronary arteries provide the entire blood supply to the myocardium. As seen in figure 2.8, the left and right coronary arteries are the first vessels

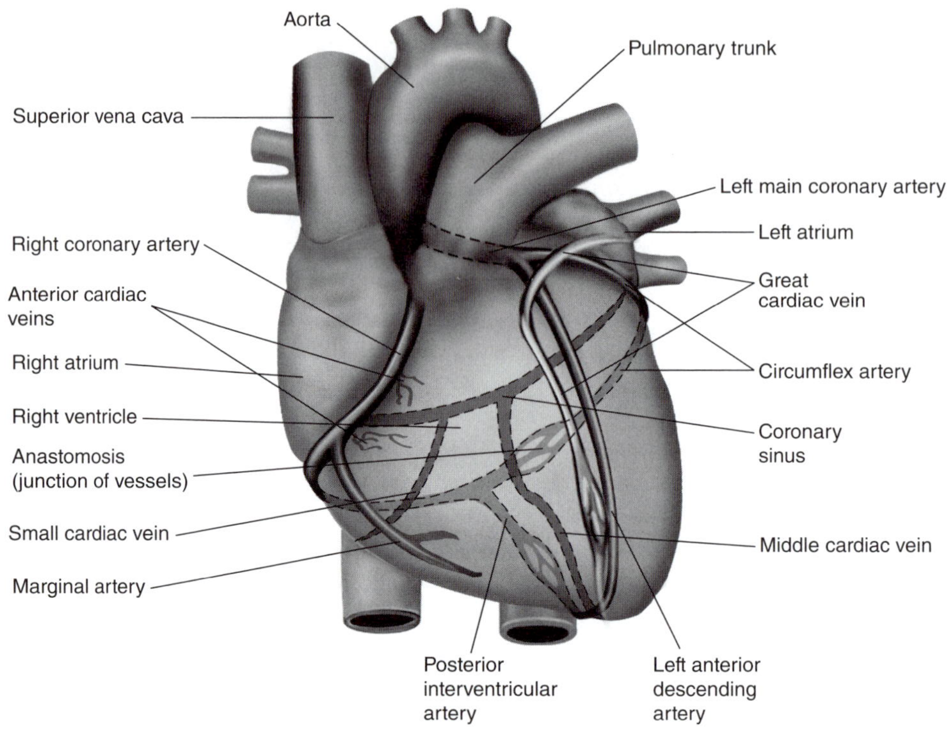

Figure 2.8 Coronary circulation, illustrating the right and left coronary arteries and their major branches.

Reprinted, by permission, from J.W. Wilmore, D.L. Costill, and W.L. Kenney, 2008, *Physiology of Exercise and Sport,* 4th ed. (Champaign, IL: Human Kinetics), 127.

that bifurcate off the root of the aorta. The right coronary artery supplies blood to the right atrium and ventricle; the left coronary artery divides near its origin into the anterior descending and circumflex arteries that together supply the majority of blood to the left atrium and ventricle. After blood travels through the coronary capillary beds, it is returned to the right atrium through the anterior coronary veins and the coronary sinus.

Coronary Blood Flow

Contraction of the myocardium produces the pressure (aortic pressure) that is necessary to perfuse the myocardium. The heart also influences blood supply to the coronary vessels by the squeezing effect of the contracting myocardium on the coronary vessels. In fact, during early systole, blood flow in large coronary vessels is actually reversed owing to the high force generated in the myocardium. Maximal left coronary blood flow occurs in early diastole when the ventricles are relaxed and compression of the coronary vessels is minimal.

Changes in coronary vessel resistance play a major role in determining coronary blood flow. As with other muscle, the diameter of the vessel is determined largely by the metabolic demands of the tissue it supplies: When the cardiac muscle is active and there are high metabolic needs, the vessels dilate in order to increase blood flow.

Sympathetic nerve stimulation also increases coronary blood flow—an effect that is the sum of several opposing forces on blood flow. Sympathetic stimulation increases the rate and force of cardiac contraction. The increased heart rate means a shorter diastolic filling period, which, along with a more forceful contraction, limits myocardial blood flow. Sympathetic nerve stimulation also tends to cause vasoconstriction in the coronary arteries. While the factors noted earlier have the effect of decreasing coronary blood flow, the increase in metabolism (resulting from a higher heart rate and more forceful contraction) more than offsets these factors so that coronary blood flow increases with sympathetic stimulation.

Myocardial Oxygen Consumption

The heart relies almost entirely on aerobic metabolism. Thus, there is a strong relationship between myocardial oxygen consumption and coronary blood flow because the production of ATP for contraction is tightly linked to the supply of oxygen. Even under resting conditions, humans extract a very high percentage (65-70%) of oxygen from the blood as it travels through the cardiac capillaries. Any increase in oxygen consumption of the myocardium, therefore, is dependent on an increase in blood flow.

Under resting conditions, the heart consumes approximately 8 to 10 ml of oxygen per 100 g of tissue each minute. As left ventricular work increases, oxygen consumption increases; the amount of oxygen consumed by the myocardium is a function of the amount of work being performed by the heart. Under normal resting conditions, the basal metabolism of the heart accounts for approximately 25% of myocardial oxygen consumption, and contraction of the heart muscle accounts for the remaining 75%. The oxygen consumed during contraction primarily reflects the need for oxygen to produce ATP for the cross-bridge cycling that occurs during muscle contraction—both the isovolumetric contraction period and the ejection period. Approximately 50% of the total myocardial oxygen consumption occurs during the isovolumetric contraction period because of the high forces that are developed in this period (even though the heart performs no external work). The amount of energy and oxygen used during this period is largely dependent on the afterload against which the heart must pump. Thus, as blood pressure increases, so does myocardial oxygen consumption. During the ejection period, external work is performed as the heart contraction expels blood from the ventricles. The oxygen consumption during this period depends on how much work is being done. The external physical work of the left ventricle in one beat is the stroke work, and it is equal to the area enclosed by the left ventricular pressure–volume loop (see figure 2.4, p. 17).

Changes in myocardial contractility affect myocardial oxygen consumption in several ways: Increased contractility affects basal metabolism, tension generation, and external work. Also, enhanced contractility necessitates additional energy to develop tension in the cross-bridges more quickly.

Heart rate is one of the major determinants of myocardial oxygen consumption, therefore, the oxygen cost of cardiac contraction must be multiplied by the number of times the heart beats each minute.

Effect of Heart Size and Pressure on Energy Cost

According to the law of Laplace, left ventricular wall tension is directly proportional to the radius of the ventricle and intraventricular pressure, and inversely related to

MEASURING MYOCARDIAL OXYGEN CONSUMPTION

Myocardial oxygen consumption is difficult to measure directly; blood flow to the myocardium must be measured, and the difference between oxygen content of the arterial and venous blood must be assessed. In other words, myocardial oxygen consumption equals blood flow × a-v O_2 difference.

Given the difficulties in obtaining these measures, it is fortunate that myocardial oxygen consumption can be easily estimated using the rate–pressure product (RPP). The RPP is calculated as (HR × SBP) / 100. The components of this equation represent the frequency of the heartbeat and the pressure that must be generated by the myocardium; thus they would be expected to be directly related to myocardial oxygen consumption, and experimental evidence confirms this hypothesis. The RPP is often used during exercise testing to monitor the work of the heart.

ventricular wall thickness. The law of Laplace applied to a spherical structure (such as a ventricle) states the following:

$$\text{Wall tension} = \frac{\text{Transmural pressure} \times \text{Vessel radius}}{2 \times \text{Wall thickness}}$$

$$T = \frac{(P \times r)}{2h}$$

On the basis of this mathematical relationship, it can be seen that an increase in ventricular wall thickness (ventricular hypertrophy) leads to a reduction in wall tension. On the other hand, an increase in ventricular cavity size (an increase in r) leads to an increase in wall tension. An increase in wall tension means that the myocardium must work harder, and this is reflected in a greater energy cost. In the case of heart failure, the ventricular cavity may be enlarged at the same time there is significant thinning of the ventricular wall, thus leading to a considerable increase in the energy required to eject blood from the ventricle (Shepherd and Vanhoutte, 1980).

MEASURING CARDIAC FUNCTION

Cardiac function is routinely assessed in hospital and clinical settings in order to diagnose disease and guide treatment. Cardiac function is also commonly measured in exercise physiology laboratories in order to understand the effect of exercise and exercise training on cardiovascular parameters. One of the most powerful and versatile tools for assessing cardiac function is a Doppler ultrasound machine (figure 2.9).

Cardiac Output

Cardiac output is a measure of the ability of the heart to pump the blood needed to meet the metabolic demands of the body. It is determined by the product of heart rate and stroke volume. Measuring heart rate is relatively easy (by palpitation, use of electronic measuring devices [such as an HR watch] or use of ECG equipment).

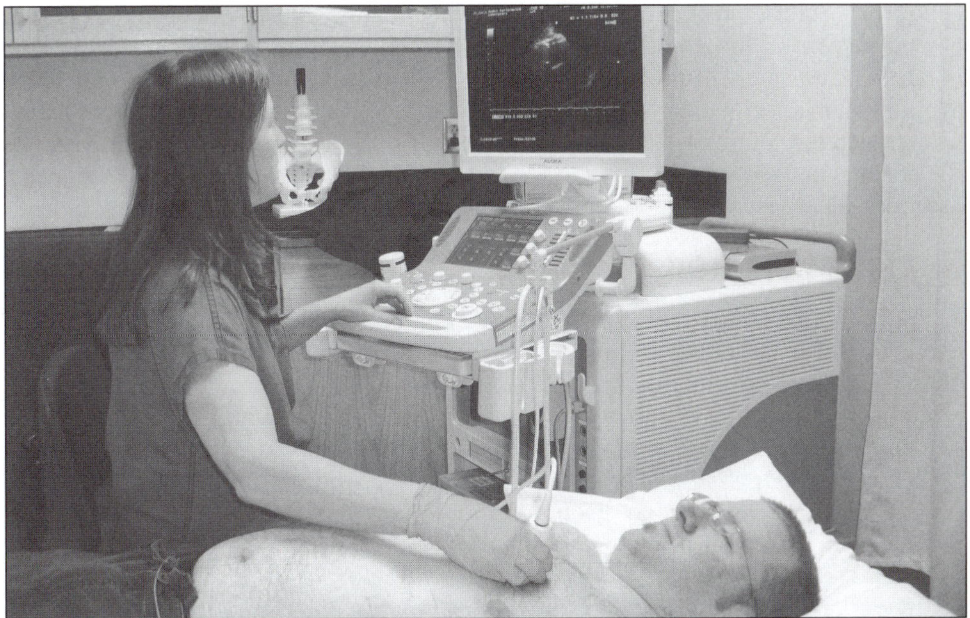

Figure 2.9 Ultrasound unit. Cardiac ultrasound is used to measure heart dimensions and assess cardiac and vascular function.

Measuring stroke volume, however, is more challenging. Stroke volume is often measured clinically using echocardiography. Under resting conditions the ventricular dimensions are imaged, and volumes are calculated based on geometric assumptions (figure 2.10). In this method, volume in the ventricle at the end of diastole and the end of systole is determined, and stroke volume is calculated as the difference between end-diastolic volume and end-systolic volume. Stroke volume is an important clinical measure because it is often used as a global measure of systolic function.

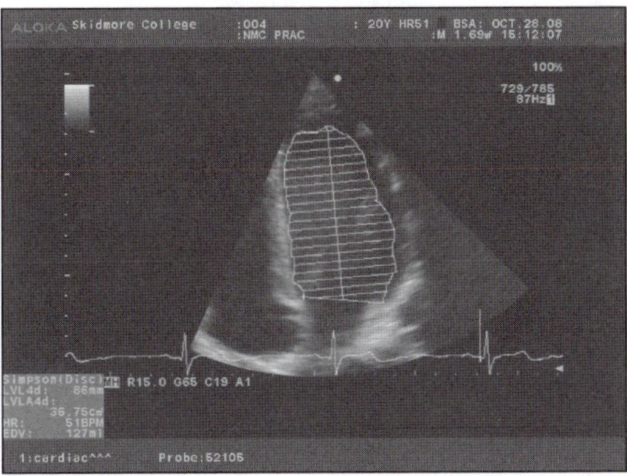

Figure 2.10 Stroke volume assessment. Cardiac dimensions are imaged during diastole and systole to provide a measure of ventricular volume. Stroke volume is the difference between end-diastolic ventricular volume and end-systolic ventricular volume.

During exercise, stroke volume must be measured by different methods because of movement artifact that limits the validity of the method just described. Under exercise conditions, it is more appropriate to calculate SV using Doppler echocardiography. Stroke volume is equal to the product of aortic cross-sectional area and the time–velocity integral of blood flowing in the ascending aorta. The cross-sectional area of the aorta is imaged using ultrasound, and the time–velocity integral of blood in the ascending aorta is measured using a Doppler probe.

Systolic Function

Systolic function is a term used to express the ability of the heart to contract forcefully enough to effectively eject blood. Several echocardiography variables can be used as an expression of systolic function (or contractility), including fractional shortening, ejection fraction, and regional wall motion analysis. Fractional shortening is the percentage change in left ventricle dimensions with each contraction (FS = [(LVED − LVES) / LVED] × 100). The most common expression of global left ventricular function is the left ventricular ejection fraction. Ejection fraction, the ratio of blood volume ejected from the heart to the volume of blood in the heart at the end of diastole, SV/EDV, is a widely used index of contractility. Regional wall motion is determined by dividing the left ventricle into 16 discrete segments, and a numeric scoring system is used to score the contractility of each segment (1 = normal; 2 = hypokinesis; 3 = akinesis; 4 = dyskinesis). A wall motion score index (WMSI) is derived by dividing the sum of the wall motion scores by the number of segments visualized during the echocardiography examination, and reflects the extent of regional wall motion abnormality (Oh, Seward, and Tajik, 2007).

Diastolic Function

Diastolic function is a measure of the ability of the myocardium to relax and allow adequate ventricular filling. It is now well appreciated that diastolic function is integral to the health and function of the heart. Many individuals with heart disease have

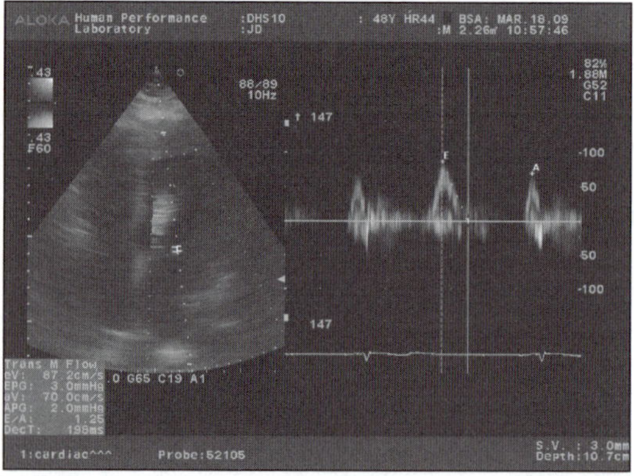

Figure 2.11 Mitral inflow velocities. The mitral inflow velocity is obtained by a pulsed wave Doppler recording with samples taken at the mitral valve. The E wave represents the inflow pattern during early diastole (when the ventricle is recoiling and relaxing). The A wave represents the inflow pattern during late filling after atrial contraction.

diastolic dysfunction with preserved systolic function. Diastolic dysfunction can be assessed by echocardiography. An enlarged left atrium reflects chronic left ventricular filling pressure. On the other hand, Doppler flow measures across the mitral valve reflect acute filling dynamics (figure 2.11). The E wave reflects early diastolic mitral inflow immediately after the opening of the valve. The A wave represents the atrial contractile component of mitral filling and is primarily influenced by left ventricle compliance and left atrial contractility. The E/A ratio also provides a valuable index of diastolic function (Lester et al., 2008).

SUMMARY

The heart is a muscular pump responsible for generating the pressure to distribute blood throughout the circulation. There are large variations in ventricular pressure and volume throughout the cardiac cycle—the alternating period of relaxation (diastole) and contraction (systole). The amount of blood ejected from the heart with each beat, stroke volume, is determined by preload, afterload, and contractility. Cardiac output, the amount of blood pumped per minute, is the product of heart rate and stroke volume. Cardiac output varies considerably to meet the metabolic needs of the individual at any given time. During exercise, cardiac output can increase five- to sevenfold, with a large portion of the increase in blood flow being directed to the working muscles.

CHAPTER 3

Cardiac Myocytes

The pumping function of the heart is directly dependent on the contraction of cardiac muscle cells, or myocytes. While the previous chapter described how the heart functions as an organ, this chapter will describe how the individual myocytes generate the force necessary for the pumping action of the heart. Not only must the myocytes contract to produce force; they must also contract in unison if the heart is to function as an effective pump. The synchronized contraction of the myocytes is dependent on a specialized conduction system and the structural and functional properties of the myocyte membrane that permit electrical signals to pass from cell to cell.

There are two primary types of cardiac muscle cells: the **contractile myocytes** that perform the mechanical work of the heart, and the modified muscle cells called **pacemaker-conduction cells** that initiate the electrical signal and conduct it through the heart. The majority of cardiac cells are contractile myocytes. The contraction of these myocytes produces the force to eject blood from the heart. The contractile myocytes will not contract unless they receive an electrical signal to do so. In contrast, cells of the conduction system do not contract. Instead, these cells serve the important function of ensuring rapid conduction of the impulses through the heart, allowing for the coordinated contractile action that is essential for the efficient pumping activity of the heart. The electrical activity associated with contractile myocytes and the electrical activity in the specialized conduction system are discussed in detail in chapter 4. This chapter is devoted to the structure and function of the contractile myocytes.

MICROSCOPIC ANATOMY OF CARDIAC MYOCYTES

Like all muscle fibers, cardiac muscle fibers (myocytes) contract in response to an electrical signal (action potential) in the cell membrane. Specific features of the cell membrane, along with specialized organelles, are responsible for transmitting the electrical signal into the interior of the cell. The structure of cardiac myocytes has many similarities with that of skeletal muscle (including many organelles and specifically the protein filaments actin and myosin). There are some important differences, however, between skeletal and cardiac muscle; cardiac muscle fibers are shorter and more branched than skeletal muscle fibers. The structure of the myocyte directly supports its function—contracting as part of a functional unit to eject blood from the chambers of the heart.

Myocytes are approximately 50 to 100 μm long and 10 to 20 μm in diameter and have a striated appearance when viewed under a microscope (figure 3.1), which results from the overlapping contractile filaments (thick and thin filaments). The cell membranes of adjacent cardiac muscle cells are attached to one another end to end by a specialized junction, the intercalated disc. Myocytes have a large centrally located nucleus and a large number of mitochondria. Other specialized organelles of the myocyte include transverse tubules (T-tubules) that carry an action potential into the interior of the cell, and sarcoplasmic reticulum that stores and releases calcium.

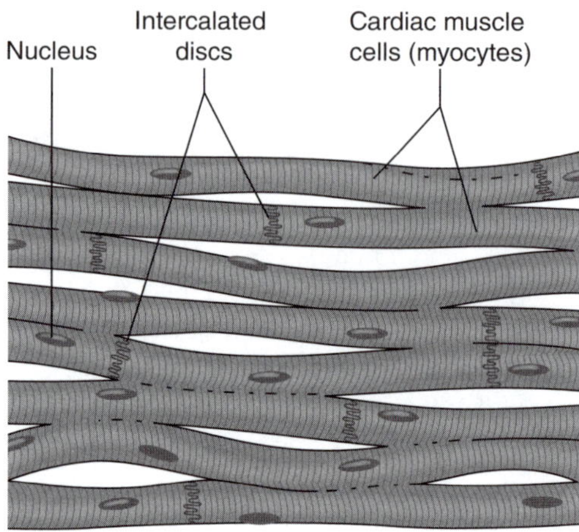

Figure 3.1 Cardiac myocytes.

Cell Membrane

As with all muscle cells, the sarcolemma of the myocyte is composed of a phospholipid bilayer. As described fully in the following chapter, an action potential is generated when there is a reversal of charge across the cell membrane. The action potential initiates muscle contraction via the sliding of myofilaments over one another. The spread of the action potential over a myocyte causes it to contract via the process of excitation–contraction coupling. At the same time, an action potential in a contractile myocyte is spread to adjacent myocytes via intercalated discs. The spreading of action potentials to adjacent myocytes ensures that all the myocytes contract at nearly the same time and that the heart functions as an effective pump.

Intercalated Discs

Intercalated discs structurally and functionally connect adjacent myocytes. An intercalated disc is an undulating double membrane separating adjacent cells that contains two primary types of junctions: gap junctions and desmosomes (figure 3.2). The gap junctions are made up of connexons, hexagonal structures that form a connection between two adjacent cells and provide a low-resistance pathway that allows the transmission of ion currents and thus electrical impulses from one myocyte to the next. Because the electrical signal (action potential) can flow from one myocyte to the next, the myocardium behaves as a functional unit—it acts as a **syncytium,** thus

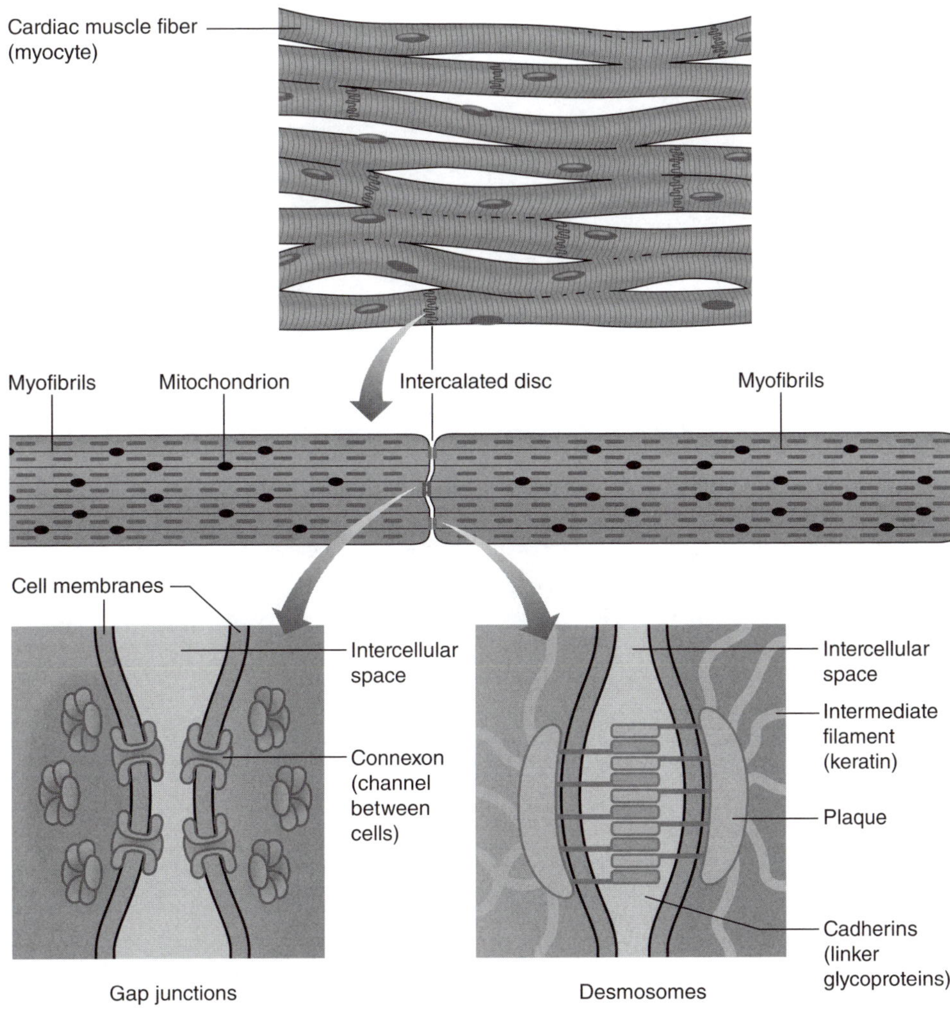

Figure 3.2 Structure of intercalated discs.

ensuring the effective pumping action of the heart. **Desmosomes** serve to bind adjacent myocytes together. There is a thickening (called a plaque) on the sarcoplasmic side of both adjacent myocytes at the site of the desmosome. The adjacent myocytes are held together by thin protein filaments (called cadherins) that extend from the plaque and interdigitate like the teeth of a zipper to hold the myocytes together. Thicker filaments (called intermediate filaments) extend from the plaque across the cell to provide additional mechanical and tensile strength to the cell. Because adjacent cells are held together by desmosomes, the myocytes are not pulled apart by the force of contraction.

Myofibrils

Myocytes contain long, cylindrical-shaped organelles called myofibrils that are themselves composed of overlapping thick and thin myofilaments (figure 3.3). It is the sliding of these filaments past one another that produces force during muscular

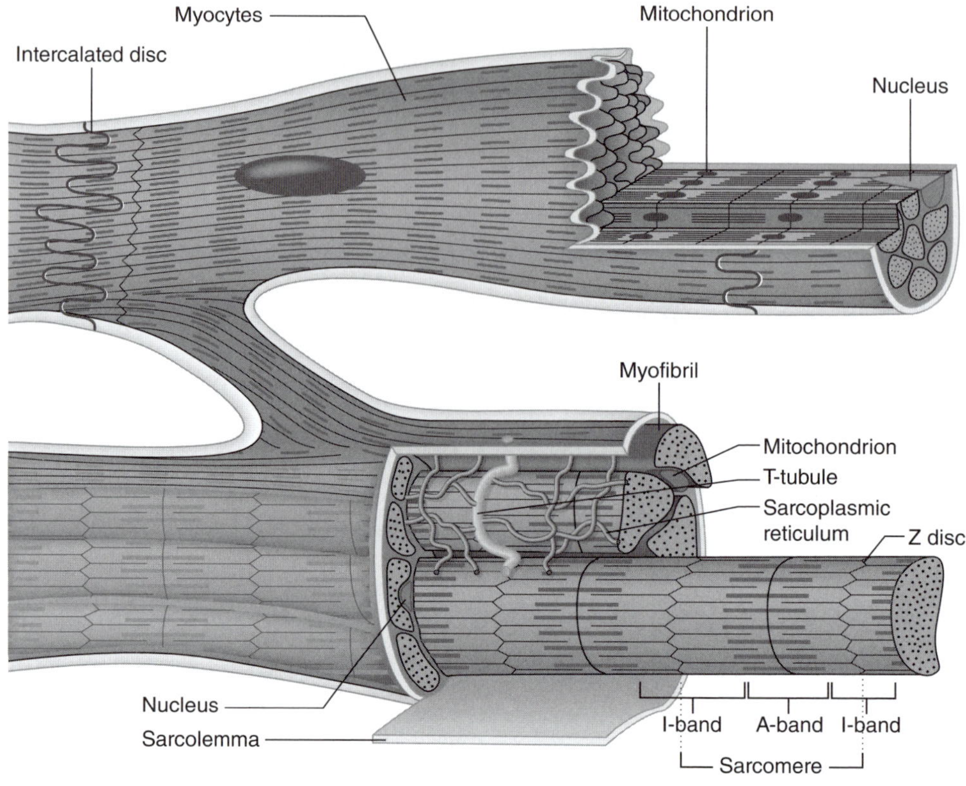

Figure 3.3 Microscopic structure of myocytes.

contraction. It is the arrangement of the sarcomeres along the myofibrils, and across the many myofibrils of the muscle cell, that gives the muscle cell its striated appearance. The myofibrils of cardiac muscle are very similar in structure, function, and appearance to those of skeletal muscle.

Each myofibril is composed of smaller functional units called sarcomeres that are oriented end to end along the myofibril. The **sarcomere** is the basic contractile unit of the myocyte. Sarcomeres are composed primarily of overlapping thick and thin filaments, although they also contain additional proteins and connective tissue that constitute the cytoskeleton of the sarcomere and contribute to the mechanical stiffness and elasticity of muscle tissue. A sarcomere extends from one Z disc to the adjacent Z disc. The Z disc that forms the partition of the sarcomere is composed of a-actinin protein. Between the Z discs are the thick and thin filaments. The thick filaments are composed of the contractile protein myosin; the thin filaments are composed primarily of the contractile protein actin. The thin filament also contains the regulatory proteins, troponin and tropomyosin, which play a regulatory role in muscle contraction.

Transverse Tubules

Transverse tubules (T-tubules) are deep invaginations in the sarcolemma at each Z disc that provide the mechanism for transmitting an electrical impulse from the sarcolemma into the myocyte. Because of their number and their ability to spread the

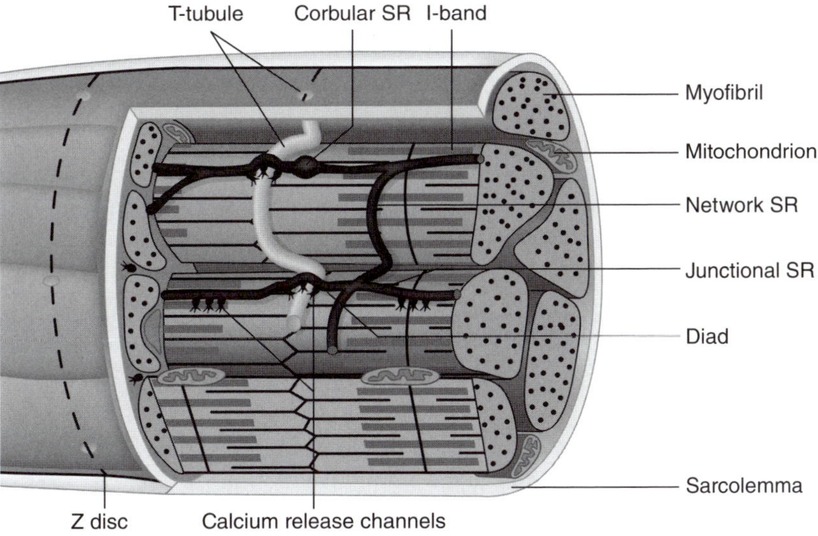

Figure 3.4 Transverse tubules (T-tubules) and sarcoplasmic reticulum.

electrical signal into the cell rapidly, the T-tubules play an important role in ensuring that all the myofibrils are activated almost simultaneously. The T-tubules are structurally and functionally connected to the sarcoplasmic reticulum, and together these organelles control intracellular calcium levels (figure 3.4).

Sarcoplasmic Reticulum

The **sarcoplasmic reticulum (SR)** is an extensive series of tubular structures within the myocyte that store and release calcium. The SR accounts for approximately 5% of cell volume and is associated with regulating intracellular calcium levels. High intracellular calcium concentrations play a critical role in controlling cardiac muscle contraction via the interaction with regulatory proteins on actin. Furthermore, myocardial relaxation is achieved primarily by the removal of calcium ions. Thus, it is important to understand how the muscle cell handles calcium, and to realize that myocytes differ from skeletal muscle in the way in which calcium is handled. The SR surrounds the myocyte and comes in close proximity to the T-tubules (figure 3.4). The SR has three distinct regions, each with a specific role in calcium handling. The junctional SR, which contains large stores of calcium, comes very close (within 15 nm) to the sarcolemma and the T-tubules. The area where the junctional SR and T-tubules nearly meet is termed a **diad.** Protein channels also extend from the junctional SR; these are alternately called **calcium release channels, calcium-induced calcium release channels,** or **ryanodine receptors.** They derive these names because they release calcium, because they release calcium when stimulated with calcium, and because they bind the drug ryanodine. In addition to releasing calcium, these receptors also bind to calcium that enters the cell through the sarcolemma or T-tubules (thus they are also sometimes called calcium-induced calcium release channels), making their placement next to the sarcolemma perfectly logical.

Corbular SR is a sac-like expansion of the SR that lies next to the I-bands and contains a high concentration of calcium. Network SR runs in parallel with the T-tubules and is primarily responsible for the uptake of calcium from the sarcoplasm after a muscle contraction. Network SR contains an abundant number of SR calcium-ATPase pumps ($SERCA_2$) to ensure that it can effectively pump calcium into the SR against its concentration gradient. In humans, approximately 75% of calcium ions are removed from the sarcoplasm during relaxation by the SR calcium-ATPase pumps (Hasenfuss, 1998). These calcium-ATPase pumps are regulated by the protein phospholamban.

The normal cycle of systole and diastole requires a precise, transient increase and decrease in the intracellular concentration of calcium ions. The SR plays an integral role in orchestrating the movement of calcium ions with each contraction and relaxation of the heart.

EXCITATION–CONTRACTION COUPLING

The sequence of events whereby an electrical stimulus in the sarcolemma leads to force generation in the myofilaments—the series of electrical, chemical, and mechanical events—is termed excitation–contraction coupling (ECC). The steps involved in ECC are summarized in figure 3.5 and the following discussion.

An action potential in the sarcolemma is spread cell to cell via gap junctions. The action potential in the sarcolemma is also carried into the interior of the cell via T-tubules (figure 3.5, a). The action potential causes an increase in calcium permeability, and calcium enters the cell though channel proteins called **L-type receptors** located in the sarcolemma and T-tubules (figure 3.5, b).

Some of the calcium that enters the cell binds to large calcium release channels located on the junctional SR (figure 3.5, c). These protein channels have T-shaped tubes in the interior that appear to be responsible for the release of calcium from the SR. Thus, the influx of extracellular calcium through the L-type calcium channel serves as a trigger for release of calcium stored in the SR through the calcium release channels—this is known as **calcium-induced calcium release** (figure 3.5, d).

The intracellular calcium binds to the regulatory protein troponin (figure 3.5, e), which is bound to another regulatory protein, tropomysin, on the thin filaments. The binding of calcium to troponin causes troponin to undergo a configurational change and pulls tropomyosin deeper into the groove along the actin strand and off of the active site on actin. This allows the myosin heads to bind to the exposed active site and the cross-bridge cycle is initiated, creating force generation and muscle shortening.

At the end of a cardiac action potential, calcium influx ceases and the SR is no longer stimulated to release calcium. Rather, **sarcoplasmic reticulum calcium-ATPase ($SERCA_2$)** pumps in the network SR actively pump calcium back into the SR (figure 3.5, f). In addition to the SR calcium pump, some (~25%) of the free calcium is expelled from the cell by the sarcolemmal sodium–calcium exchanger. As the level of free cytosolic calcium drops, calcium dissociates from troponin, and the troponin-tropomyosin complex resumes its blocking position on actin—leading to relaxation, or on the whole muscle level, diastole.

The SR calcium-ATPase pumps that pump calcium back into the network SR are regulated by the protein phospholamban. **Phospholamban** stimulates reuptake of calcium into the SR and thus enhances myocardial relaxation. Norepinephrine and epinephrine (resulting from sympathetic nervous stimulation) cause the phosphorylation and hence increased activity of phospholamban, enhancing myocyte relaxation. Thus, sympathetic nervous simulation increases the myocardial contractility (inotropic action), the rate of contraction (chronotropic action), and the rate of relaxation (lusitropic action). That is, the heart contracts and relaxes more quickly under sympathetic nerve stimulation.

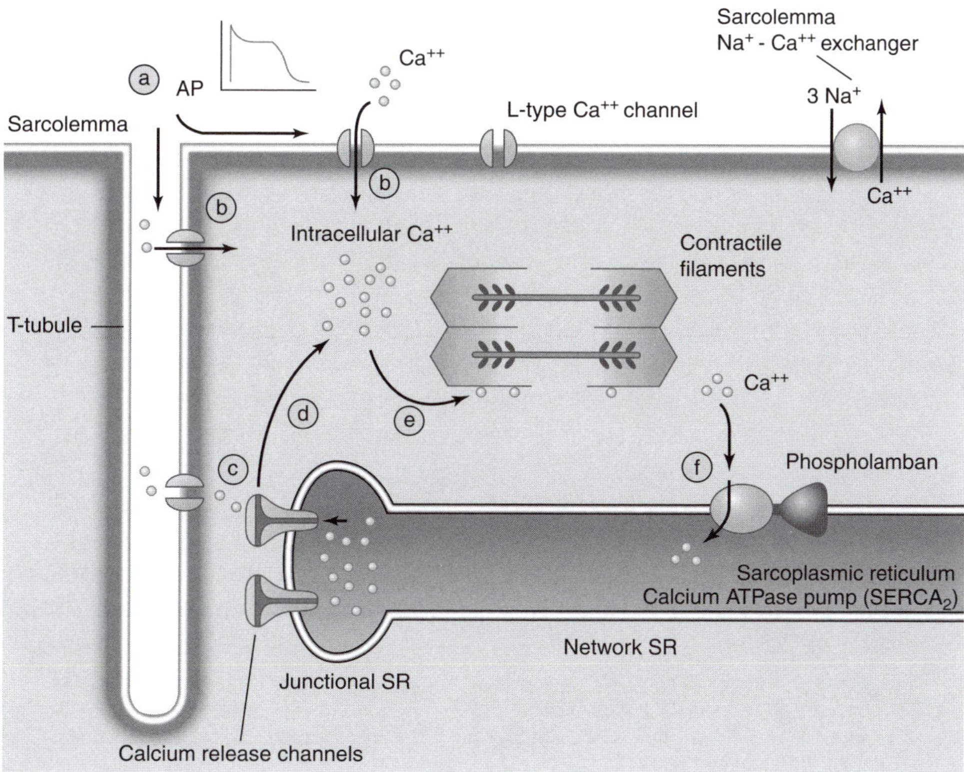

Figure 3.5 Sequence of events in excitation–contraction coupling. An action potential is carried into the interior of the cell via T-tubules (a). Calcium enters the cell through L-type receptors (b). Calcium binds to calcium release channels on junctional sarcoplasmic reticulum (SR) (c), causing the release of calcium from the junctional SR (d). Intracellular calcium binds to the regulatory protein troponin, leading to cross-bridge formation (e). At the end of a cardiac action potential, SR calcium-ATPase pumps in the network SR actively pump calcium back into the SR, and the sarcoplasmic sodium–calcium exchanger ejects calcium from the cell (f).

MECHANISMS OF CONTRACTION

Like skeletal muscle cells, cardiac muscle cells contract when the myofilaments slide over one another toward the center of the sarcomere, causing shortening of the sarcomere and of the muscle fiber. That is to say, shortening of the sarcomere is explained by the sliding filament theory of muscle contraction. Specifically, the thin filaments slide into the spaces between the thick filaments, causing the I-band to shorten while the A-band remains unchanged.

Role of Calcium

Calcium plays a critical role in cardiac muscle contraction because it regulates the position of tropomyosin. In the resting state, tropomyosin blocks the active site on actin, and the myosin heads cannot form a cross-bridge with actin. A rise in intracellular calcium concentration, however, causes the active sites on actin to become exposed, allowing cross-bridge formation to occur. This is accomplished by intracellular calcium binding to troponin (which is bound to tropomyosin). The binding

MYOCARDIAL STUNNING: WHAT IS THE ROLE OF CALCIUM?

Myocardial stunning refers to a decrease in myocardial function following a period of ischemia and reperfusion. Ischemia can cause structural changes in myocytes that are reversed with the restoration of normal blood flow. A number of studies in marathon and ultra-endurance athletes suggest that exercise may lead to myocardial stunning or exercise-induced cardiac fatigue. Whether this observation represents true myocardial stunning or a "pseudo stunning" is not clear, but it is the topic of great research interest.

One of the primary theories to explain myocardial stunning is the "calcium overload" theory. This theory suggests that disturbed myocyte calcium hemostasis, caused by prolonged, severe exercise, leads to decreased calcium responsiveness, apoptosis, and proteolysis of contractile proteins.

The calcium theory is not the only theory to explain the transient decrease in myocardial function following exercise (myocardial stunning), and it is likely that this phenomenon is a multifaceted entity and that several mechanisms may be involved. The reader is referred to several articles to explore the evidence of myocardial stunning with exercise and its possible mechanisms.

Dawson, E., et al. 2003. Does the human heart fatigue subsequent to prolonged exercise? *Sports Medicine* 33:365-380.

Kusuoka, H., and E. Marban. 1992. Cellular mechanisms of myocardial stunning. *Annual Review of Physiology* 54:243-256.

Scott, J., and D. Warburton. 2008. Mechanisms underpinning exercise-induced changes in left ventricular function. *Medicine and Science in Sports and Exercise* 40(8):1400-1407.

of calcium to troponin causes the troponin to undergo a configurational change that pulls tropomyosin from the blocking position on the active sites of actin and thereby exposes the active site to permit myosin binding.

The removal of calcium from the sarcoplasm is also essential to relaxation of the myocardium. Calcium is removed by SR calcium-ATPase (~75%) and sarcolemma sodium–calcium exchangers (~25%). Factors that limit calcium uptake into the SR or removal from the cell impair myocardial relaxation and can contribute to diastolic dysfunction.

Force of Contraction

The force of contraction in a cardiac myocyte is determined by the number of cross-bridges formed during contraction. The number of cross-bridges that are formed during contraction is determined primarily by two factors: the degree of stretch of the myocyte, and the intracellular calcium concentration.

Resting Muscle Length

As with skeletal muscle, there is a definite relationship between the resting length of a myocyte and the tension it develops. A cardiac myocyte develops maximal force when the cell begins its contraction with a resting length of 2 to 2.4 μm. At this length, there is maximal overlap between the thick and thin filaments so that a maximal number of cross-bridges can be made. For the cell to acquire this optimal resting length, it needs to be stretched. The subsequent stretch of cardiac muscle leads to increased force of contraction. This principle is evidenced in the heart muscle collectively by the Frank-Starling law of the heart—increased preload leads to increased force of contraction (see figure 2.5, p. 20). The increase in the stretch of the heart is provided by the filling of the ventricles, which then stretches the myocytes, thus increasing preload.

Intracellular Calcium Concentration

In cardiac muscle, the number of cross-bridges that are formed, and hence the strength of contraction, is directly proportional to intracelluar calcium concentration. Anything that increases calcium concentrations will increase the number of cross-bridges formed and enhance the force of contraction. Norepinephrine released from sympathetic nerve fibers is one of the major factors that increase calcium concentration under stress and exercise conditions.

Figure 3.6 illustrates the influence of calcium concentration on the tension that a cardiac muscle can generate. As calcium concentration increases, force of contraction increases with a steep increase in tension developed within the physiological range of calcium. For any given concentration of calcium, however, the tension generated by the muscle cell is greater if the sarcomere started at a longer resting length. The increased force generated when the sarcomeres are stretched is the result of optimal overlap of the contractile filaments and increased sensitivity to calcium in the stretched position. Figure 3.6 depicts the effect of muscle length and calcium concentration on force within a cardiac myocyte. Figure 2.6 (on p. 20) depicts the effects of ventricular filling (which stretches the myocardium) and sympathetic nerve stimulation (which increases intracellular calcium) at the level of the heart organ. As would be expected, these figures are very similar.

Several mechanisms can raise cytosolic calcium concentration and thus increase the force of contraction. For example, epinephrine and norepinephrine increase calcium

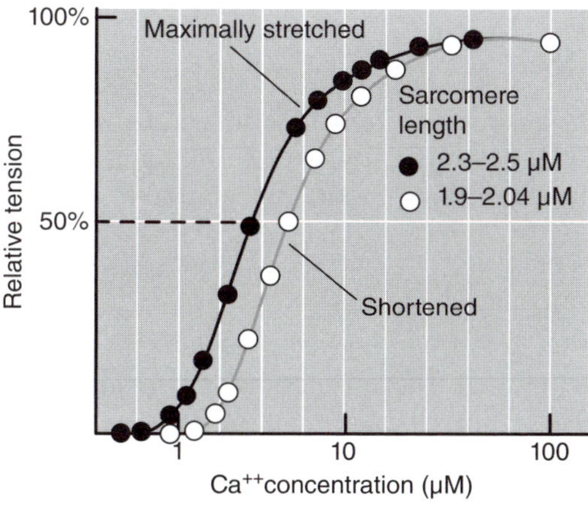

Figure 3.6 Effect of resting muscle length and calcium concentration on muscle tension.

Reprinted, by permission, from M.G. Hibberd and B.R. Jewell, 1982, "Calcium- and length-dependent force production in rat ventricular muscle," *Journal of Physiology* 329: 527-540.

entry into the cell by phosphorylation of calcium channels via a cyclicAMP(cAMP)-dependent protein kinase. Additionally, epinephrine and norepinephrine enhance myocardial contractility by increasing the sensitivity of the contractile elements to calcium. An increase in extracellular concentration of calcium or a decrease in the sodium gradient across the cell membrane can also result in an increase in intracellular concentration of calcium. In contrast, force of contraction is diminished by factors that lead to lower intracellular calcium levels, including a lower extracellular calcium concentration, an increase in the sodium gradient across the sarcolemma, or calcium channel agonists that prevent calcium from entering the cell.

METABOLIC REQUIREMENTS

Myocytes depend on the production of adenosine triphosphate (ATP) to provide the chemical energy for the cross-bridge cycling and other cellular work. The vast majority of this ATP is produced through aerobic metabolism.

ATP Requirements

Myocytes have a very high metabolic rate because cells are always contracting; the myocytes contract one to three times per second throughout life. Adenosine triphosphate provides the energy for muscle contraction. An ATP molecule is used by each myosin head that binds to actin during each cycle of the cross-bridging cycle. As described in the section on cross-bridge cycling, the ATP binds to the myosin head, causes the actin and myosin to detach, and then is hydrolyzed to provide the energy to change the myosin head to its activated configuration. Adenosine triphosphate is also necessary for the maintenance of ATP-dependent pumps that operate during contraction and relaxation.

In order to supply the ATP that is constantly needed to support contraction, the myocytes rely on aerobic ATP production. Hence, myocytes have a high mitochondrial density (accounting for 30% to 35% of cell volume) and rely heavily on an abundant supply of oxygen to support aerobic metabolism (see "Coronary Circulation," chapter 2). The need to produce ATP, and hence oxygen demand, increases dramatically when heart rate and contractility increase.

Energy Sources

Myocytes can utilize several different fuel substrates to produce ATP, including glucose, fatty acids, ketone bodies, pyruvate, and lactate. The primary fuel source at any point is largely determined by the fuel source that is in greatest supply. In a resting state, following an overnight fast, approximately 60% of the energy is produced through the oxidation of fatty acids and 40% from carbohydrate. Following a high-carbohydrate meal, the myocytes rely almost exclusively on carbohydrates as a fuel source. During exercise, the heart uses lactate to produce ATP. The uptake of lactate by the heart increases as a function of arterial lactate levels (Gertz et al., 1988; Massie et al., 1994). Thus, as exercise intensity increases and lactate levels in the blood increase, the heart relies more heavily on lactate oxidation to produce ATP.

SUMMARY

Contraction of cardiac myocytes generates the force necessary to eject blood from the heart. Structurally, there are many similarities between cardiac muscle fibers and skeletal muscle fibers. However, there are also important differences; most notably, cardiac muscle cells are shorter and more branched than skeletal muscle cells, and cardiac muscle cells have specialized gap junctions between adjacent cells that allow electrochemical signals to pass between adjacent cells. Like skeletal muscle, cardiac muscle fibers contract when the thick fibers slide over the thin fibers, causing a shortening of the sarcomere. Excitation–contraction coupling explains the sequence of events by which an action potential in the membrane leads to shortening of the myofilaments. In cardiac muscle fibers, an action potential increases the cell's permeability to calcium. As extracellular calcium enters the cell, some of it binds to calcium release channels on the SR, causing the release of intracellular (within the SR) calcium stores. The high cytosolic calcium binds to troponin, which undergoes a configurational change and removes tropomyosin from its blocking position on actin. This permits the myosin heads to bind to actin, and cross-bridge cycling is initiated. The force of myocyte contraction is dependent on resting length of the cell and intracellular calcium concentration.

CHAPTER 4

Electrical Activity of the Heart

For the heart to contract, an electrical event must occur first and must stimulate the muscle fibers to shorten. These electrical events are very important, as under normal circumstances they make the heart contract in a rhythmical manner. Disturbances in the electrical system of the heart can cause dysrhythmias that may become life threatening. This chapter examines the factors that contribute to the electrical events in the heart, discusses how the electrical signal is transmitted throughout the heart, and explains how the electrical activity of the heart is measured. The chapter also examines how heart rate is controlled and how this is measured.

ION BASIS OF ELECTRICAL ACTIVITY

Many human cells, but especially those of neurons, skeletal muscle, and heart muscle, have an electrical charge compared to their surroundings. This charge is caused by separation of ion pairs, leading to a difference in the ion concentration inside compared to outside the cell. The cell membrane functions as a gatekeeper, allowing some ions to move through it and stopping or greatly reducing the flow of other ions. A cardiac muscle cell has a greater extracellular than intracellular concentration of Na^+, whereas the intracellular K^+ concentration is much greater than the extracellular concentration. Since ions have a natural tendency to diffuse across the membrane to create equal concentrations on the two sides, it is clear that the cell membrane is not freely permeable to Na^+ and K^+.

The separation of ion pairs creates an electrical force, and the balance between the electrical and chemical forces is called the **electrochemical equilibrium potential (E_m)**. This potential can be described by the Nernst equation as

$$E_m = -61.5 \times \log \frac{\text{intracellular } [ion]}{\text{extracellular } [ion]}$$

To calculate the E_m for potassium, simply substitute K^+ in the equation as follows:

$$E_{K+} = -61.5 \times \log \frac{\text{intracellular } [K^+]}{\text{extracellular } [K^+]}$$

Because the concentrations of potassium and sodium ions on the inside and the outside of the cell differ vastly, there is a difference between the electrical charge inside of the cell membrane and that outside of it. This difference in charge is referred to as the *membrane potential.* The resting membrane potential is approximately −90 mV in myocardial cells.

RESTING MEMBRANE POTENTIAL

The rate of ion transfer across the cell membrane is determined by the concentration difference and the permeability of the membrane to each specific ion. The permeability to an ion is in turn determined by the opening and closing of ion gates or channels. These gates are specific to each ion. Under normal circumstances there is always some "leakage" of ions across the cell membrane, since the inherent electrochemical forces work toward establishing an equilibrium. Thus, there is always some K^+ crossing from inside to outside the cell, while the reverse is true for Na^+. This means that the membrane potential would be gradually altered if the ion balance were not restored. In fact, the inward movement of Na^+ would eventually cause membrane depolarization unless these ions were transported back to outside the cell. The ions are moved outside the cell through active transport, a form of a metabolic sodium-potassium pump, involving the membrane enzyme Na^+, K^+-ATPase. As with any pump, energy is required in order for it to work, and the energy is provided by the phosphorylation of adenosine triphosphate (ATP). The active transport of K^+ back into and Na^+ out of the cell maintains the resting membrane potential at close to −90 mV.

ACTION POTENTIAL

Just as with skeletal muscle, for a myocardial muscle fiber to contract, an action potential must be generated. A typical ventricular action potential is depicted in figure 4.1. The action potential looks slightly different depending on whether it is from a conducting cell or a contracting cell (shown in figure 4.3, p. 47), and the following description applies to the contracting cell. There are five distinct phases of the cardiac action potential, labeled in the figure as phases 0 through 4 (phases are labeled 0 through 4 by convention). In general, the myocardial action potential is characterized by a rapid depolarization (0), followed by a slight overshoot (1); a delay or plateau in the membrane potential, referred to as the refractory period (2), which then proceeds to a rapid repolarization (also part of the refractory period) (3); and a return to resting membrane potential (4). The myocardial action potential lasts much longer than that of the skeletal muscle, and the delay before rapid depolarization causes a refractory period during which another depolarization is not possible. Figure 4.1 shows a comparison of a typical skeletal muscle action potential compared to the typical action potential from a ventricular myocyte. The lengthening of the myocyte action potential is caused by Ca^{++} entry into the cell, which does not happen in the skeletal muscle. The specific events causing the myocardial action potential are described next.

- Phase 0—Depolarization: Fast, voltage-gated sodium channels, also called m gates, on the cell membrane open and allow small amounts of Na^+ to enter the cell.

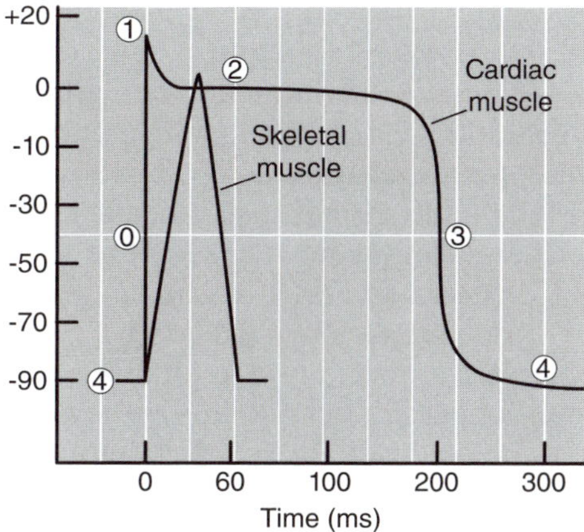

Figure 4.1 Action potential of cardiac and skeletal muscle. The action potential of a myocyte is vastly different from that of a skeletal muscle cell, primarily because of the difference in the refractory period.

This causes a very rapid depolarization. As the membrane potential gets close to 0, the entry of Na^+ into the cell is slowed. This channel remains open for only a short period of time (a few 10,000ths of a second) and then closes. This results in rapid depolarization and actually causes the cell membrane potential to become positive for a short period of time, called the overshoot.

- Late phase 0—The overshoot: As the membrane potential approaches 0, sodium inactivation gates start to close, further decreasing the amount of Na^+ entering the cell. However, the substantial concentration gradient allows some Na^+ to continue to enter the cell, causing the overshoot. As the membrane potential reaches 20 to 30 mV, the inactivation gates close completely and no more Na^+ enters the cell, effectively stopping continuation of the overshoot.

- Phase 1—Early repolarization: The fast sodium channels have closed, and the potassium channels now open, allowing K^+ to leave the cell. The repolarization of the cell membrane starts immediately after the overshoot has stopped, and is caused by an efflux of K^+ from the cell. This causes the initial partial repolarization of the cell membrane. However, soon after the onset of depolarization, the permeability to K^+ also decreases substantially; thus the initial repolarization is fast but very short.

- Phase 2—The plateau (delayed repolarization): The permeability to K^+ is similar to the permeability of Ca^{++}, which is drastically increased by the opening of Ca^{++} channels. This allows Ca^{++} to enter the cell. The increased Ca^{++} influx is offset by the K^+ efflux, which creates the plateau and delays repolarization. The inward flux of Ca^{++} also makes these ions available for excitation–contraction coupling as discussed in chapter 3.

- Phase 3—Rapid repolarization: The Ca^{++} channels close, preventing further entry of Ca^{++} into the cell. At the same time, K^+ channels remain open and K^+ permeability is increased, causing K^+ to leave the cell. This rapidly restores the cell membrane potential to resting levels.

- Phase 4—Resting membrane potential: The resting membrane potential is maintained by the "pump" through the action of Na^+, K^+-ATPase as already discussed. Ca^{++} is removed from the cell by a Na^+–Ca^{++} exchanger and by an active transport Ca^{++} pump.

Repolarization

As shown in figure 4.1, the duration of the action potential of a cardiac myocyte is as much as 200 times longer than that of the action potential in skeletal muscle. The major difference is the plateau created by the Ca^{++} influx. Another important difference between cardiac muscle and skeletal muscle cells is that the delay in repolarization creates a refractory period in cardiac muscle cells. The refractory period prevents the possibility of another depolarization. The refractory period has two components: an absolute or effective refractory period and a relative refractory period. It is possible for another depolarization to occur as a result of a stronger than normal signal during the relative refractory period, but not during the absolute refractory period. Furthermore, a depolarization during the relative refractory period has the possibility of disturbing the normal pattern of rhythmic depolarizations in the heart, and thus may cause significant dysrhythmias. Therefore, the relative refractory period has also been referred to as the vulnerable period. For example, as demonstrated in figure 4.2a, skeletal muscle has the capability to produce what is called treppe or summation of contraction until tetanus. This is not possible in the heart, since the refractory period

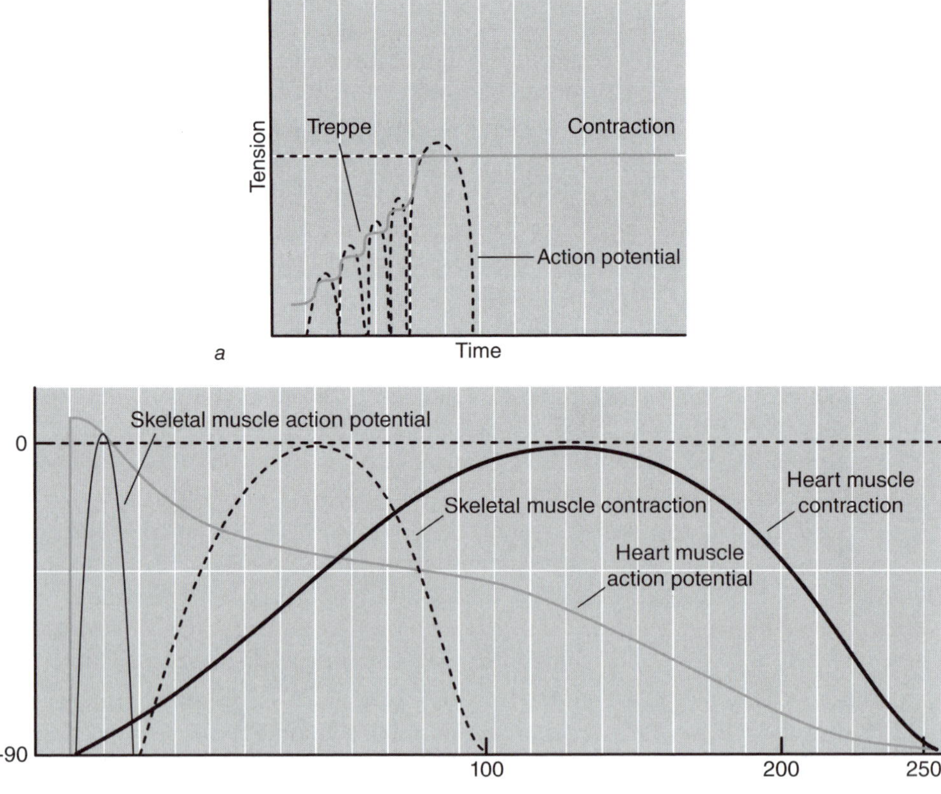

Figure 4.2 The relationship between the action potential and muscle cell contraction. The action potential occurs prior to muscle fiber contraction. Whereas the action potential in skeletal muscle is complete before the initiation of contraction, the myocyte cell contraction occurs primarily during the refractory period of the action potential. Thus summation of impulses can occur in skeletal muscle (treppe), but not in cardiac muscle.

prevents summation of contraction and tetanus. The myocardial refractory period is necessary to allow the ventricles to fill with blood before the next contraction. It is also easy to understand why a sustained isometric contraction (tetanus) of the heart would not be beneficial, since it would prevent the heart from pumping blood. Thus, the specialized action potential of cardiac myocytes is a sophisticated physiologic design to facilitate cardiac output.

The action potential precedes the force development of the myocardial muscle cell, which helps to facilitate the coordination of contraction with ventricular filling. This is also shown in figure 4.2. In skeletal muscle, the action potential ends as the muscle starts to contract and produce force. Thus a second action potential can now cause summation or treppe. The situation is different in cardiac muscle, as force development reaches its peak during the absolute refractory period; thus an additional action potential cannot be produced to cause summation of contraction force in cardiac muscle fibers. The cardiac muscle force returns to baseline almost simultaneously with the action potential. Thus, the cardiac muscle cell is ready for another contraction after it has relaxed and force development has ceased. This particular characteristic is important, as fast muscle relaxation characteristics allow for better filling of the ventricles.

Action Potential Variances Within the Heart

The action potential differs between different sites in the myocardium. For instance, the ventricular action potential has a shorter refractory period and a greater upstroke during phase 0 compared to the action potential of cells in the conduction system such as the sinoatrial node. This is shown in figure 4.3. In addition, some specialized myocardial cells can generate their own action potential without any external input, such as a signal from the nervous system. In fact, the heart will generate its own electrical activity and beat even if it is not connected to the nervous system or if it

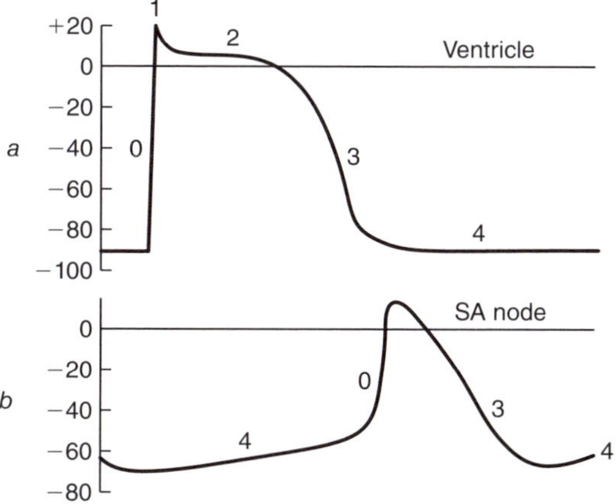

Figure 4.3 The action potential of a *(a)* myocyte compared to the *(b)* sinoatrial (SA) node. The action potential in the SA node has a "wandering" baseline that eventually crosses the threshold of contraction, and a very short refractory period compared to the actin potential in a ventricular myocyte.

Reprinted, by permission, from P. Brubaker, L. Kaminsky, and M. Whaley, 2010, *Coronary artery disease: Essentials of prevention and rehabilitation programs* (Champaign, IL: Human Kinetics), 115.

is removed from the body (as long as blood and oxygen supply is adequate). This is termed the autorhythmicity of the heart.

CONDUCTION SYSTEM OF THE HEART

To facilitate organized conduction, the heart has a specialized conduction system. This system consists of specialized cardiac muscle cells, which function almost like a nervous system within the heart but are not neural tissue. The heart's conduction system ensures that conduction of rhythmical impulses is rapid throughout the heart, and that the conduction is organized to facilitate the heart's pumping action. For example, the atria contract before the ventricles, allowing blood to flow from the atria and fill the ventricles before ventricular contraction. Furthermore, the apex of the heart contracts first, allowing blood to be pushed toward the aorta. This coordinated contractile action is essential for organized pumping activity and maintenance of cardiac output.

The conduction system of the heart is depicted later in figure 4.5 (p. 50). Under normal circumstances, the electrical impulse originates in the sinoatrial (SA) node, located in the right atrium. The SA node fibers are autorhythmic and can thus generate their own action potential, and the SA node is usually known as the pacemaker of the heart. Since the SA node is directly connected to atrial muscle fibers, the impulse is spread directly and immediately throughout the atria. The depolarization wave is further facilitated by the internodal pathways, which carry the impulse throughout the atria to the atrioventricular (AV) node.

The AV node is located in the right atrium, close to the junction with the right ventricle. The major function of the AV node is to delay the signal to prevent the impulse from traveling to the ventricles too rapidly. This delay allows the atria to contract and allows time for the ventricles to fill before the beginning of ventricular contraction. The delay is approximately 0.06 to 0.10 s. The cause of the delay is probably related to a combination of very small fibers in the AV node and a lower number of gap junctions, creating resistance to conduction. The conduction from the atria to the ventricles can occur only through the AV node, because there is a fibrous barrier between the atria and the ventricles that prevents transmission of an electrical impulse through any means other than the AV node. The AV node is also designed for one-way conduction (conduction from the atria to the ventricles only), which prevents reentry of impulses from the ventricles back to the atria.

The impulse travels from the AV node to the AV bundle (also called the common bundle or the bundle of His). From the AV bundle, the impulse is transmitted along the interventricular septum through the left and right bundle branches. The left and right bundles carry the impulse to the apex of the heart, at which point they split into smaller branches in a progressive fashion. These smaller branches are called Purkinje fibers. The Purkinje fibers turn back around and curl back toward the base of the heart. The transmission through the Purkinje fibers is very rapid, reaching speeds of up to 4 m/s, which is about 150 times greater than the speed of transmission through the AV node. The impulse spends only 0.03 s being transmitted from the bundle branches to the end of the Purkinje fibers. The Purkinje fibers also have an action potential very similar to that of other ventricular fibers, with a prolonged refractory period. This serves to protect the ventricles from premature contractions, as many premature or early depolarizations from the atria are blocked by the Purkinje fibers.

The Purkinje fibers do not innervate each cardiac muscle cell; in fact, most ventricular cells do not come in contact with these fibers. Instead the action potential is spread from cell to cell by the ventricular muscle fibers themselves. This is very different from what occurs in skeletal muscle, in which each muscle fiber must be innervated by a neuron. The impulse is transmitted from cell to cell via the gap junctions at the intercalated discs (see chapter 3). The gap junctions between cardiac cells are normally open, allowing ion transfer from cell to cell resulting in depolarization. However, the impulse has a considerably lower velocity than in the Purkinje fibers, traveling at speeds up to 0.5 m/s—about six times slower than observed in the Purkinje fibers.

The nature of impulse transmission and cardiac muscle depolarization results in a coordinated wave of depolarization, which in turn produces a coordinated sequence of atrial and ventricular contractions. The coordination of contraction is further helped by the arrangement of the ventricular muscle fibers; they are arranged in a double spiral around the heart, separated by fibrous tissue. Because of this arrangement, the apex and the base of the heart are pulled closer together during a contraction, almost in a screwing or wringing motion, facilitating ejection of blood into the aorta.

AUTORHYTHMICITY OF CONDUCTION CELLS

The action potential of a conducting myocardial cell is decidedly different from that of a ventricular myocyte. Whereas an electrical event in a ventricular myocyte results in contraction of the muscle fiber, an action potential in a conduction cell results in the spread of the electrical signal throughout the heart so that the contraction of myocytes is coordinated. As shown in figure 4.4, the resting membrane potential of pacemaker cells is only −60 mV, and it is not stable. Instead, there is a gradual decrease in voltage from −60 mV to −40 mV, at which point the cell depolarizes. The point of depolarization, −40 mV, is referred to as the threshold. However, the depolarization is not due to opening of fast Na^+ channels, but instead to an opening of slower Ca^{++} channels. Thus, the upstroke of the depolarization is slower. As depicted in figure 4.4, the slow change in membrane potential from −60 to −40 mV is primarily caused by an opening of channels permeable to both Na^+ and K^+, called I_f channels. These channels are more permeable to Na^+ than to K^+, resulting in a slow depolarization. Close to the threshold point of −40 mV, the I_f channels gradually close, and instead, some of

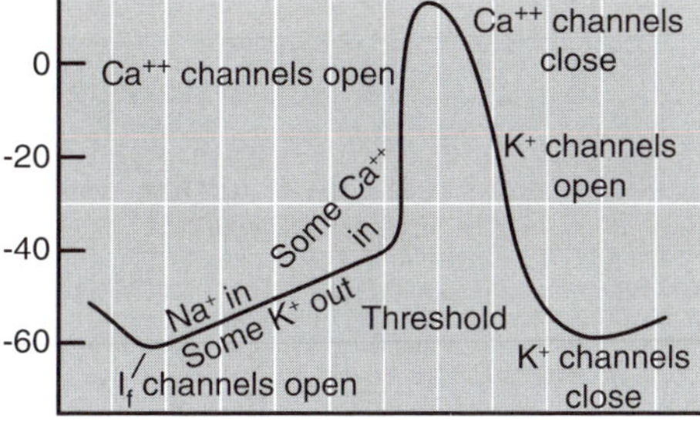

Figure 4.4 The action potential of a conducting cell. The autorhythmicity of pacemaker cells is due to a continual "leakage" of sodium ions into the cell, with some leakage of potassium out of the cell. This causes a gradual increase in baseline membrane potential, until the threshold is reached and depolarization occurs. The depolarization is caused by an influx of calcium ions, not sodium ions as in the ventricular myocyte.

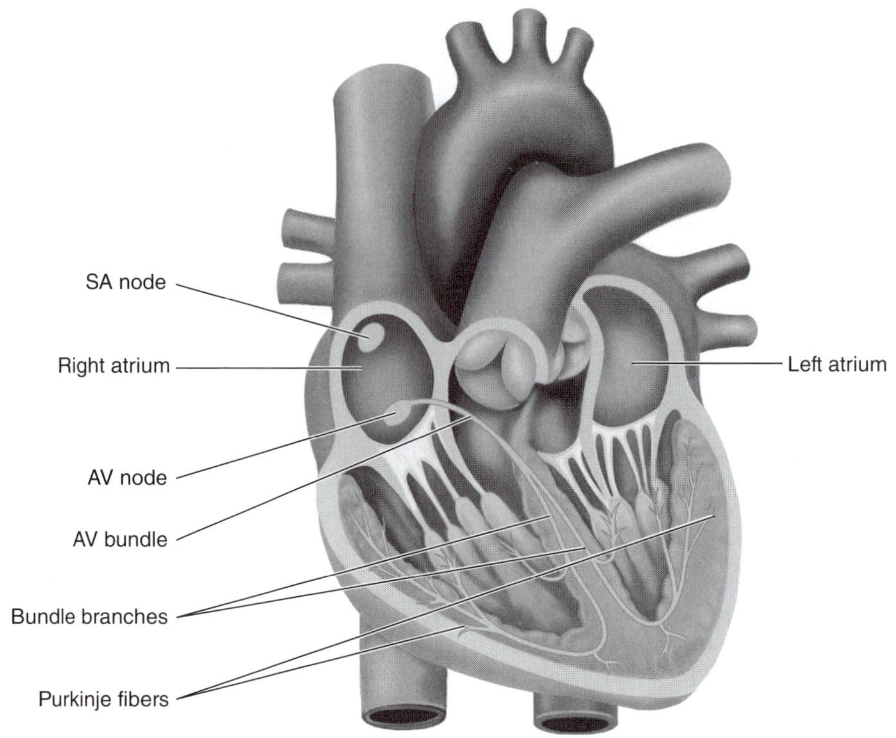

Figure 4.5 The conduction system of the heart. The impulse is generated in the sinoatrial node and is then transmitted throughout the atria and collected in the atrioventricular node; it then continues through the bundles to the Purkinje fibers and the individual muscle cells.

Reprinted, by permission, from J.W. Wilmore, D.L. Costill, and W.L. Kenney, 2008, *Physiology of Exercise and Sport,* 4th ed. (Champaign, IL: Human Kinetics), 128.

the Ca^{++} channels open. At threshold, the Ca^{++} channels are wide open, allowing Ca^{++} entry into the cell, creating the steep depolarization phase of the action potential.

At the peak of depolarization, the Ca^{++} channels close and K^+ channels open, resulting in rapid repolarization. Unlike what occurs with the ventricular myocyte action potential, there is no plateau, since the Ca^{++} channels have closed and the K^+ channels are now wide open. When the membrane potential reaches −60 mV, the K^+ channels close and the I_f channels open again, starting the process all over. This automated process is continual, throughout a person's life. Because of the inherent autorhythmicity of the heart, the heart can function without any external neural input. As long as the heart receives blood and nutrients, it can continue to beat. Thus a heart can be removed from the body and continue to beat for some time, as long as blood flow is maintained. A heart transplant is a case in which the heart continues to beat within the body but without any neural input from the central nervous system.

PACEMAKERS OF THE HEART

Under normal conditions, the SA node functions as the pacemaker of the heart. The SA node usually sets a rate of 60 to 100 depolarizations per minute, or a heart rate of

60 to 100 beats/min if each depolarization is conducted through the heart and results in a contraction. However, other areas of the heart also have autorhythmic tissue, including the AV node and the Purkinje fibers. The AV node will set a rate between 40 and 60 beats/min, while the Purkinje fibers will set a rate of 15 to 40 beats/min. The AV node and Purkinje fibers can be thought of as backup systems in case the SA node fails. They do not interfere with the SA node because the rate of the SA node discharge is much higher, causing depolarization of both the AV node and the Purkinje fibers before their own autorhythmicity can take over. However, if the SA node fails, then the AV node will have time to reach its self-excitatory threshold, creating an action potential that will then be conducted throughout the conduction system. However, atrial contraction may not be well coordinated with ventricular contraction since the impulse now travels in the "wrong" direction. If both the SA node and the AV node fail, the Purkinje fibers will have time to reach their own self-excitatory threshold, causing a discharge throughout the ventricles. Obviously, if both the SA and AV nodes fail, this is a serious medical condition that can be fatal. Even if life is sustained, at a heart rate of 15 to 40 beats/min, cardiac output is at best substantially reduced and a pacemaker usually needs to be implanted.

It is possible for the heart to develop what is called an ectopic pacemaker. This happens when some part of the heart tissue develops a discharge that is more rapid than that of the SA node. The AV node and the Purkinje fibers can become ectopic pacemakers, causing an abnormal sequence of contraction. It is also possible for the heart muscle fibers to develop into ectopic pacemakers. Ectopic pacemakers can deleteriously affect both heart rhythm, resulting in a dysrhythmia, and the pumping action of the heart, decreasing the cardiac output.

CONTROL OF HEART RATE

The heart is richly innervated by the autonomic nervous system. The parasympathetic system innervates primarily the SA and AV nodes through the vagus nerves. There is some vagus innervation to the atria but very little directly to ventricular myocytes. The sympathetic system is distributed to most parts of the heart, including the SA and AV nodes, but with a rich innervation to the ventricular muscle. Under normal circumstances in healthy individuals, the SA node is under tonic influence of both the parasympathetic and sympathetic systems; however, the parasympathetic system predominates at rest. This has been shown through the creation of a vagal block (blocking of vagal activity using the drug atropine), which increases heart rate by approximately 30 to 40 beats/min. Conversely, blocking the sympathetic system (using a beta blocker such as propranolol) decreases resting heart rate by only 10 beats/min. When both the sympathetic and parasympathetic system are blocked, heart rate is typically between 100 and 120 beats/min, depending on the age and health of the person. This heart rate is called intrinsic heart rate, or the rate at which the heart beats if not innervated by the autonomic nervous system. Intrinsic heart rate is seen in heart transplants, as the nerves are cut during the transplantation; these patients usually have resting heart rates above 100 beats/min.

When the vagus nerve fires, acetylcholine is released from the nerve endings. This causes the membrane potential of autorhythmic cardiac cells (such as the SA node) to change. The resting membrane potential is decreased to around −70 mV because of an increased permeability to K^+, creating a state of hyperpolarization. This slows

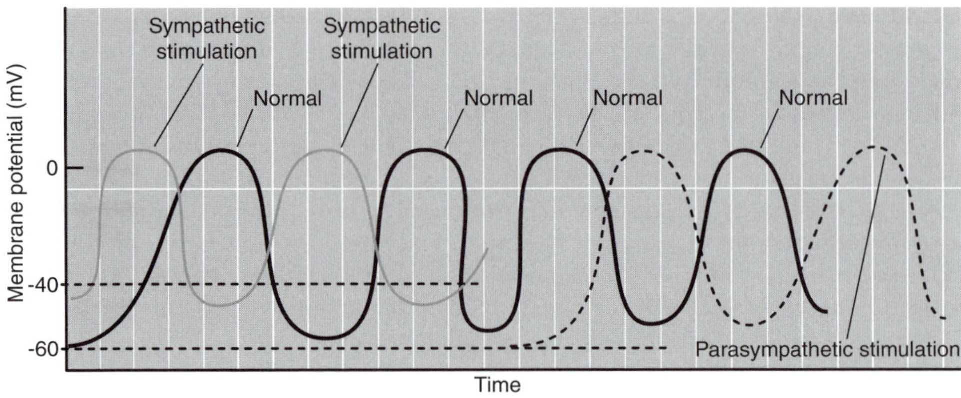

Figure 4.6 The effect of neural stimulation on membrane potential. Sympathetic stimulation hypopolarizes the resting membrane potential of the sinoatrial node; thus, the time needed between depolarizations decreases, producing an increase in heart rate. Parasympathetic stimulation hyperpolarizes the membrane, having the opposite effect.

the rate of depolarization due to the leakage of Na^+ and thus prolongs the time to reach the threshold, resulting in a decreased heart rate (figure 4.6). Furthermore, the conduction speed to the ventricles is decreased by a decrease in the excitability of the AV node fibers.

Conversely, sympathetic stimulation results in the release of norepinephrine from the nerve endings. This causes increased permeability to Na^+ and Ca^{++}, creating a more positive resting membrane potential; and the depolarization threshold is reached sooner (figure 4.6), increasing the heart rate. Sympathetic stimulation also increases the rate of conduction throughout the heart and increases the force of contraction. Although increases or decreases in heart rate can be caused by reciprocal changes in sympathetic and parasympathetic stimulation (e.g., an increase in heart rate can be caused by a decrease in vagal firing and an increase in sympathetic stimulation), either of the autonomic divisions can also act alone without reciprocal action by the other.

Another important aspect of autonomic control of heart rate is the differential speed at which sympathetic and parasympathetic stimulation affect heart rate. When

BETA BLOCKERS

Beta blockers are medications very commonly used to treat hypertension. They block the beta-adrenergic receptors, thus reducing the sympathetic influence on both heart rate and the strength of ventricular contraction. At rest, beta blockers reduce heart rate by only about 10 beats/min because the heart is primarily under vagal control at rest. Thus, reducing the sympathetic influence has a relatively small effect. However, during exercise, when the sympathetic influence becomes more important, beta blockers have a greater effect on heart rate. In fact, it is not unusual for maximal heart rate to be reduced by 60 to 80 beats/min during beta blockade. This also reduces maximal cardiac output and may contribute to reduced maximal work capacity.

the vagus nerve is stimulated, the heart rate response is immediate; that is, heart rate decreases instantaneously. When vagal stimulation ceases, the heart rate also returns to initial level instantaneously. This immediate effect is probably due to a direct coupling of muscarinic receptors with the K^+ channels, allowing for immediate changes in K^+ efflux. Conversely, sympathetic stimulation causes a more gradual increase in rate and a more gradual return to initial levels as illustrated in figure 4.7. This is probably related to the greater amount of norepinephrine required to cause a change in heart rate, compared to acetylcholine. Also, norepinephrine is not directly coupled with any of the ion channels involved in the action potential, but operates through a second messenger system, which requires more time. Thus, when a rapid increase in heart rate is required, for instance at the initiation of exercise, this is initially caused by vagal withdrawal, followed by sympathetic stimulation. At the cessation of exercise, the speed at which heart rate returns to baseline depends on the amount of vagal reactivation and the amount of residual sympathetic stimulation, including the amount of available norepinephrine. Therefore, heart rate recovery following a substantial sympathetic challenge, such as exercise, usually takes longer than the initial increase in heart rate.

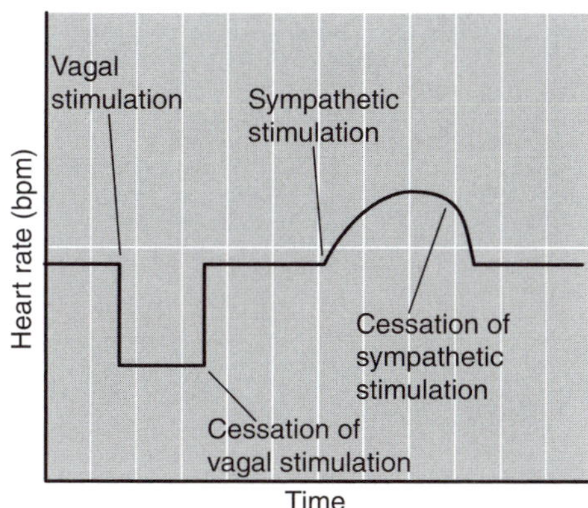

Figure 4.7 The effect of neural stimulation on heart rate. Vagal stimulation creates an immediate heart rate response, decreasing heart rate during stimulation. When stimulation ceases, the heart rate returns to resting levels immediately. Sympathetic stimulation increases heart rate in a more gradual fashion, and heart rate also decreases gradually after cessation of sympathetic stimulation.

BRAIN- AND RECEPTOR-MEDIATED HEART RATE CONTROL MECHANISMS

Various regions in the brain can increase or decrease heart rate when stimulated. Specific cortical centers are involved in causing increases in heart rate in response to both emotional and physical stress. The hypothalamus is involved in altering heart rate in response to environmental stressors such as temperature. Central command has been identified as a contributor to the initial increase in heart rate with some types of exercise, but the contribution is relatively minor.

The baroreceptors, located in the carotid sinuses and in the aortic arch, affect heart rate in response to changes in blood pressure. Although the main function of the baroreceptors is to control blood pressure, they achieve this partly by changes in heart rate. An increase in blood pressure causes a reflex decrease in heart rate, whereas a decrease in blood pressure causes an increase in heart rate. The effect is not equal for all pressure changes, as the relationship between blood pressure and heart rate is sphygmoidal in nature (figure 4.8). The baroreceptors are most responsive to changes in the steep portion of the curve; thus a smaller change in blood pressure will cause a larger change in heart rate in the responsive range (the steep portion of the curve) of the baroreceptor reflex. In the response range, reciprocal changes in vagal and sympathetic activity are responsible for the changes in heart rate. Thus, the baroreceptors cause changes in heart rate by altering parasympathetic and sympathetic stimulation to the SA node.

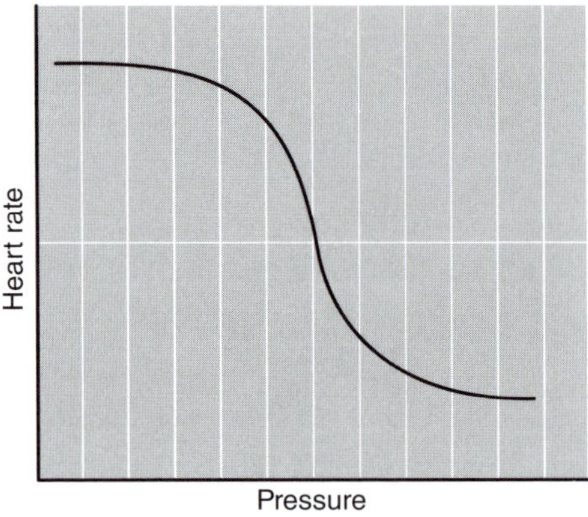

Figure 4.8 The baroreceptor sensitivity showing the heart response to a change in pressure. The ability of a change in pressure to alter heart rate has an effective range. At the lower and higher ends of the curve, a change in pressure has a very small effect on heart rate.

There are cardiopulmonary receptors in the venous side of circulation and in both atria that respond to atrial stretch. Most of these receptors are in the right atrium, close to the vena cava. An increase in blood flow (such as an increase in venous return) stretches these receptors and causes an increase in heart rate. This is called the Bainbridge reflex. However, changes in blood volume can cause confusing changes in heart rate as both the Bainbridge and baroreceptor reflexes are activated, with opposing effects. It appears that the increase in heart rate following loss of blood volume is caused by vagal withdrawal, and not by changes in sympathetic stimulation.

Respiration also influences heart rate. In most individuals, heart rate varies rhythmically with respiratory rate. Normally, heart rate increases during inspiration and deceases during expiration. Sympathetic nerves are activated during inspiration, whereas vagal activity increases during expiration. However, the primary influence of respiration is through vagal activity. The influence of respiration on vagal activity is probably caused by differential rates of removal of acetylcholine versus norepinephrine. Acetylcholine (released from vagal nerve endings) is removed very rapidly. In contrast, removal of norepinephrine (released by sympathetic nerve endings) takes longer, so norepinephrine has a more constant effect (tonic effect). Consequently, the respiratory variation in heart rate is primarily a function of vagal activity. This is evidenced in individuals with enhanced vagal tone, who have large respiratory variations in heart rate—a common finding in well-trained endurance athletes.

Peripheral and arterial chemoreceptors can also affect heart rate, but this influence is relatively minor under normal circumstances. The chemoreceptors have greater influences on respiration, as ventilation is substantially increased (in both depth and rate) when the arterial chemoreceptors are stimulated, but heart rate changes very little. The primary effect of stimulation of the chemoreceptor is a decrease in heart rate, as a result of excitation of the medullary vagal center in the brain. However, this can be confusing, since an increase in respiration has an inhibitory effect on the medullary vagal center, causing a increase in heart rate. In normal healthy individuals, chemoreceptors have minimal influence on heart rate.

HEART RATE VARIABILITY

Heart rate variability (HRV) refers to the beat-to-beat variability of the R-R interval. Although we typically measure heart rate as a single number, this is actually an average of several beats. In fact, the beat-to-beat variability can be substantial, as shown in figure 4.9. Heart rate variability has clinical utility, as a low HRV is associated with mortality following a myocardial infarction (Task Force, 1996); but it can also be used to provide physiologic insight regarding cardiovascular control (Akselrod et al., 1981).

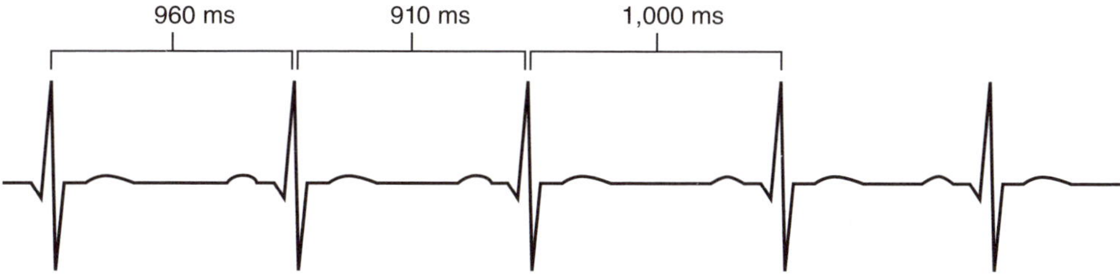

Figure 4.9 Variability of beat-to-beat heart rate. The period between heartbeats varies from beat to beat; thus heart rate also varies beat to beat.

Heart rate variability is measured via either short-term or long-term recordings of the electrocardiogram (ECG). Long-term recordings are usually obtained over a 24 h period using Holter monitoring. Although short-term recordings vary in length, 5 min recordings have been suggested as a standard measurement length (Task Force, 1996). The ECG recordings are converted to tachograms by plotting the R-R interval over the number of beats, as shown in figure 4.10*a*. The data are then analyzed using a time domain analysis or frequency analyses. Common time domain variables include the standard deviation of normal R-R intervals (SDNN) and the square root of the mean squared differences of successive normal R-R intervals (RMSSD). SDNN is a measure of total variability, and RMSSD is a measure of high-frequency variations, or vagal modulation.

Frequency analyses are done through evaluation of the spectral power of the R-R variability in relation to the frequency. Very-low-frequency (VLF) oscillations are difficult to interpret and have little physiological meaning when obtained from short-term recordings. High-frequency (HF) oscillations are typically centered around 0.2

Hz in a 0.15 to 0.4 Hz frequency band. High-frequency oscillations are mediated by the parasympathetic influence on the SA node (Task Force, 1996; Jalife and Michaels, 1994). Since vagus nerve activity has an immediate effect on the R-R interval (increases R-R interval) and cessation of vagus nerve activity has a similar immediate effect (decrease in the R-R interval), vagal activity is characterized by HF power. Thus, the greater the HF power, the greater the influence of vagal activity on the SA node.

Low-frequency (LF) oscillations are considered a marker of either sympathetic modulations on the SA node (Malliani, Lombardi, and Pagani, 1991) or a combination of vagal and sympathetic influences (Akselrod et al., 1981). Since sympathetic stimulation of the SA node results in a gradual decrease in the R-R interval while cessation of sympathetic stimulation results in a similar gradual increase in the R-R interval, this results in LF oscillations and an increase in LF power. However, this may apply only to normalized LF power (LF power normalized to the total HRV power, minus the VLF power). It is possible that normalized LF and HF power provide reasonable estimates of vagal and sympathetic modulation, respectively, but the interpretation of LF power is still controversial. Instead, the LF/HF ratio is commonly evaluated as a measure of the sympathovagal balance, with an increase in the ratio signifying a more sympathetic influence (Task Force, 1996). However, the relationship between the parasympathetic and sympathetic systems is not always reciprocal. Therefore, it may be more appropriate to think of the LF/HF ratio as an index of sympathovagal dominance. An example of a frequency analysis with accompanying power spectra is shown in figure 4.10*b*.

Heart rate variability analysis provides insight regarding autonomic control of the heart. It is common to evaluate the HRV response to a stressor, such as exercise, to evaluate cardiovascular autonomic control. For instance, the HRV response to isometric handgrip exercise is commonly evaluated, and a typical response is shown in figure 4.11. The HF power is reduced during handgrip, indicating vagal withdrawal, but returns during recovery (vagal reactivation). In contrast, the LF/HF ratio is increased during the handgrip exercise, indicating greater sympathetic dominance, but then returns to baseline during recovery (sympathetic withdrawal). Contrast this

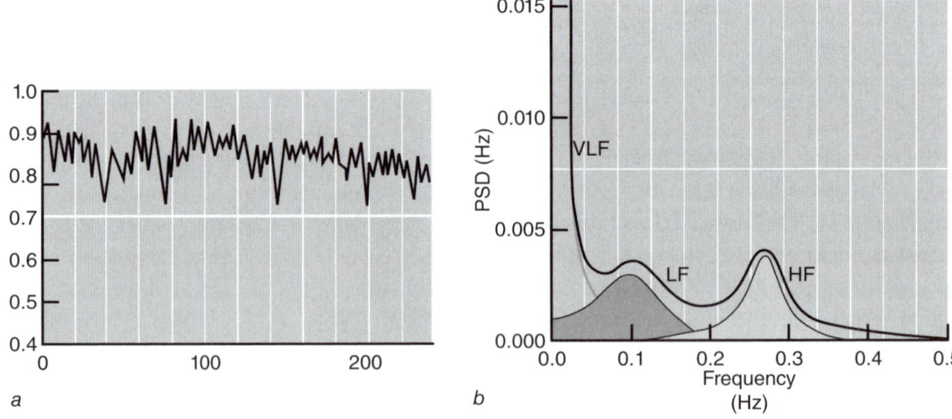

Figure 4.10 A tachogram of heart rate variability and the subsequent spectral analysis of the tachogram. *(a)* A tachogram is created by plotting the interbeat time interval against the number of beats. *(b)* The tachogram is then analyzed using spectral analysis, creating the spectral power curves. The area under the curve is calculated in each spectral band.

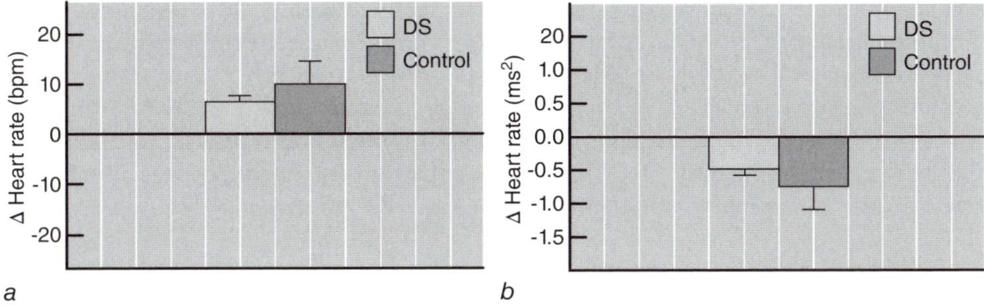

Figure 4.11 Heart rate and heart rate variability response to an isometric handgrip task. High-frequency spectral power (parasympathetic modulation) decreases during isometric handgrip exercise (vagal withdrawal), leading to an increase in heart rate. A healthy control subject (shaded bars) exhibits larger changes in vagal withdrawal compared to a person with autonomic dysfunction (open bars).

to a person with autonomic dysregulation, also shown in figure 4.11. This person does not decrease HF during the handgrip exercise and LF/HF does not increase much, and as a result, heart rate does not increase appropriately during the exercise.

Heart rate variability can also be evaluated in response to standard endurance exercise as long as the recordings are made during steady state. Figure 4.12 shows

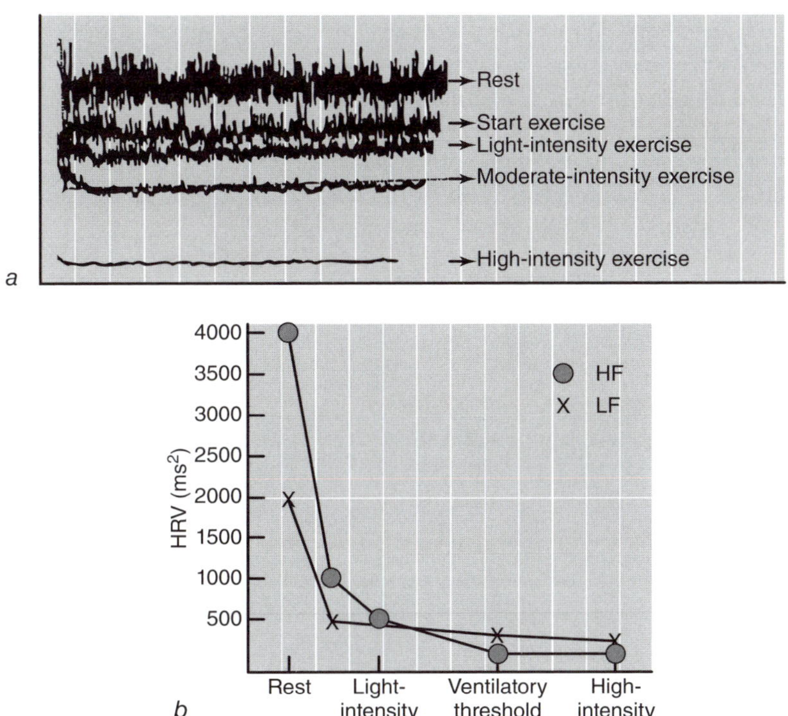

Figure 4.12 Heart rate variability response from rest to high-intensity cycle exercise. The tachogram *(a)* illustrates that heart rate variability decreases from rest to exercise and continues to decrease as exercise intensity increases. Spectral analysis of the tachogram *(b)* illustrates that both HF and LF power decrease.

a typical response to increasing intensities of cycling exercise. It is clear that vagal withdrawal occurs immediately with the start of exercise, and then continues until the ventilatory threshold. At this point, HF modulation is virtually nonexistent (indicting very little vagal influence at this exercise intensity), but the LF/HF ratio still increases, denoting further sympathetic dominance. This is also accompanied by a drastic reduction in total HRV power, showing that the interpretation of exercise HRV responses is highly dependent on the total HRV power response.

SUMMARY

The myocardial action potential is substantially different from the action potential in skeletal muscle, primarily through the prolonged repolarization causing the refractory period in cardiac cells. The action potential also varies between myocardial cells, depending on location and purpose of the cell. The main difference is in the conducting cells, which possess a great deal of autorhythmicity, allowing the heart to contract without neural innervation. Furthermore, depolarization of conducting cells is caused by an influx of calcium ions, not an influx of sodium ions as in the nonconducting cells. The signal is conducted throughout the heart by specialized conducting cells, starting in the SA node. The signal is conducted throughout the atria, and then collected and delayed in the AV node, before continuing through the main bundle and the left and right bundle branches. The conducting cells do not innervate each individual myocyte; instead the signal is spread from cell to cell through gap junctions and intercalated discs. The heart is normally primarily under vagal control at rest, but sympathetic fibers also innervate both the SA node and the ventricles. During exercise, the initial increase in heart rate is caused by vagal withdrawal followed by more substantial contributions from sympathetic stimulation. There are several potential control mechanisms of heart rate, including central command, arterial baroreceptors, cardiopulmonary receptors, peripheral and arterial chemoreceptors, and ventilation. The vagal modulation of the heart can be estimated noninvasively using the HF portion of the frequency domain of HRV.

CHAPTER 5

The Electrocardiogram

The electrical activity of the heart, produced by the ion changes leading to action potentials (discussed in chapter 4), can be recorded via the placement of electrodes on the surface of the skin. This recording is called an electrocardiogram (ECG), and the ECG provides a picture of the electrical activity of the heart. The first ECG was recorded in the late 1800s, in 1887 to be precise, and has since become a very valuable clinical tool used to understand and diagnose various aspects of abnormal heart function. Willem Einthoven, a Dutch physiologist, is considered the father of ECG. Although the technology used to collect ECG recordings has obviously changed in the last 120 years, the basics of the ECG and its interpretation have changed very little during this time.

THE ECG TRACING

The ECG tracing is an average of all the action potentials in the heart; thus it looks very different from any individual action potential. However, the various parts of the ECG provide valuable information about depolarization and repolarization of the myocardium. These electrical events precede and provide the stimulus for the mechanical contraction and relaxation of the heart. The names of the various waves, complexes, segments, and intervals of one complete cardiac cycle are shown in figure 5.1. It is important to remember that the ECG has a size (vertical deflections) and a time (horizontal component) component. In general, a greater deflection from baseline (the height of the wave) denotes a stronger electrical signal; and the horizontal (the width) portion of a wave, segment, or interval depicts the time of that portion of the ECG (Stein, 1992b). The height of the waves (voltage) is measured in millivolts (mV), and 10 mm equals 1 mV. In fact, most ECG machines produce a calibration box, which is usually 1 mV (10 mm) high. Most of the time, the chart speed of the ECG is standardized to 25 mm/s. Thus, each 1 mm square represents a 0.04 s time interval, and each large square (5 mm) represents a 0.2 s time interval. Thus, if a wave is 3 mm wide, it lasts for 0.12 s. The meaning of each of the ECG components is explained next.

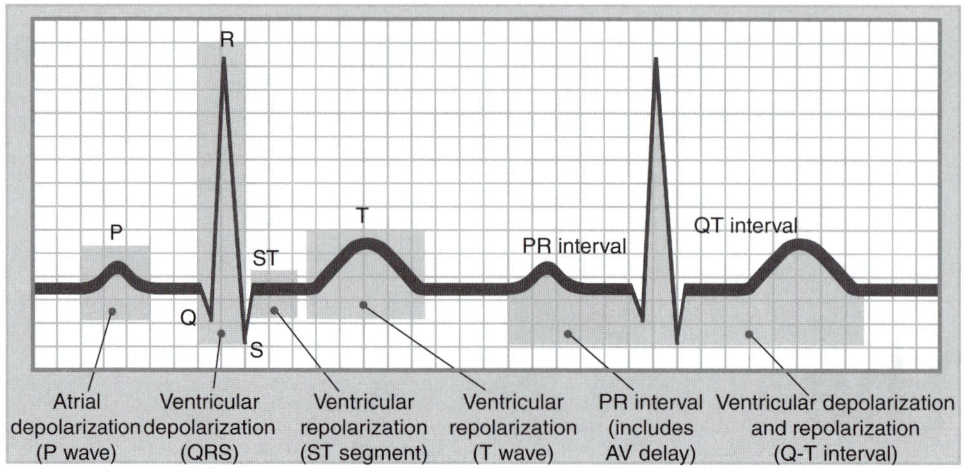

Figure 5.1 The components of the ECG complex.

Reprinted, by permission, from J.W. Wilmore, D.L. Costill, and W.L. Kenney, 2008, *Physiology of Exercise and Sport,* 4th ed. (Champaign, IL: Human Kinetics), 131.

Waves

The ECG tracing is an average of all the action potentials in the heart, and as such is an electrical representation of the myocardial contraction and relaxation. The various waves, intervals, and segments denote specific events during the cardiac cycle, helping us understand how the heart works from an electrical perspective. Waves can represent both depolarization (contraction) and repolarization (relaxation) of the heart.

- P wave. The first wave in the cardiac cycle, the P wave depicts the depolarization of the atria. The direction (up or down), size of the deflection (up or down), configuration, and length (duration) of the P wave are evaluated and provide information on atrial function. P waves are usually less than 2.5 mm high, have a rounded contour, and should not be peaked or notched. A normal P wave duration is between 0.06 and 0.11 s.
- The Q, R, and S waves. The Q wave is the first negative (below baseline) deflection (in most leads) after the P wave. The R wave is the first positive (above baseline) deflection after the P wave, and the S wave is the second negative deflection below baseline after the P wave. Together these waves make up the QRS complex and depict the depolarization of the ventricles. As with the P wave, the size, configuration, and duration of the waves and the complex as a whole are evaluated. The Q wave should be less than 25% in height (below baseline) of the R wave and less than 0.04 s in duration. The R wave and S wave should be sharply peaked and have no notches. The sum of the S wave in lead V1 or V2 (see p. 61 for an explanation of leads) and the R wave in lead V5 or V6 should be less than 35 mm. Overall, the duration of the QRS complex should be less than 0.12 s (3 mm) (if it is too long, this indicates a block in the impulse transmission). It is common for the QRS complex not to have all three waves. The Q wave is often very small or not present; and it is also common for the S wave to be absent, sometimes in conjunction with an absent Q wave. Nevertheless, the complex is referred to as the QRS complex even if only an R wave is present.

• T wave. The T wave depicts ventricular repolarization. It is the upward deflection following the QRS. It can vary in length and size but is usually rounded and modest in height. The T wave should generally go in the same direction as the R wave. Deviations in the T wave may be indicative of myocardial ischemia, ventricular hypertrophy, or electrolyte disturbances.

• There is no recorded wave for atrial repolarization. The atrial repolarization wave is hidden in the QRS complex.

Segments and Intervals

Segments and intervals are isoelectric (in the normal heart) portions of the ECG. They denote either a delay (atrioventricular [AV] node) of the impulse, or repolarization.

• The P-R segment—from the end of the P wave to the beginning of the Q wave (or R wave if Q is absent). This segment represents the delay in the impulse in the AV node, before it continues down the bundles to depolarize the ventricles.

• The P-R interval—from the beginning of the P wave to the beginning of the Q wave. This represents the atrial depolarization and the delay in the AV node. The P-R interval is normally between 0.12 and 0.2 s. A shorter interval may denote a preexcitation syndrome, and a longer interval may denote a block in the AV node.

• The Q-T interval—from the beginning of the Q wave to the end of the T wave. This interval denotes the depolarization and repolarization of the ventricles. The Q-T interval will vary with heart rate, as a higher heart rate decreases the Q-T interval. Both short and long Q-T interval syndromes tend be associated with genetic variations that are linked to higher risk of sudden death, presumably from cardiac dysrhythmias.

• The ST segment—from the end of the QRS complex to the beginning of the T wave. This segment denotes the early repolarization of the ventricles. The ST segment is evaluated as isoelectric or for deviations up or down from baseline. A significant deviation is indicative of myocardial ischemia.

• There may be a U wave after the T wave; however, this is very uncommon and not usually observed. The U wave (if it is present) may represent the repolarization of the papillary muscles.

MEASURING THE ECG

The ECG is measured through the use of surface electrodes connected to an ECG machine, which is really a sophisticated galvanometer. The placement of the electrodes is standardized, which is important for uniformity and interpretation. In general, the ECG leads can be thought of as limb leads or chest leads. As shown in figure 5.3, the limb leads are attached to the arms and legs, usually close to the wrists and ankles. The chest leads are attached to the chest area (see figure 5.2) in standardized locations (Guyton and Hall, 2000).

The standard limb leads are named leads I, II, and III, and these are the leads originally used by Einthoven. These leads are bipolar, which means they use two electrodes, one originating electrode and one sensing electrode. Thus, an electrode is not the same as a lead. Although there are a total of four limb electrodes, one on each arm and one on each leg (figure 5.3), only three electrodes are used for leads.

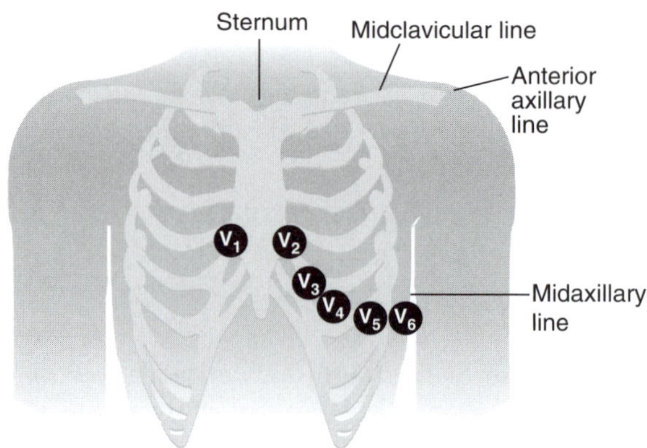

Figure 5.2 Chest electrode placement for the 12-lead ECG electrode placements.

Reprinted, by permission, from V. Heyward, 2010, *Advanced fitness assessment and exercise prescription*, 6th ed. (Champaign, IL: Human Kinetics), 36.

The electrode on the right leg is used as a ground and is not part of the lead system. Lead I has the originating electrode on the right arm and the sensing electrode on the left arm (shown in figure 5.3). Lead II also originates from the right arm, but the sensing electrode is on the left leg. Lead III originates from the left arm, and the sensing electrode is on the left leg. It is helpful to think of the sensing electrode as the detecting electrode, or in a sense a camera, looking at the electrical signal coming from the originating electrode. If the sensing electrode detects the signal coming toward it, then the QRS will be positive (deflection above the baseline or isoelectric line). If the signal is going away from the sensing electrode, then the QRS will be negative (deflection below the baseline). Combining the vectors of all three leads enables determination of the mean vector, or the combined sum of all electrical activity of the heart in the frontal plane. The sum of the vectors in a normal heart is assumed to be in the center of a triangle formed by the three leads, called Einthoven's triangle. The normal mean vector is also shown in figure 5.3. In the normal ECG, the QRS (and also both the P and T waves) will be positive (deflections above the baseline) but the deflection in lead II will be greatest, since the normal mean conduction is very close to the alignment of lead II, in the direction from the right shoulder to the left foot (Stein, 1992b).

There are also three unipolar limb leads, using the same electrodes as leads I, II, and III. These leads do not have an originating electrode per se, only a sensing electrode, hence the term unipolar. The origin for these leads is assumed to be the center of the chest. These leads are called augment leads. AVR is the right augment lead, with the sensing electrode on the right arm. Thus, AVR is negative (QRS deflection below baseline) in the normal ECG, since the mean signal travels away from the sensing electrode. AVL uses the left arm as the sensing electrode, and AVF uses the left leg as the sensing electrode; thus both of these leads are positive in the normal ECG. Combining all six leads in what is called a hexaxial reference system allows determination of the mean axis of the ECG in the frontal plane. The hexaxial reference system is depicted in figure 5.4. A normal mean axis is located primarily in the left quadrant. A quick way to determine if the axis is normal is to evaluate the QRS

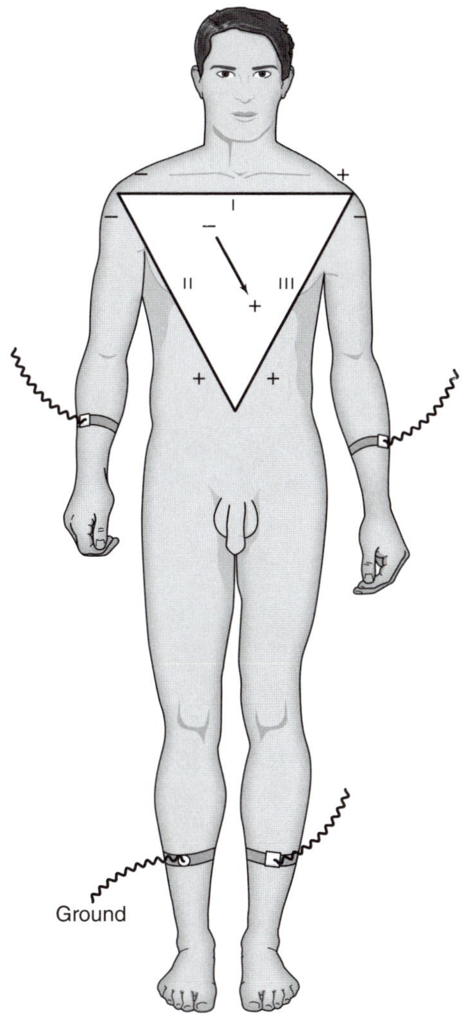

Figure 5.3 Einthoven's triangle, showing the direction of each lead. There will also be an electrode on the right leg, but it serves as a ground electrode and is not used to create a lead.

in lead I and AVF. If the QRS is positive in both leads, then the axis is normal. The following steps allow a more precise determination:

1. Determine the most isoelectric lead. This is the lead where the sum of the positive and negative deflections of the QRS are closest to zero. For example, if a lead had a 3 mm positive deflection and a 2 mm negative deflection, then the overall deflection of that lead would be +1 mm. If no other lead came closer to zero, then this would be the most isoelectric lead.

2. The axis will be 90 degrees from the most isoelectric lead. Thus, if lead I was the most isoelectric lead, the axis would be either −90 or +90 degrees.

3. Examine lead I and AVF to determine which quadrant the axis is in. In the preceding example, if AVF is positive, then the axis is +90 degrees. If AVF is

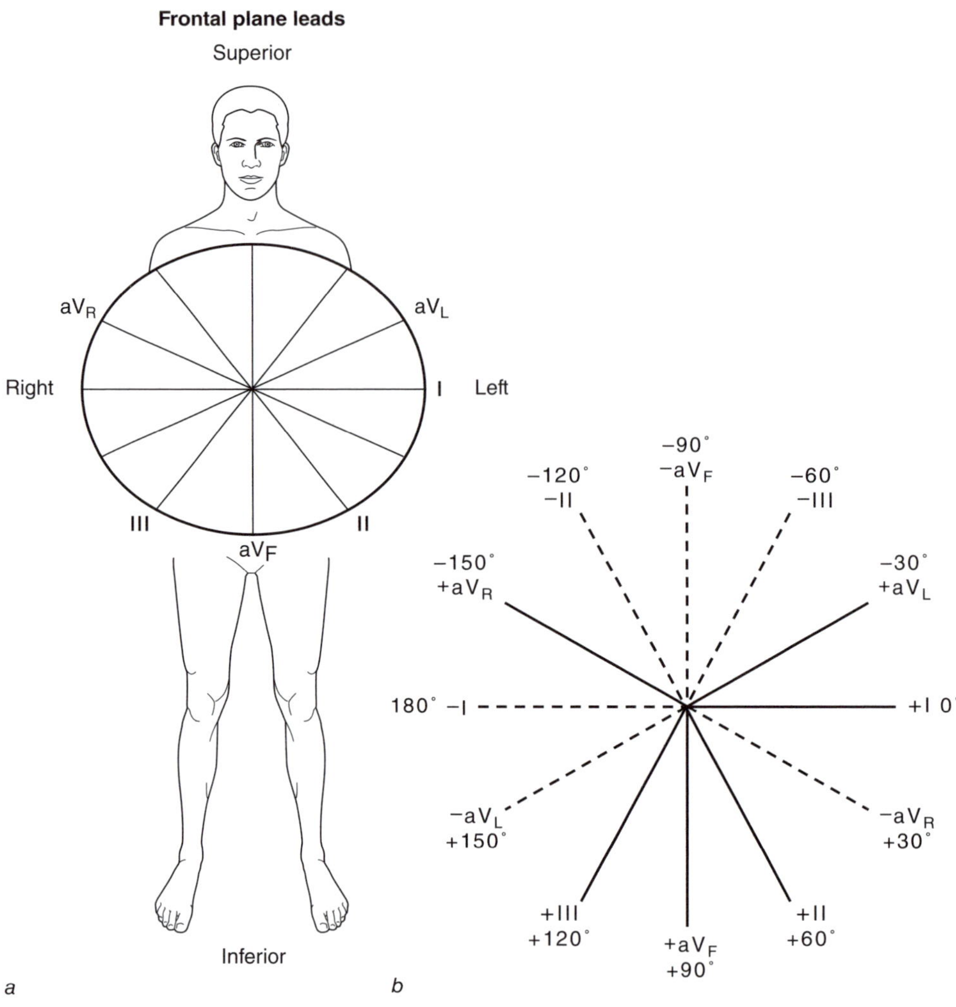

Figure 5.4 The hexaxial reference system used to calculate the axis. See text for details.

Reprinted, by permission, from G. Whyte and S. Sharma, 2010, *Practical ECG for exercise science and sports medicine* (Champaign, IL: Human Kinetics), 43,49.

negative, the axis is −90 degrees. An abnormal axis can indicate that conduction through the heart is not progressing in a normal manner and may be related to heart disease.

The precordial or chest leads (location depicted in figure 5.2, p. 62) are also unipolar leads, but provide an "ice pick" view of the heart in the anterior–posterior plane, as opposed to the frontal plane. These leads are denoted V1 through V6. Together with the six limb leads, they make up the standardized 12-lead ECG. The following conventions are used for placement of the precordial leads:

- V1: Located in the fourth intercostal space just to the right of the sternum.
- V2: Located in the fourth intercostal space just to the left of the sternum (usually directly across from V1).

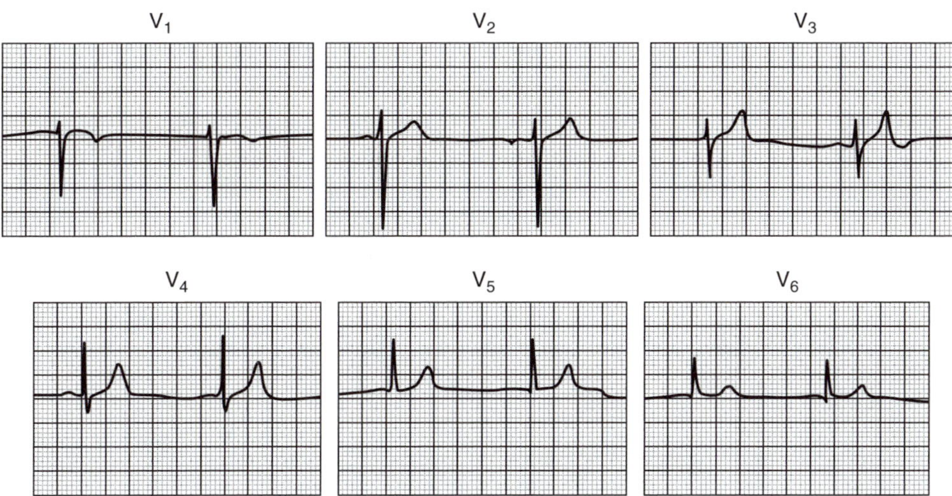

Figure 5.5 The progression of the ECG on the chest (V) leads. The R wave is small in the right V leads (V1 and V2) but increases in leads V4 through V6 with concomitant decreases in the S wave.

- V3: In between V2 and V4.
- V4: Located in the fifth intercostal space, in the midclavicular line.
- V5: Located in the fifth intercostal space at the anterior axillary line (think of the chest as a box—V5 is located on the line between the front and sides of the box). Since the intercostal space often curls upward a great deal, V5 is normally placed in line with V4.
- V6: Located on the midaxillary line (in the middle of the side of the box), at the same level as V4 and V5.

An example of a normal progression of the chest leads is shown in figure 5.5. V1 is primarily negative, and V4 through V6 are primarily positive. V1 and V2 are sometimes referred to as the right chest leads, whereas V5 and V6 are referred to as the lateral chest leads. Although a mean vector could be determined from the chest leads, this is not usually done. Altogether, from the 10 electrodes used (the electrode on the right leg is a ground electrode and is not used by any lead), 12 leads are produced (six limb leads and six chest leads), providing 12 different views of the heart.

MEASURING HEART RATE

Since the ECG paper speed is standardized and each millimeter represents 0.04 s, it is possible to calculate heart rate from an ECG tracing. There are several methods for doing this; only two of the most common methods are discussed here. One method uses a 6 s strip (or a 3 s strip) from the ECG (Stein, 1992b; Thaler, 1988) and the following steps to calculate heart rate:

1. Determine the 6 s time period by counting 30 large boxes—each large box is 5 mm or 0.2 s (see figure 5.6).
2. Count the number of QRS complexes in the 6 s time period.

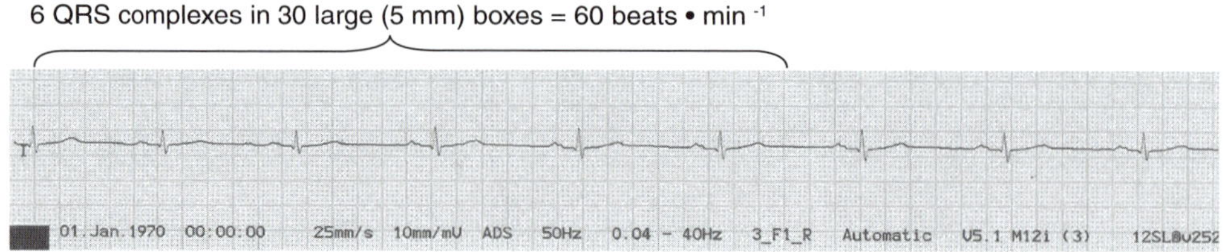

Figure 5.6 Calculating heart rate from the ECG. Each large box is 0.2 s in duration; thus 30 boxes constitutes 6 s. The number of QRS complexes in the 6 s is counted and multiplied by 10 to calculate the heart rate in beats per minute.

Reprinted, by permission, from G. Whyte and S. Sharma, 2010, *Practical ECG for exercise science and sports medicine* (Champaign, IL: Human Kinetics), 45.

3. Multiply the number of QRS complexes in the 6 s time period by 10 to derive heart rate. In the example provided in figure 5.6, there are six QRS complexes during the 6 s time period; thus the heart rate is 60 beats/min.

Sometimes a faster method of heart rate determination is required; this is especially useful when a person is exercising. The following method is very quick (after some practice).

1. Find a QRS complex that falls right on a big box line. If the next QRS falls on the next big box line, then the heart rate will be 300 beats/min. Obviously this would not be normal; thus such a finding would be unlikely. Therefore, the following steps are needed.

2. Find the next QRS complex after the first one on a big box line—if it is two large boxes after the first one, the heart rate is 150 beats/min. Simply divide the number of big boxes between the QRS intervals into 300; 300 divided by 2 yields a heart rate of 150. If there were three large boxes between the QRS complexes, then heart rate would be 100 beats/min and so on. This is depicted in figure 5.7. It is likely that the QRS complexes will not fall exactly on the two big boxes, but heart rate can easily be estimated by dividing the difference between the heart rate of the big boxes around the QRS by 5. For instance, if the second QRS falls between the big box denoting 100 and 75. Thus, the difference is 25 beats; each millimeter in between then denotes 5

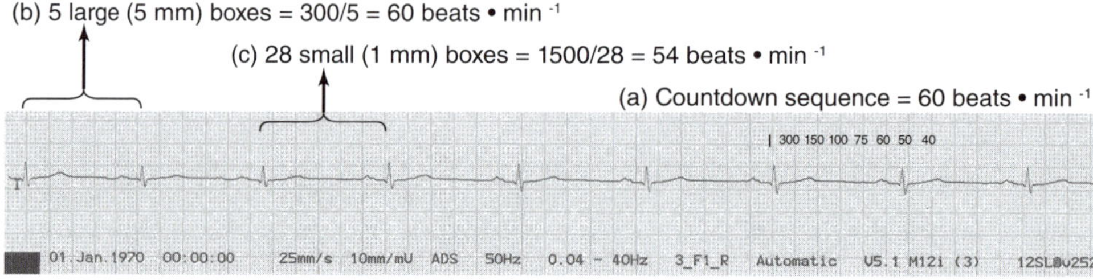

Figure 5.7 Simplified procedure for heart rate determination. Count the number of big boxes between R waves and divide into 300 to calculate the heart rate.

Reprinted, by permission, from G. Whyte and S. Sharma, 2010, *Practical ECG for exercise science and sports medicine* (Champaign, IL: Human Kinetics), 44.

beats/min. In the example, the QRS is 2 mm away from the large box line, denoting a heart rate of 75 beats/min. Adding 10 beats to 75 yields a heart rate of 85 beats/min. Although this method is not 100% accurate, in most instances it will provide heart rate measurements that are very close to the real heart rate.

CARDIAC RHYTHMS

A normal cardiac rhythm is called sinus rhythm because the heart is paced by the sinoatrial (SA) node. This means that the electrical signal originates in the SA node and is conducted normally throughout the heart. A normal sinus rhythm is denoted by a regular rhythm, with normal P waves and QRS complexes throughout. All the P waves and QRS complexes also look the same from beat to beat, with a heart rate between 60 to 100 beats/min, as shown in figure 5.8*a*. Since sympathetic stimulation can increase heart rate (for instance during exercise) but the signal still originates from the SA node and is conducted normally, it is possible to have a sinus rhythm with a heart rate above 100 beats/min. This is called sinus tachycardia (figure 5.8*b*). If the heart rate is lower than 60 beats/min but originates from the sinus node and is normally conducted, it is termed sinus bradycardia (figure 5.8*c*). It is not unusual

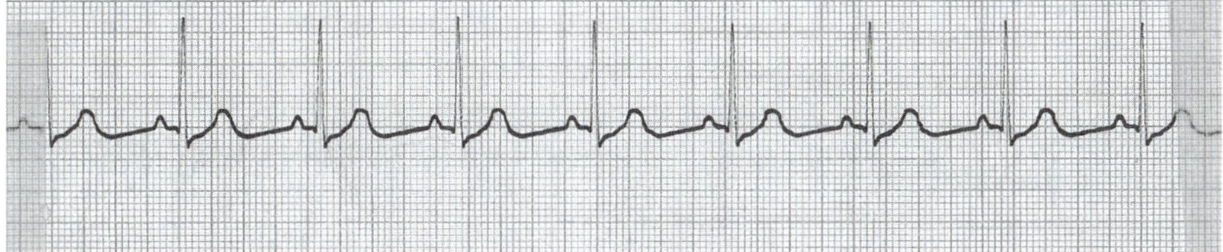

a

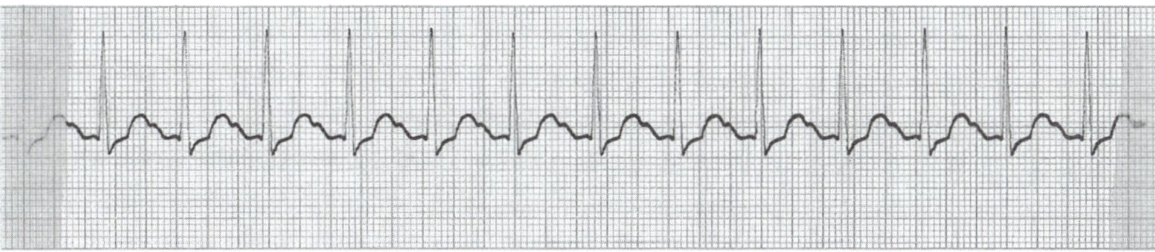

b

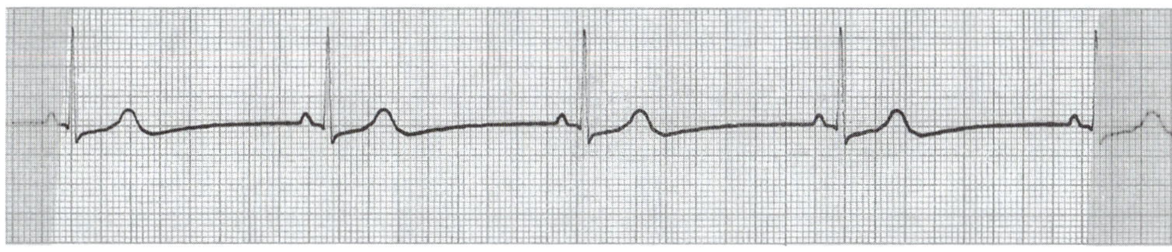

c

Figure 5.8 *(a)* Normal sinus rhythm, *(b)* sinus tachycardia, and *(c)* and sinus bradycardia. All rhythms originate from the sinus node and have normal P waves and QRS complexes, and the rhythm is regular. The only difference is the heart rate.

Reprinted, by permission, from P. Brubaker, L. Kaminsky, and M. Whaley, 2010, *Coronary artery disease: Essentials of prevention and rehabilitation programs* (Champaign, IL: Human Kinetics), 114, 129.

to have a slight variation in cardiac rhythm that still originates from the SA node, because respiration affects sinus node discharge. As explained in chapter 4, during expiration, vagal activity increases and thus heart rate slows down. This is sometimes called respiratory cardiac dysrhythmia or sinus dysrhythmia, but it is a common occurrence in healthy people (Guyton and Hall, 2000).

Ectopic Beats

An **ectopic beat** is a beat that does not originate from the SA node. Thus, the impulse to contract is initiated someplace else in the heart. Ectopic beats can originate from the AV node, from bundles, or from the heart tissue itself. An ectopic beat that appears before it is expected is called a premature beat. Premature beats originating from the atria are called atrial premature beats, those that originate from the AV node are called premature junctional beats, and those originating from the ventricles are called premature ventricular beats. Sometimes it is difficult to differentiate if an ectopic beat originates from the atria or the AV node; thus these ectopic beats are sometimes referred to as supraventricular beats, denoting their origin above the level of the ventricles.

The ECG in figure 5.9*a* shows a premature atrial beat (PAC). The R-R interval is short between the PAC and the beat preceding it, much shorter than expected and shorter than the R-R intervals between the other beats. The short R-R interval between the preceding beat and the PAC is what identifies this as a premature beat. The P wave appeared early, followed by a normal QRS complex; thus even though the atria produced a premature depolarization, this was normally conducted through the ventricles. Furthermore, the P wave of the premature beat looks different than the other P waves. This different configuration is due to the different location of the origin of the depolarization. Since the SA did not cause this depolarization, the speed of depolarization through the atria is altered, leading to the difference in

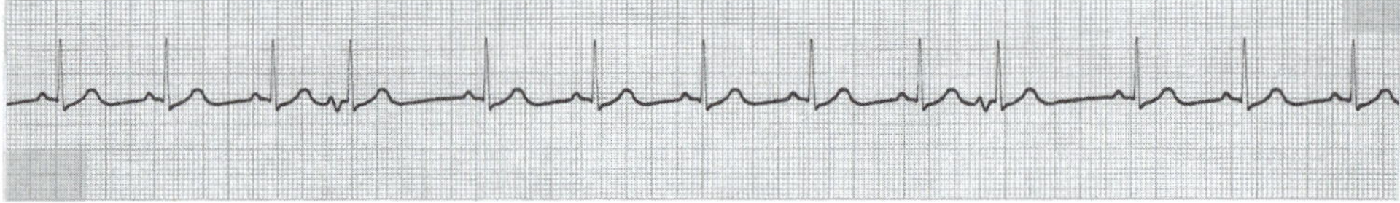

a

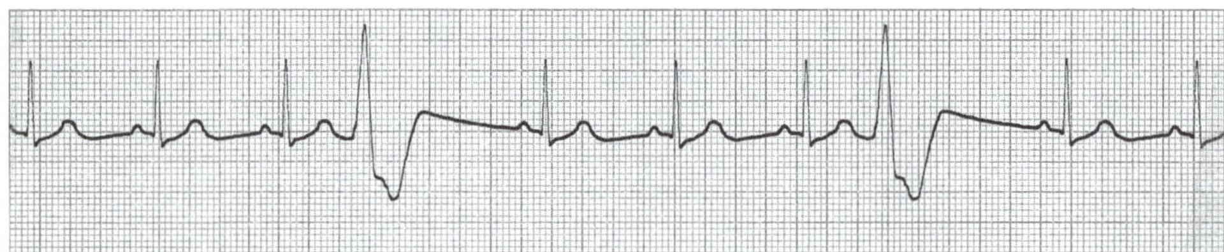

b

Figure 5.9 Comparing *(a)* a premature atrial to a *(b)* premature ventricular contraction. The premature ventricular beat is wide and bizarre, occurs early, and has no P wave. The premature atrial contraction is characterized by an early P wave that looks different from the other P waves, while the QRS is normal.

Reprinted, by permission, from P. Brubaker, L. Kaminsky, and M. Whaley, 2010, *Coronary artery disease: Essentials of prevention and rehabilitation programs* (Champaign, IL: Human Kinetics), 129, 132.

P wave appearance. If the ectopic beat originates from the AV node, the P wave will often be negative (when a positive deflection is expected), or it will be missing altogether.

Now compare the ECG in figure 5.9*a* with the one in figure 5.9*b*. The ECG in figure 5.9*b* illustrates a premature ventricular beat, indicating that the beat originated in the ventricles. Because the depolarization of this beat does not follow the normal conduction pattern, the configuration of the QRS complex and the T waves looks very different from the normal QRS segment. Typically, premature ventricular contractions (PVC) are easy to pick out because of their wide, bizarre QRS complex, which looks very different from the normal QRS complexes, and because the T wave is also in the opposite direction of the QRS segment. In addition, the PVC is not preceded by a P wave. The PVC is also followed by a compensatory pause; thus the time between the PVC and the next beat is long. The compensatory pause is caused by retrograde conduction of the beat in the conduction system, preventing the conduction of the next P wave because the conduction system is still refractory. The second P wave is conducted normally. This causes the R-R interval between the normally conducted QRS complex preceding the PVC and the normally conducted QRS after the PVC to be two normal R-R intervals long. When multiple PVCs occur that all look alike, they all come from the same ectopic focus and are called unifocal PVCs. If multiple PVCs occur that do not look alike, this denotes that they originate from different foci; these are called multifocal PVCs. Premature beats occur in most people and are not necessarily associated with problems or disease. Stress, caffeine, and other lifestyle factors are all associated with premature beats.

Atrial Tachycardias

Sometimes an ectopic focus can produce a **tachycardic rhythm** in which all of the beats are from the ectopic focus. Atrial tachycardia is characterized by short or long "runs" of multiple atrial contractions in a row. The onset of atrial tachycardias is often sudden; thus they are called paroxysmal atrial tachycardias. An example of atrial tachycardia is shown in figure 5.10*a*. The QRS is usually normal, and it is often difficult to discern any P waves. Hence it is impossible to tell if the tachycardia originates from the atria or the AV node. Therefore, these tachycardias are often referred to as supraventricular tachycardias. Depending on the ventricular rate, supraventricular tachycardia can cause light-headedness or loss of consciousness because a very high heart rate prevents adequate ventricular filling between beats. This leads to reduced stroke volume and cardiac output.

Sometimes the atria depolarizes at a very rapid rate, but in a regular rhythm. Since the rate is very high, usually 250 to 300 beats/min, each atrial depolarization is not conducted through to the ventricle. Instead, every two to five beats are usually conducted through, as shown in figure 5.10*b*. This is called atrial flutter. This dysrhythmia is characterized by the flutter waves (P waves) that create a sawtooth pattern.

Fibrillation denotes an irregular type of contraction in which the atrial muscle fibers do not contract in an organized, unified manner. Instead the atria "fibrillate"; this type of contraction does not produce any type of blood flow to the ventricles as a result of atrial contraction. On the ECG, this shows up as irregular waves between the QRS complexes, and no P waves are discernable (see figure 5.10*c*). The QRS rhythm is usually quite irregular as well. Both atrial flutter and atrial fibrillation occur in various types of heart disease, but these dysrhythmias are usually not fatal. It is, in

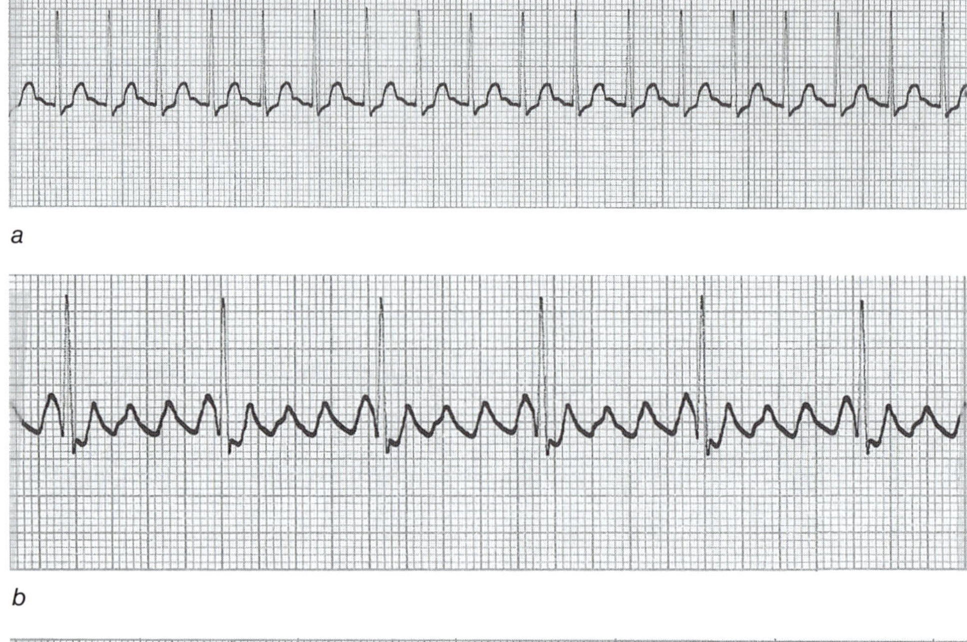

a

b

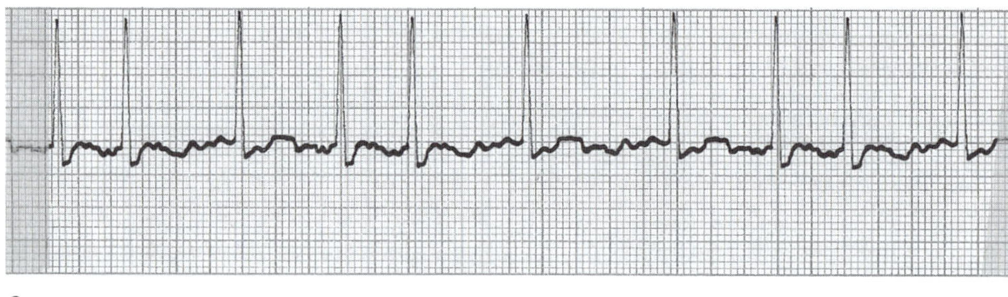

c

Figure 5.10 Examples of atrial dysrhythmias. *(a)* Paroxysmal atrial tachycardia, *(b)* atrial flutter, *(c)* atrial fibrillation.

Reprinted, by permission, from P. Brubaker, L. Kaminsky, and M. Whaley, 2010, *Coronary artery disease: Essentials of prevention and rehabilitation programs* (Champaign, IL: Human Kinetics), 130.

fact, quite common for people to live for many decades with these dysrhythmias; and most have a normal life, including physical activity. Because the blood is not ejected from the atria in an organized manner, the risk of blood clotting is higher in these individuals, and they are often prescribed anticoagulatory medications.

Ventricular Tachycardias

Ventricular tachycardias are, unlike the atrial tachycardias, life threatening. Ventricular tachycardia, shown in figure 5.11*a,* is essentially several consecutive PVCs. The QRS complexes are wide and bizarre because the ectopic focus originates from the ventricles, which distinguishes this dysrhythmia from supraventricular tachycardia. Ventricular tachycardia is often sudden in onset. At high ventricular rates, dizziness or loss of consciousness is common. Ventricular tachycardia is a very dangerous

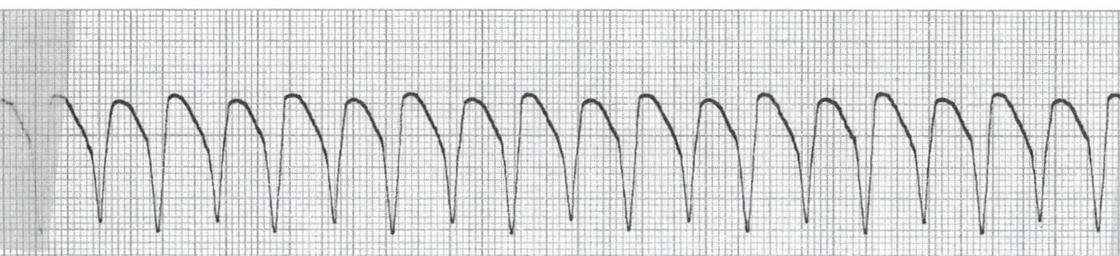

a

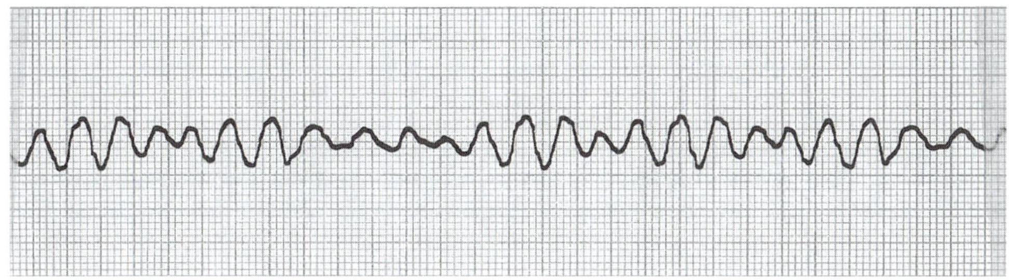

b

Figure 5.11 Examples of ventricular dysrhythmias. *(a)* Ventricular tachycardia, *(b)* ventricular fibrillation.

Reprinted, by permission, from P. Brubaker, L. Kaminsky, and M. Whaley, 2010, *Coronary artery disease: Essentials of prevention and rehabilitation programs* (Champaign, IL: Human Kinetics), 133.

dysrhythmia because it can deteriorate into **ventricular fibrillation,** which is usually lethal without medical intervention.

Ventricular fibrillation, shown in figure 5.11*b,* is a lethal dysrhythmia in which the muscle fibers of the ventricles are fibrillating. Since no organized muscle activity is present, no blood is ejected from the heart; thus there is no cardiac output. This is manifested on the ECG as irregular fluctuations, of fibrillation waves, but there are no discernable QRS complexes or T or P waves. A person with ventricular fibrillation will lose consciousness within a few seconds and usually needs to be defibrillated (be given an electrical shock) in order to survive. The defibrillation works by simultaneously depolarizing all the cardiac cells, which "resets" the system. Hopefully, after the electrical charge has been delivered and all the cells depolarized, the internal pacemaker (SA node) will take over and set the pace again. Ideally, the SA node starts a new impulse before any other impulses are generated elsewhere in the ventricle. This is possible because the fairly long refractory period of ventricular cells allows the signal to be generated by the SA node and transmitted normally through the conduction system before an action potential is generated by the ventricular myocyte itself (Stein, 1992a).

CONDUCTION BLOCKS

A block in the normal conduction of impulse transmission either delays the signal transmission or completely blocks it, causing characteristic changes in the ECG. Blocks can occur in the AV node or in the bundles below the AV node. Blocks can be caused by heart disease, ischemia, medications, and in some instances exercise training. This section describes AV and bundle branch blocks.

Atrioventricular Blocks

Atrioventricular blocks (AV blocks) occur when there is either a slowing of or a complete block of conduction at the level of the AV node. These delays or blocks can be readily evaluated from the ECG. AV blocks may be of little clinical significance (first degree block) or can be of substantial clinical importance (third degree block) (Stein, 1992a).

- First-degree AV block: This is a delay of the signal, usually in the AV node or through the bundles. It is defined by a prolongation of the PR interval, to greater than 0.2 s (>5 mm).
- Second-degree AV block: There are two types of second-degree AV blocks, Mobitz 1 (also called a Wenkeback) and Mobitz 2. In both cases, the ventricles do not respond to an atrial conduction, resulting in a P wave that is not followed by a QRS complex. In Mobitz 1, the PR interval becomes gradually prolonged, to the point that a QRS is not conducted. After the ventricular beat is "dropped," the cycle starts over

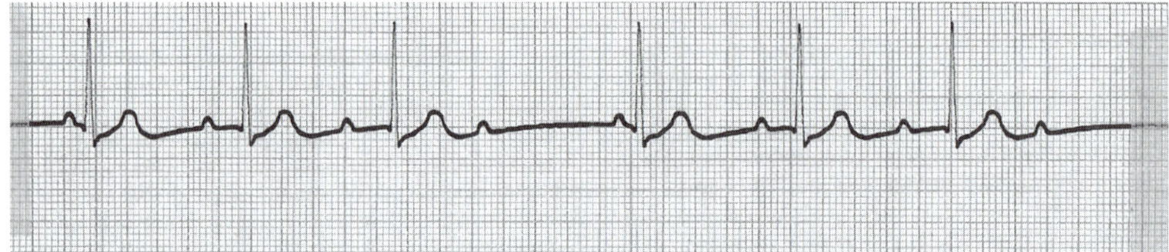

a

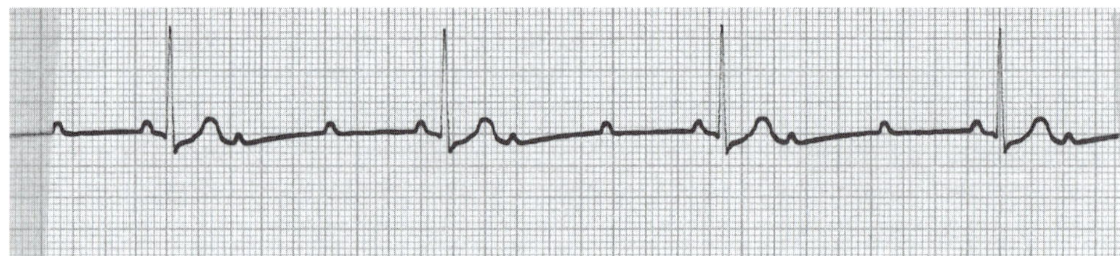

b

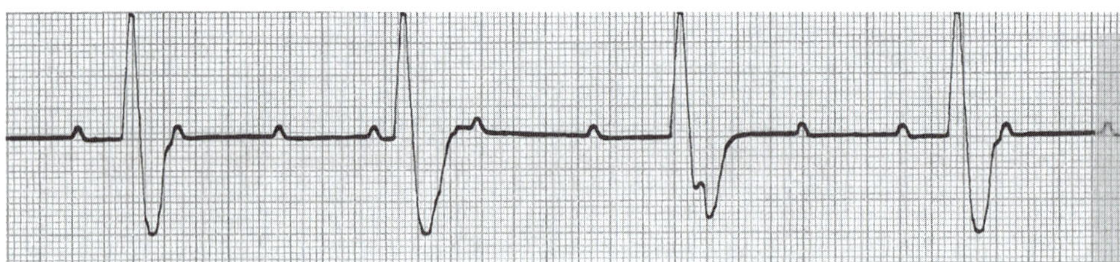

c

Figure 5.12 ECG showing second- and third-degree AV blocks. *(a)* Mobitz 1, *(b)* Mobitz 2, *(c)* example of a third-degree AV block.

Reprinted, by permission, from P. Brubaker, L. Kaminsky, and M. Whaley, 2010, *Coronary artery disease: Essentials of prevention and rehabilitation programs* (Champaign, IL: Human Kinetics), 135, 136.

again. This is shown in figure 5.12*a*. The block is usually in the AV node in Mobitz 1. In Mobitz 2, QRS complexes are periodically "dropped," and only a P wave will be evident. The PR interval is prolonged, and P waves seem to drop unexpectedly (figure 5.12*b*). The key difference in the ECG between Mobitz 1 and Mobitz 2 is that the PR interval is regular in Mobitz 2 (in beats with a QRS). The site of the block is below the AV node, and Mobitz 2 is a more serious condition that may proceed to complete AV block.

- Third-degree AV block: Third-degree AV block is a complete AV block. This means that the atria and the ventricles beat independently of each other, and there is no relationship between P waves and QRS complexes as the two are completely dissociated. This is shown in figure 5.12*c*. In third-degree AV block, the signal does not get through the AV node; hence the ventricular and atrial pacemakers are different. The QRS may be wide or normal, depending on where the ventricular pacemaker is (in the AV node or the ventricles). The ventricular rate is always less than the atrial rate. Third-degree AV block is a very serious condition and usually requires installation of a pacemaker.

Bundle Branch Blocks

In bundle branch blocks, conduction to either the right or left bundle is interfered with. This causes a wide QRS and may make the ventricles beat asynchronously. Bundle branch blocks can be caused by abnormalities in the bundles or in the myocardium.

- Left bundle branch block (LBBB): In LBBB, the signal is also blocked and the impulse is forced to travel outside the standard conduction pattern, which causes a delay and prolongation of the QRS complex. This affects both the initial and final QRS vectors, including repolarization vectors, affecting both the T wave and the ST segment. LBBB is characterized by (1) QRS prolongation (≥0.12 s); (2) negative QRS complexes in V1 and V2; (3) positive QRS complexes in V5 and V6, often notched; (4) no small normal Q waves in leads I, AVL, V5, or V6 (this may hinder the ability to recognize ECG changes associated with a myocardial infarction); and (5) resulting repolarization abnormalities that produce ST-segment and T wave dispersion in the opposite direction of the QRS. This causes ST-segment depression in leads such as leads I, II, V5, and V6. As a result, ischemic changes cannot be evaluated from the ECG during exercise in people with LBBB. The ECG changes associated with LBBB are shown in figure 5.13*a*.

- Right bundle branch block (RBBB): In RBBB, the activation of the right ventricle occurs after that of the left. This causes the terminal QRS vector to be shifted to the right. RBBB exhibits four characteristics. (1) The QRS complex is prolonged (≥0.12 s); (2) there is an S wave in lead I, caused by a terminal QRS vector that is shifted to right, producing the S wave; (3) The terminal right shift of the QRS vector also produces a second R wave in lead V1, called an R'; (4) RBBB causes repolarization abnormalities. The T wave is in the opposite direction of the terminal deflection of the QRS complex. Thus the T wave is negative in V1 and positive in lead I. The ECG changes in RBBB are shown in figure 5.13*b*.

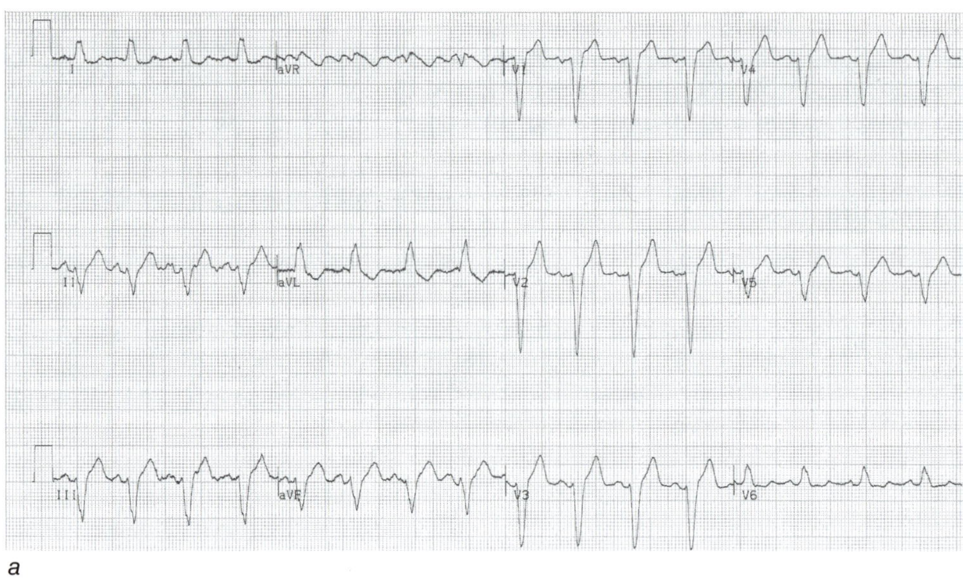

a

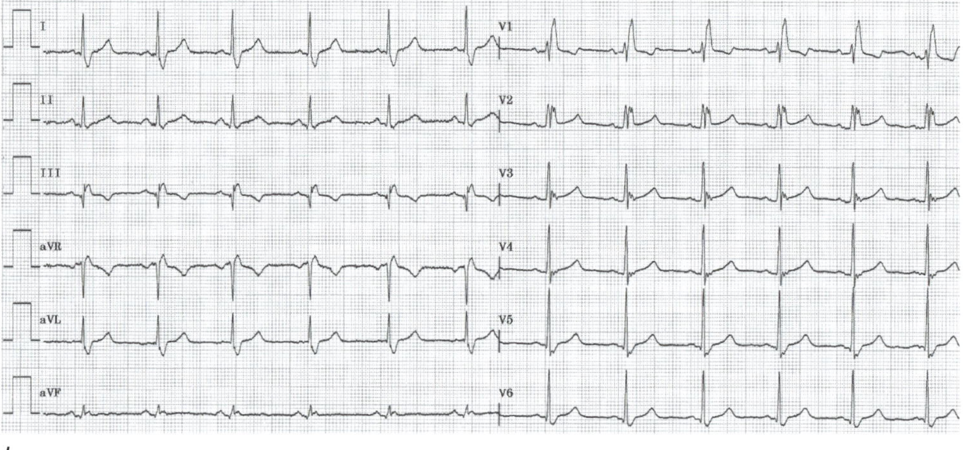

b

Figure 5.13 Examples of *(a)* left and *(b)* right bundle branch blocks.

Reprinted, by permission, from P. Brubaker, L. Kaminsky, and M. Whaley, 2010, *Coronary artery disease: Essentials of prevention and rehabilitation programs* (Champaign, IL: Human Kinetics), 139, 140.

VENTRICULAR HYPERTROPHY

Ventricular hypertrophy or enlargement can be ascertained using ECG recordings; however, the gold standard for determining ventricular hypertrophy is cardiac ultrasound. The ECG cannot distinguish between muscle hypertrophy (wall hypertrophy) and chamber dilation. Nevertheless, left ventricular hypertrophy (LVH) is associated with increased mortality and morbidity.

Left Ventricular Hypertrophy

Left ventricular hypertrophy is commonly caused by hypertension and various forms of heart disease. It results in increased voltage on the ECG and may also produce repolarization abnormalities termed a strain pattern. There are many different criteria for determining LVH from the ECG, but one commonly used—and easy to use—set of criteria is the Scott criteria. Use of the Scott criteria requires an evaluation of the QRS amplitude in various leads, and LVH is considered present if one or more of the following criteria are met:

1. The S wave in V1 or V2 + the R wave in V5 or V6 is equal to or greater than 35 mm.
2. The R wave in V5 or V6 is equal to or greater than 26 mm.
3. The S wave + the R wave in any precordial lead is equal to or greater than 45 mm.
4. The R wave in lead I + the S wave in lead III is equal to or greater than 25 mm.
5. The R wave in aVL is equal to or greater than 7.5 mm.
6. The R wave in aVF is equal to or greater than 20 mm.
7. The S wave in aVR is equal to or greater than 14 mm.

In addition, the presence of a strain pattern would strengthen the determination of LVH. If LVH is present, this may also increase the likelihood of false positive ECG changes in response to a stress test.

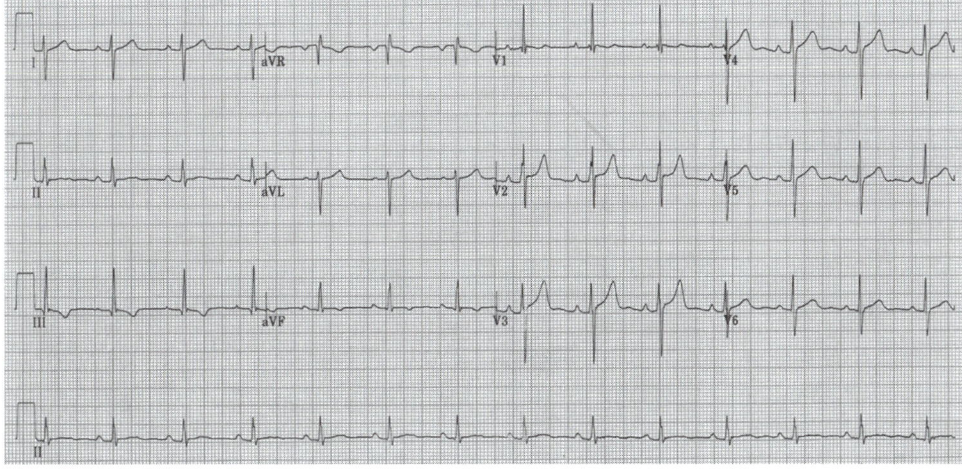

Figure 5.14 Example of left ventricular hypertrophy.

Reprinted, by permission, from G. Whyte and S. Sharma, 2010, *Practical ECG for exercise science and sports medicine* (Champaign, IL: Human Kinetics), 82.

Right Ventricular Hypertrophy

Right ventricular hypertrophy (RVH) is less common than LVH, and is usually caused by congenital heart disease, pulmonary disease, or certain types of heart valve problems. The following are typical ECG manifestations of RVH:

1. Right axis deviation is present.
2. A negative QRS is present in lead I (required for right axis deviation).
3. The QRS is positive in V1, and the R wave is greater than 7 mm. Repolarization abnormalities are consistent with a strain pattern in V1 or V2 or in AVF.

An example of an ECG manifestation of RVH is shown in figure 5.15.

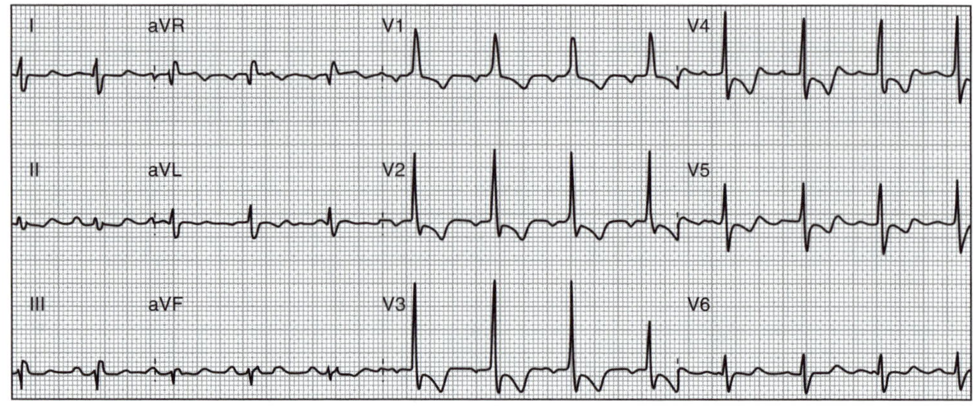

Figure 5.15 Example of right ventricular hypertrophy.

Reprinted, by permission, from P. Brubaker, L. Kaminsky, and M. Whaley, 2010, *Coronary artery disease: Essentials of prevention and rehabilitation programs* (Champaign, IL: Human Kinetics), 143.

ST-SEGMENT CHANGES (ISCHEMIA)

The ECG can be a very useful tool to evaluate ischemic responses. It is commonly used as part of a graded exercise test, the so-called stress test. This test evaluates if the heart can respond to increasing exercise stress in an appropriate manner. If a person has atherosclerotic plaque, blood flow to the myocardium is reduced, creating an ischemic response. This may not be evident at rest; but as exercise intensity increases, the heart needs to increase blood flow in order to do the extra work. If blood flow cannot increase appropriately, as when atherosclerotic plaque is present, an ischemic response may ensue. It is often manifested as ST-segment changes on the ECG, but may also be accompanied by other dysrhythmias and chest pain.

The ST segment is evaluated as a change from the isoelectric line. This involves locating the J point (the junction of the S wave and the ST segment) and then evaluating the ST segment 80 ms (2 mm) after the J point, in relation to the isoelectric line. Figure 5.16 shows an example. The amount of displacement from the isoelectric line is evaluated, as well as the configuration of the ST segment. For a horizontal or downsloping ST segment, a 1 mm ST-segment depression is considered significant, whereas for an upsloping ST segment, a 1.5 mm ST-segment depression is significant. The greater the displacement, the more significant the ischemic response. An example of a typical ischemic response to exercise is shown in figure 5.17. It is common for the resting ST segment to be normal but for the J point to drop farther below baseline as exercise becomes more intense. As shown in figure 5.17, an ischemic response is often characterized by a progressive drop in the ST segment, sometimes also accompanied by a change in configuration.

MYOCARDIAL INFARCTION

When blood flow to a region of the heart is blocked, the tissue beyond the blockage becomes ischemic. If the ischemic condition persists (blockage remains), the tissue becomes injured and eventually infarcts. An acute event (infarct in progress) may

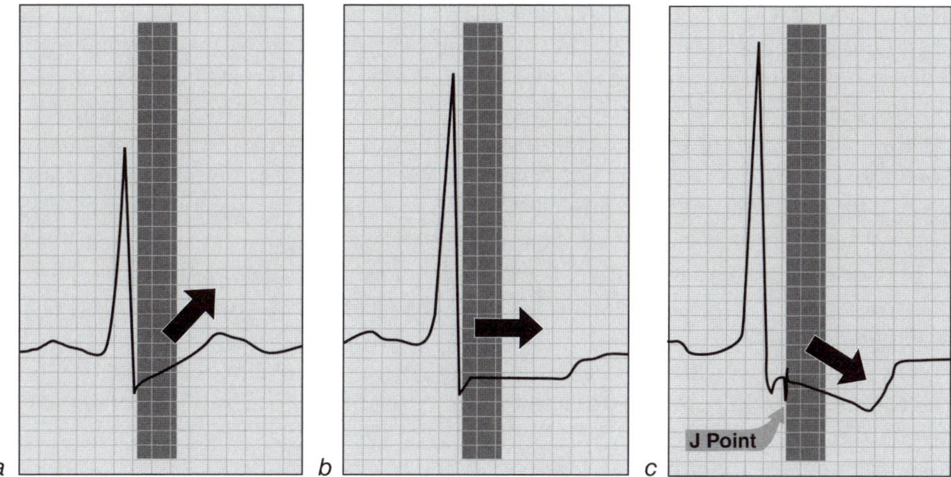

Figure 5.16 ST-segment determination and different ST-segment configurations. Identify the J point and evaluate the amount of ST-segment displacement 2 mm (0.08 s) after the J point. The segment can be upsloping *(a)*, horizontal *(b)*, or downsloping *(c)*.

Reprinted, by permission, from P. Brubaker, L. Kaminsky, and M. Whaley, 2010, *Coronary artery disease: Essentials of prevention and rehabilitation programs* (Champaign, IL: Human Kinetics), 126.

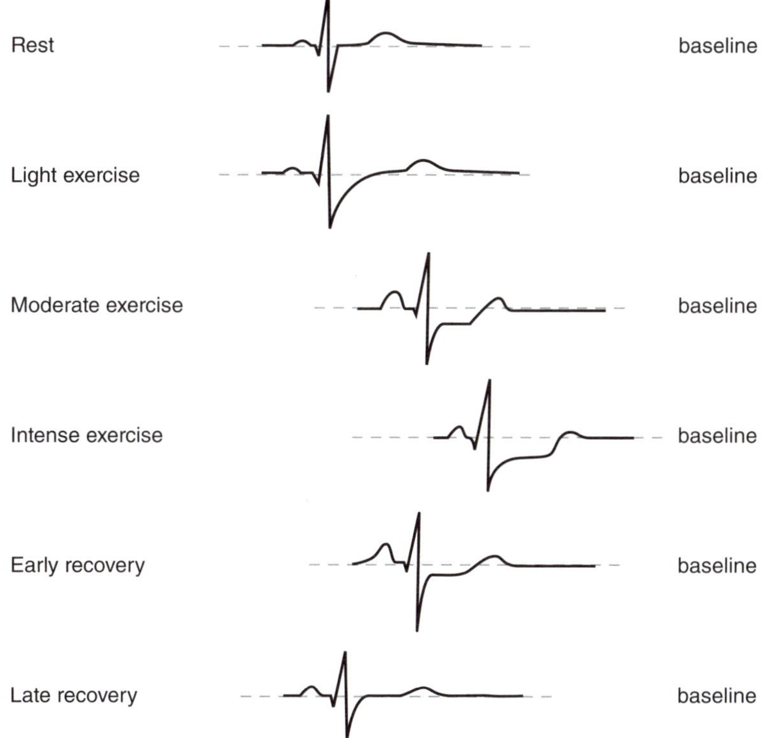

Figure 5.17 ST-segment changes with an ischemic exercise response. The ST segment gradually becomes more depressed with increasing exercise intensities, and then returns to baseline in recovery.

produce the ECG changes shown in figure 5.18. The early ECG changes to blood flow obstruction are manifested by a severe ST-segment elevation, followed by an upright T wave. If the condition persists, the ST-segment elevation persists but becomes more concave, and the T wave becomes inverted. At this point Q waves may or may not develop. In general, Q waves develop several hours to several days following blood flow obstruction, but they may not develop at all (called a non-Q wave infarction). A significant Q wave is at least 1 mm wide (0.04 s in duration) or one-third of the height of the R wave. Several days after the infarction, the ST segment normalizes; but the T wave often remains inverted, and the Q wave remains as well. After several weeks (or months) the T wave also normalizes, but the Q wave remains (in the case of a Q wave infarction).

The ECG changes may also help to locate where in the heart the infarction took place. The following general guidelines are helpful for providing a location of the myocardial infarction based on ECG changes in the leads affected (development of ST-segment elevation, inverted T waves, and Q waves):

- Lateral infarction: Leads I, AVL, and V4 through V6
- Anterior infarction: Leads V1 through V4
- Inferior infarction: Leads II, III, and AVF
- Posterior infarction: Leads V1 and V2—prominent R wave and ST-segment depression

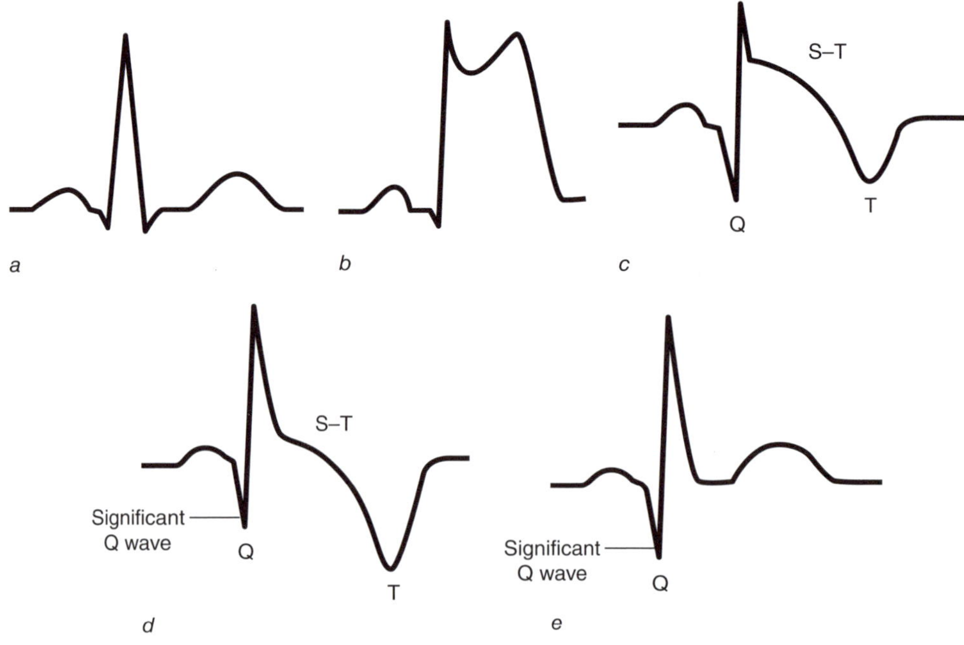

Figure 5.18 The ECG changes associated with myocardial infarction, progressing from obstruction of blood flow to injury and finally to infarction. *(a)* baseline, *(b)* hours following blood flow obstruction, *(c)* hours to days following obstruction, *(d)* days to weeks following obstruction, and *(e)* weeks to months following obstruction.

TEST CONSIDERATIONS

Resting ECG recordings should always be made with the patient in a supine position (electrodes should also be attached with the patient supine) using the traditional limb lead placements as shown in figure 5.2. However, exercise testing is not conducive to limb leads on the arms and legs; thus those leads are usually moved to the torso in what is called the Mason-Likar placement (figure 5.19). Because this electrode configuration changes the axis and amplitude of the ECG, a resting ECG using traditional electrode placements should always be obtained first.

It is recommended (American College of Sports Medicine, 2000) that ECG measurements be obtained at rest in the supine and exercise postures before the initiation of exercise. During exercise, a 12-lead ECG should be obtained in the last 15 s of each exercise stage and immediately upon cessation of exercise. It is also recommended that a three-lead ECG be continuously monitored during the test. Since the left precordial leads are the most sensitive leads (V5 and V6) for detecting ST-segment changes, it is highly recommended that V5 or V6 be one of the three leads used for continuous monitoring during the exercise test. Finally, ECG recordings should be obtained every 1 to 2 min in recovery for at least 5 min, or until exercise-induced changes return to baseline.

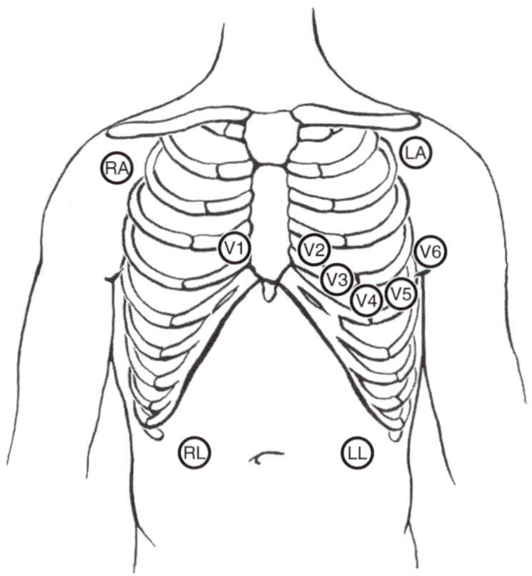

Figure 5.19 The Mason-Likar electrode placement used during exercise testing.

Reprinted, by permission, from P. Brubaker, L. Kaminsky, and M. Whaley, 2010, *Coronary artery disease: Essentials of prevention and rehabilitation programs* (Champaign, IL: Human Kinetics), 117.

COMMON ECG CHANGES IN ATHLETES

It is not unusual for athletes to exhibit an "abnormal" ECG due to the effect of exercise training on the anatomy and autonomic control of the heart. Most of the changes are present in athletes who participate in endurance events; they are more common in males than in females, and in individuals of African decent (Corrado et al., 2009,

2010). Up to 80% of highly trained athletes show some evidence of ECG changes. Most of these changes can be considered physiologic adaptations to exercise training and do not necessarily indicate any pathology or concern. However, it is important to understand which ECG abnormalities may be physiologic and which are likely to be pathologic. A joint statement by the European Association of Cardiovascular Prevention and Rehabilitation and the European Society of Cardiology provides the following guidelines (Corrado et al., 2010).

The most common training-induced ECG changes in athletes are sinus bradycardia, first-degree AV block, second-degree AV block (Mobitz type 1 only), incomplete RBBB (QRS duration <0.12 s), early repolarization, and isolated QRS voltage criteria for LVH. It is generally agreed that a high degree of endurance training increases parasympathetic tone and decreases sympathetic control of the heart. These changes are likely responsible for the high prevalence of sinus bradycardia, first-degree AV block, and Mobitz type 1 in athletes. However, more serious conditions, such as Mobitz 2 or third-degree AV block, as well as sick sinus syndrome, are indicative of pathologic changes that need medical attention.

Training-induced increases in wall thickness and chamber size are the likely causes of isolated QRS voltage criteria for LVH and for incomplete RBBB. The increased LV size causes higher QRS amplitude, and the increased RV size causes the conduction delay responsible for the incomplete RBBB. However, other criteria for LVH, such as a strain pattern or enlarged atrial size, are rarely present in athletes; and those changes are considered pathologic. Furthermore, a full RBBB or LBBB is not training induced and should also be considered pathological.

Early repolarization is present in 50% to 80% of highly trained athletes. It is characterized by a J point elevation of at least 0.1 mV (1 mm) from baseline, which is often associated with a slurring or notching of the terminal QRS complex. It is most often found in the left precordial leads. It is generally believed that autonomic influences induced by training cause early repolarization, since a slowing of the

DIFFERENTIATING PHYSIOLOGIC AND PATHOLOGIC HEART RATES IN AN ATHLETE

A 20-year-old male athlete was having episodes of dizziness during intense exercise training bouts and was referred for further evaluation, even though he had had a normal preseason physical. The resting ECG showed LVH by voltage criteria, and was accompanied by a significant strain pattern in the lateral precordial leads. A 24 h continuous ECG recording (Holter monitoring) showed several episodes of 20 to 30 s of ventricular tachycardia. Since these findings are not consistent with training-induced ECG changes but probably indicate underlying heart disease, the patient was referred for an echocardiogram. The ultrasound evaluation showed severe thickening of the ventricular septum, consistent with hypertrophic cardiomyopathy (HCM). Further follow-up testing confirmed the diagnosis of HCM. Since HCM is the most common cause of sudden cardiac death in young athletes, the patient was advised to discontinue competitive athletics and was started on medication to prevent potentially lethal dysrhythmias such as ventricular tachycardia.

heart rate exaggerates the ST-segment elevation. However, ST-segment depression or ST-segment changes indicative of a strain pattern are not common and should be considered pathological.

SUMMARY

The ECG provides an electrical representation of atrial and ventricular contraction and relaxation. The standardized 12-lead ECG is used to detect abnormalities that may be related to disease or other causes. The ECG is an excellent tool for detecting cardiac dysrhythmias, alterations in conduction patterns of the heart, signs of ischemia and myocardial infarction, and evidence for left or right ventricular remodeling indicating hypertrophy. The ECG also provides a very accurate measure of heart rate. Furthermore, the ECG is an excellent tool for evaluating changes during an exercise test (most often done to detect latent atherosclerotic disease), and an exercise stress test is often the first test used to screen for suspected heart disease. Highly trained endurance athletes, primarily males, often exhibit an "abnormal" ECG; but most of the time these changes are considered normal variants induced by high-level training. It is important to distinguish these "normal variants" from more serious problems caused by underlying disease.

CHAPTER 6

Hemodynamics and Peripheral Circulation

Blood flow through the vascular system of the human body is determined by a series of physical properties and forces. The relationships among these properties and forces explains where blood flows and the amount of blood flow in any particular tissue. Hemodynamics refers to the study of blood flow. Hemodynamics explains blood flow by applying simple principles of physics to a very complex system. Such an approach allows for an explanation of blood flow that is not different from the explanation of water flow through any plumbing system. This works very well if one assumes that blood is a simple fluid like water (which it is not, since blood contains cells and particles that create an entirely different viscosity than that of water) and that the flow is primarily laminar through rigid tubes. Such a simple explanation is not entirely accurate in the human cardiovascular system for several reasons: The vascular system is made of pliable tubing with many branch points; blood is not a simple fluid; and flow is often turbulent. Nevertheless, knowledge of the principles of hemodynamics provides a basic understanding of vascular flow and the interrelationships between pressure, flow, and vessel size.

THE PRESSURE DIFFERENTIAL

Since the human circulation is a closed-loop system, physical factors determine the rate and amount of blood flow through the system. There are two primary determinants of blood flow: the pressure differential and the resistance to flow (Badeer and Hicks, 1992; Berne and Levy, 2001). This relationship can be expressed with the formula:

$$\text{Flow} = \frac{\text{Pressure differential}}{\text{Resistance to flow}}$$

The influence of the pressure differential on blood flow is shown in figure 6.1. Assuming that the resistance to flow is held constant, the rate of blood flow in a tube is determined by the difference in pressure between the ends of the tube. If the pressure is 100 mmHg at one end of the tube and 50 mmHg at the other, the fluid (blood) will flow from the high-pressure end to the low-pressure end. If the pressure at the low-pressure end is lowered to 0 mmHg, then the rate of flow will increase, since the pressure differential has increased. Conversely, if the pressure at both ends is 100 mmHg, then there will be no flow since there is no pressure differential, despite the generally high pressure in the system. In the human circulation, the high-pressure end is the aorta, with a mean arterial pressure (MAP) of about 100 mmHg; and the low-pressure end is the right atrium, with a pressure of close to 0 mmHg (0-4 mmHg). Thus, blood will flow from the aorta through the systemic circulation and back into the right atrium for recirculation.

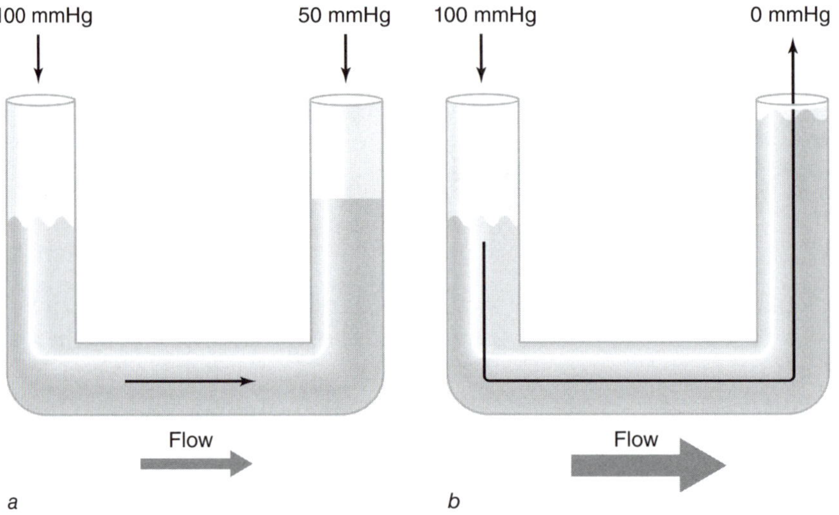

Figure 6.1 The effect of pressure differential on flow. The pressure differential between the tubes determines the rate of flow, with all other factors being equal. The flow increases as the pressure differential increases.

FLOW VELOCITY

The velocity of blood flow (v) is directly related to the amount of flow ($\dot{Q}$) and the cross-sectional area of the system (A), and can be expressed as

$$v = \dot{Q} / A.$$

In a closed system like the human vascular system, if the flow remains constant, then the velocity of the flow will vary directly in proportion to the cross-sectional area of the vascular bed. This is shown in figure 1.4 (p. 8). The velocity of flow is highest in the aorta and then diminishes as the cross-sectional area of the arterial tree increases, with the lowest velocity across the capillary beds. The velocity then starts to gradually increase, as the total cross-sectional area of the venous system decreases, and finally terminates in the vena cava. Consequently, in the example shown in figure 6.1, since there was no change in cross-sectional area, not only will blood flow increase with an increase in pressure gradient, but the velocity of blood flow will also increase. This concept is demonstrated in figure 6.2. Using a garden hose, the velocity of flow is adequate to spray water several feet. However, if the same volume of flow were to be put into a fire hose, a mere trickle would come out, since the increase in cross-sectional area would have decreased the velocity of the flow (Little, 1985).

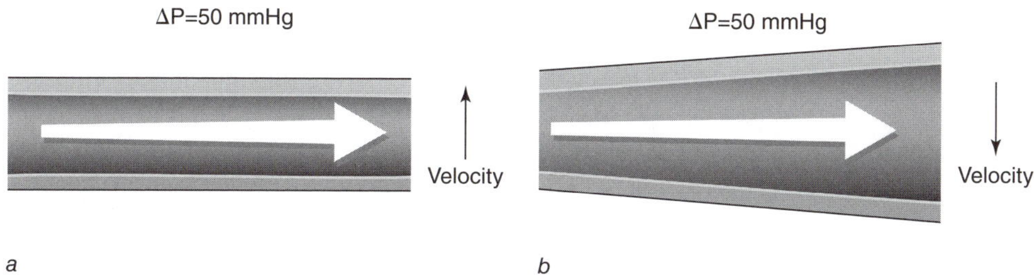

Figure 6.2 Pressure differential and flow velocity. *(a)* A change in pressure differential without a change in cross-sectional area of the tube produces an increase in flow velocity. *(b)* However, if the cross-sectional area is also increased, the flow velocity will decrease if the overall flow is constant. This is similar to what would happen if the water flow in a garden hose were placed into a fire hose.

POISEUILLE'S LAW

The mathematical relationship between pressure, flow, and resistance has been named **Poiseuille's law** after a French physiologist who carefully studied blood flow at the end of the 19th century. He used cylindrical glass tubes and simple fluids (like water) to simulate blood flow under carefully controlled conditions. Thus, this law applies to conditions of steady or constant flow (not pulsatile flow). This law also applies to laminar flow, where the fluid moves in parallel to the axis of the tube; however, the velocities of the flow vary by position in the tube. The layer closest to the wall is motionless, while the peak velocity is obtained along the axis of the tube (middle of the tube in this case). Under these conditions, Poiseuille's law is written as

$$\dot{Q} = \frac{\pi(P_1 - P_2)r^4}{8\eta\ell}$$

where P_1 is the inflow pressure and P_2 is the outflow pressure; r is the radius of the vessel; η is the viscosity of the fluid; and ℓ is the length of the tubes.

From the equation, it is easy to see that the pressure differential is essential for blood flow, but the radius of the vessel is even more important. At rest, the pressure differential is the main determinant of overall blood flow, whereas the vessel radius of the arteries determines where blood is distributed throughout the body. For example, the MAP in the aorta is close to 100 mmHg, whereas the pressure in the vena cava as blood enters the right atrium is close to 0. Thus, the pressure differential ($\Delta P = P_1 - P_2$) is close to the aortic pressure of 100 mmHg. This is the main "driver" of the overall circulation. The vessel radius changes in various segments throughout the body and determines how blood is distributed to the various tissues. However, the impact of blood viscosity and the length of the blood vessel should not be minimized, as these can make important contributions to blood flow. Change in overall blood flow, as when a person starts to exercise, is primarily a function of changes in pressure differential as aortic blood pressure increases with exercise while the large venous pressure remains close to 0. Vessel vasodilation/vasoconstriction determines how much blood the working muscles will receive compared to a nonworking tissue such as the intestinal system (Marieb, 2004). Nevertheless, changes in blood viscosity, which can occur with dehydration, will also affect blood flow during exercise.

Resistance

Resistance to blood flow can be described by Ohm's law. In an electrical circuit, the resistance R is equal to the ratio of the electromotive force, or the ratio of voltage drop (E) and the current flow (I), as follows:

$$R = E / I$$

For a fluid, such as blood, this can be expressed as $R = \Delta P / Q$. By rearranging and substituting, the following hydraulic resistance equation is derived from Poiseuille's law:

$$R = \frac{8\eta\ell}{\pi r^4}$$

From this equation it is clear that resistance depends on the size of the blood vessel and the viscosity of the blood. Since the radius of the vessel is raised to the fourth power, this is the principal determinant of resistance to flow through an individual vessel (Germann and Stanfield, 2002). In the human circulation, the greatest resistance to flow is provided by the arterioles, as illustrated in figure 1.4 (p. 8). The greatest pressure drop across the vascular system occurs in the arterioles. Considering that total flow is not altered and that the various components of the vascular system are in series, the resistance in the arterioles is proportional to the resistance in the other components of the system. This can be expressed as Ra/Rt, where Ra is the resistance in the arterioles and Rt is the resistance in the rest of the system (Guyton and Hall, 2000). Since $R = \Delta P / \dot{Q}$ and $\dot{Q}$ does not change (that is, $\dot{Q}a = \dot{Q}t$), $Ra / Rt = \Delta Pa / \Delta Pt$. Consequently, the ratio of the pressure drop in the arterioles and the rest of the system is equal to the ratio of the resistance in the arterioles and the rest of the system; thus the large drop in pressure in the arterioles is caused by the increase in resistance in these vessels. This is also facilitated by the anatomical structure of the arterioles, since the relatively large amount of smooth muscle in the arteriolar wall allows for potent vasoconstrictive properties, and only a small change in vessel lumen is required for a large change in resistance.

Another feature that affects resistance is anatomical arrangement of blood vessels. Going from the aorta to the capillaries, the vessels are arranged in series. This means that the same blood that enters the aorta also travels through the arteries, arterioles, capillaries, and so on until it is returned through the vena cava. Following the concept of hydraulic resistance, this means that under steady state flow, the resistance in any part of the system is additive to the resistance in the other parts of the system, and the total resistance is then the sum of all the individual resistances. Thus,

$$\text{Total resistance} = R1 + R2 + R3 + R4, \text{ and so on.}$$

This is depicted in figure 6.3*a*. It is important to remember that the resistance of the different portions of the vascular systems varies a great deal, but the overall resistance depends on the resistance in each individual component (Berne and Levy, 2001).

Even though the overall vasculature is arranged in series, within a particular portion of the system the vessels are arranged in parallel. This is especially prevalent at the capillary level, as shown in figure 6.3*b*. This means that blood that passes through one capillary will pass through only that particular capillary, and will not pass through any other capillaries. This arrangement also has a large effect on resistance. Consider that under steady state conditions, the inflow and outflow pressures are the same for all vessels, and the total blood flow is the sum of the flow through each individual vessel:

$$\text{Total } \dot{Q} = \dot{Q}1 + \dot{Q}2 + \dot{Q}3 + \dot{Q}4, \text{ and so on}$$

Since $R = \Delta P / Q$, this also means that the reciprocal of the total resistance (1/R) can be expressed as the sum of the reciprocal of the individual resistances:

$$1 / \text{Total resistance} = 1 / R1 + 1 / R2 + 1 / R3 + 1 / R4$$

If the resistances were all of equal magnitude, that is $R1 = R2 = R3 = R4$, then this could be restated as $1 / Rt = 4 / R1$. In this example, assuming that the resistance in all of the capillaries was the same, the total resistance would be a quarter of the individual resistance:

$$\text{Total resistance} = R1 / 4$$

This illustrates an important characteristic of parallel arrangements of blood vessels, where the total resistance is always less than any individual resistance. It explains why the resistance across a capillary bed is very low even though the resistance in an individual capillary may actually be higher than the resistance in an individual upstream vessel. Furthermore, since the velocity of flow through the capillaries is fairly slow, only a small pressure differential is required to maintain capillary perfusion (Mulvany and Aalkjaer, 1990).

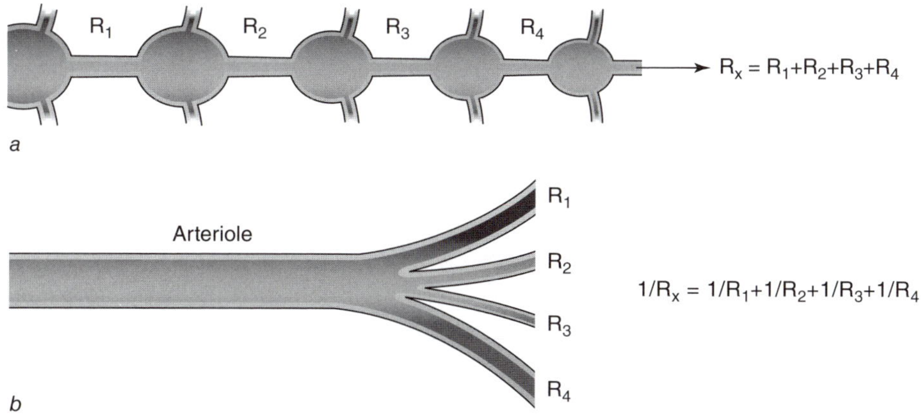

Figure 6.3 The effect of blood vessel arrangement on resistance to flow. *(a)* For blood vessels arranged in series, the total resistance is the sum of each individual resistance. *(b)* If vessels are arranged in parallel, the total resistance is less than the resistance in any individual vessel.

Viscosity

Viscosity refers to the friction between fluid layers as they slide past each other in conditions of laminar flow. This friction affects flow by impeding the movement of the individual fluid layers. The viscosity is measured as the ratio between shear stress and shear rate. Shear stress is loss of energy due to internal resistance, or the force applied to a segment of the laminar flow, divided by the area in contact with the fluid. Shear stress then is the relative velocity of the adjacent fluid. As the coefficient of viscosity increases, it takes more energy to move the fluid. The viscosity is measured in poise, which equals 1 dyne second/cm^2. However, direct measurement is very difficult, so viscosity is often measured as relative viscosity, where the fluid in question is compared to water. A higher relative number indicates a more viscous fluid. For example, human blood has a relative viscosity of 3 to 4, but human plasma has a lower viscosity of around 1.8 (Little, 1985).

The viscosity of blood is also related to the diameter of the tube, called the Fahraeus-Lindqvist effect. In general, in conditions of laminar flow, the red blood cells are concentrated in the center of the vessel, leaving a relatively cell-free layer of plasma near the vessel wall. This layer becomes proportionally greater as the size of the vessel lumen decreases, creating the appearance that the viscosity is reduced (see figure 6.4). It is likely that this effect decreases the resistance to flow in the arterioles compared to larger arteries, but not in the capillaries. In the capillaries, the red blood cells flow through the vessel lumen one at a time, thus preventing continuity of plasma flow (since the plasma is now located between the red blood cells, not between the cells and the vessel wall).

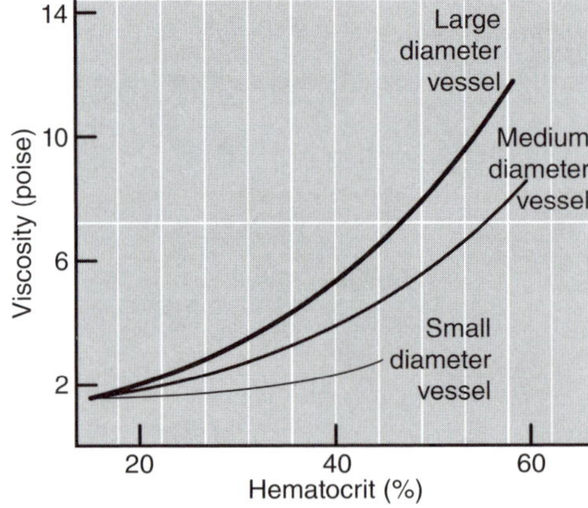

Figure 6.4 Effect of hematocrit concentration on viscosity in different-size blood vessels. The hematocrit concentration has little effect on blood viscosity in smaller vessels, but has a substantial effect in larger vessels.

In blood, viscosity is altered primarily by via alterations in the hematocrit (percentage of blood that is composed of red blood cells [RBC]). In the normal hematocrit range, a change in the hematocrit level produces an equal change in viscosity. As hematocrits increase above 50%, the viscosity increases in a nonlinear manner. This has implications for exercise, because dehydration increases hematocrit levels (as plasma is lost), thus increasing viscosity. This leads to an increase in the energy required to produce flow, thus increasing the work of the heart (Little, 1985).

BLOOD FLOW

The laws of hemodynamics described in the previous section are general guidelines for understanding human blood flow. But because blood is a somewhat viscous fluid and the arteries are not rigid tubes but flexible and capable of changing diameter, there are several other factors that affect blood flow.

Laminar Flow

Laminar flow describes what is also called streamlined flow. When flow is laminar, we can think of layers of blood moving in the same direction in a vessel, but at different speeds. The layer closest to the vessel wall is actually stationary, and its motion is prohibited by cohesive attraction between the blood and the vessel wall. The next layer slides by at a slow velocity, and then the velocity of each subsequent layer increases until the greatest velocity is reached in the center of the vessel. This is illustrated in figure 6.5.

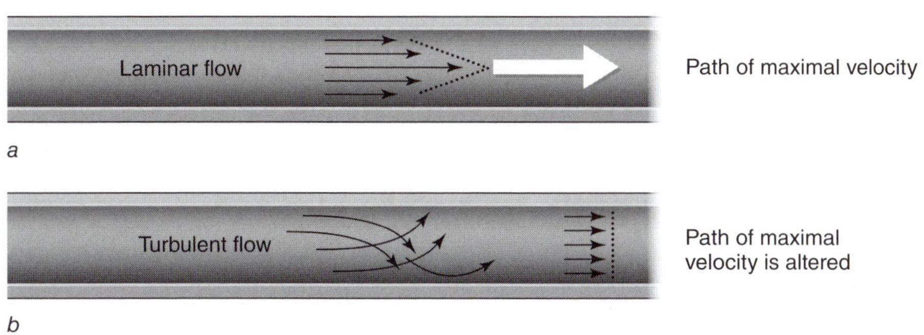

Figure 6.5 Laminar versus turbulent flow. *(a)* In laminar flow, the velocity of flow next the vessel wall is close to zero. The velocity of subsequent layers of blood is increased until the maximal velocity is reached in the center of the vessel. *(b)* During turbulent flow, the different layers of blood cross paths, creating more resistance to flow and altering the path of maximal velocity.

Turbulent Flow

Turbulent flow is primarily dependent on the velocity of flow and the size of the vessel. As the velocity increases, at some critical point the flow will start to develop swirls and vortices; thus the various layers of fluid are no longer moving linearly (see figure 6.5). The point where turbulent flow will develop can be predicted from the Reynold's number, which is calculated from the following equation:

$$Rn = vd\rho / \eta$$

Rn is the Reynolds number (which is dimensionless); v is the velocity of the blood flow; d is the diameter of the blood vessel; ρ is rho or the density; and η is the viscosity of the blood. From this equation it is easy to see that turbulent flow is more common in larger than in smaller vessels, and the higher the velocity of flow the greater the likelihood of turbulent flow. Consequently, turbulent flow is common in the aorta, close to the valve, but virtually nonexistent is smaller arterioles.

During turbulent flow, all molecules in the blood travel at the same speed in relation to the vessel wall. This is different from laminar flow, where the speed is greatest in the center of the vessel and gradually decreases closer to the vessel wall. As a result, the flow becomes proportional to the square root of ΔP, instead of being directly proportional to ΔP as during laminar flow. This leads to a much greater energy requirement during turbulent flow, increasing the load on the heart. Thus, in conditions of turbulent flow, the heart has to work harder to produce the same stroke volume compared to laminar flow conditions (Berne and Levy, 2001).

Shear Stress

When blood flows through a vessel, it pulls the endothelial lining of the vessel with it, creating what is called a shear stress on the vessel wall. This force is parallel to the surface of the vessel wall, in effect "dragging" the wall in the direction of the flow. Shear stress can be calculated by the following equation, where τ is the shear stress, η is viscosity, $\dot{Q}$ is blood flow, and r is the radius of the vessel:

$$\tau = 4\eta\dot{Q} / \pi r^3$$

From this equation, it is clear that shear stress is greater when the blood flow or viscosity increases. Shear stress is also greater in smaller vessels (assuming constant flow and viscosity), since the radius plays an important role. Shear stress is important because it stimulates physiological processes related to vasodilation, as discussed in chapter 7.

Conversion of Pulsatile to Steady Flow

As the ventricles contract, the blood entering the aorta is pulsatile in nature, as most of the stroke volume is ejected very rapidly and no blood enters the aorta following the completion of systole. Since systole composes only 33% of the cardiac cycle, no blood actually enters the aorta during the majority of the cardiac cycle. However, the capillaries require steady flow for continual exchange of gases and nutrients. The arterial system gradually converts this pulsatile flow to a steady flow by using stored elastic energy to act as secondary pumps during diastole.

The aorta is very elastic, which allows it to stretch when blood is ejected from the left ventricle. During diastole, the elastic recoil of the aorta serves as a secondary pump, pushing blood down the arterial tree. This is essentially the same principle that applies to a rubber band. After a rubber band has been stretched, if released it will return to its initial length as shown in figure 6.6.

This elastic property of the arterial system functions as a hydraulic filter, ensuring that as blood travels from large to smaller arteries, the pulsatile flow is gradually dampened and converted to a steady flow at the capillary level. This is illustrated in figure 1.4 (see page 8), where the pulsatile nature of blood flow is decreased concomitantly with the decrease in flow velocity and an increase in the cross-sectional area of the vasculature.

Compliant arteries also ensure that the workload on the heart is minimized. If the hydraulic filter were perfect, that is 100% effective, then there would be no pressure variations between systole and diastole in any of the arterial circulation. As seen in figure 1.4, this is not the case; however, compliant arteries do ensure that the pulsatile

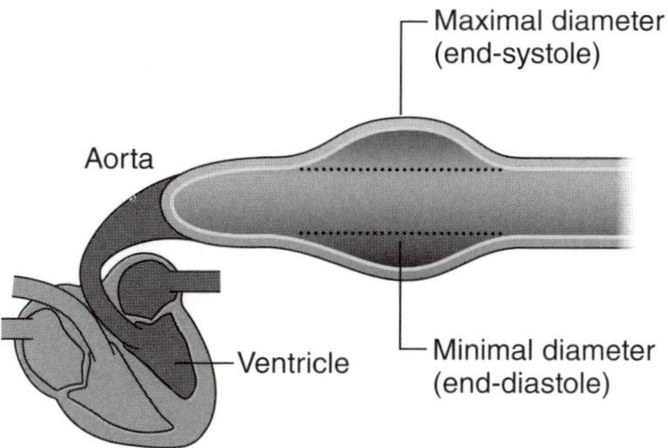

Figure 6.6 Arterial compliance assisting blood flow as secondary pump. The aorta stretches to accommodate blood flow during ventricular systole. It returns to its initial length during ventricular diastole; thus the aorta functions as a "secondary pump," assisting blood flow.

flow is eventually converted to steady flow with no pulsatile component (Remington and O'Brien, 1970). But if the arteries become stiff, as happens with aging and with various types of diseases such as heart disease, this creates two major negative effects. First, the heart has to work harder since the volume of blood now enters a smaller area (the aorta will not distend as much). This leads to higher velocity of flow, increasing the chance of turbulent flow, imposing additional work on the heart. In an attempt to conserve energy, the stroke volume will then decrease, limiting cardiac output. Thus, stiff arteries may actually contribute to limitations to exercise capacity. Second, the pulse pressure (the difference between systolic and diastolic pressure) will increase, contributing to an increase in pulsatility that extends farther down the arterial tree, thus increasing the likelihood of turbulent flow in comparatively smaller arteries. This increases the resistance to flow, and in turn further increases the blood pressure and the work of the heart (Mulvany and Aalkjaer, 1990). The effect of age on arterial stiffness is illustrated in figure 6.7. In fact, arterial stiffness increases in a linear manner from age 5 throughout life; thus the heart of an older person needs to work much harder to produce the same cardiac output as that of a younger person.

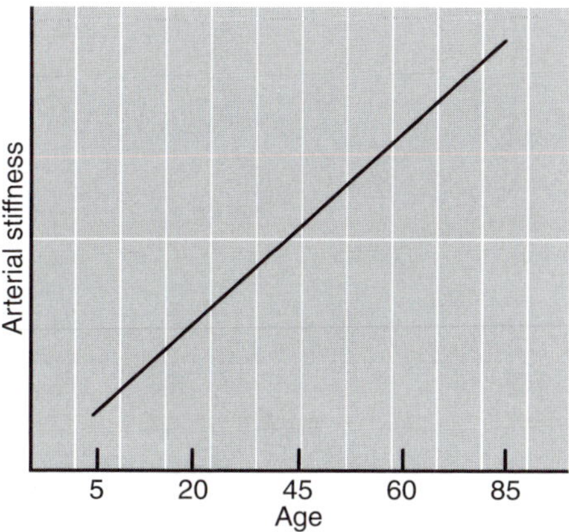

Figure 6.7 Arterial stiffness and age. Arterial stiffness increases linearly with age (after age 5) in the absence of disease. Arterial disease can increase the slope of arterial stiffness at any age.

ARTERIAL BLOOD PRESSURE

Blood pressure refers to the pressure of the blood in the arteries. Systolic blood pressure (SBP) refers to the pressure in arteries during ventricular systole, and diastolic blood pressure (DBP) refers to the pressure during ventricular diastole. The average blood pressure, mean arterial blood pressure (MAP), is also important because it is the "driving pressure"; thus MAP determines the amount of blood flow through the body (Guyton and Hall, 2000). Normal blood pressure for an adult is a pressure below 120/80 (Chobanian et al., 2003). However, ideal blood pressure is a pressure lower than 115/75. High blood pressure, or hypertension, has negative health consequences and is defined as a blood pressure above 140/90, but pressures of 120-139/80-89 are considered prehypertensive and are associated with increased risk of developing cardiovascular disease. In fact, increased risk for future cardiovascular disease starts with blood pressures higher than 115/75 (Chobanian et al., 2003).

Arterial pressure can be measured by inserting a pressure transducer directly into the artery. This produces a pressure curve, as shown in figure 6.8. A way to calculate the driving pressure (MAP) from this pressure curve is to measure the area under the curve and divide by the time from the beginning of systole to the end of diastole. However, this is a cumbersome method and not used in practical settings. Instead, MAP is determined using the following formula:

$$MAP = 1/3\ (SBP - DBP) + DBP$$

Mean arterial pressure is not simply an average of SBP + DBP because the heart spends much more time in diastole than in systole; thus the pressure during systole lasts for only a short time. Since the relationship between systole and diastole changes during exercise (relatively less time is spent in diastole as heart rate increases), it is suggested that MAP during exercise be calculated as $MAP = 1/2\ (SBP - DBP) + DBP$.

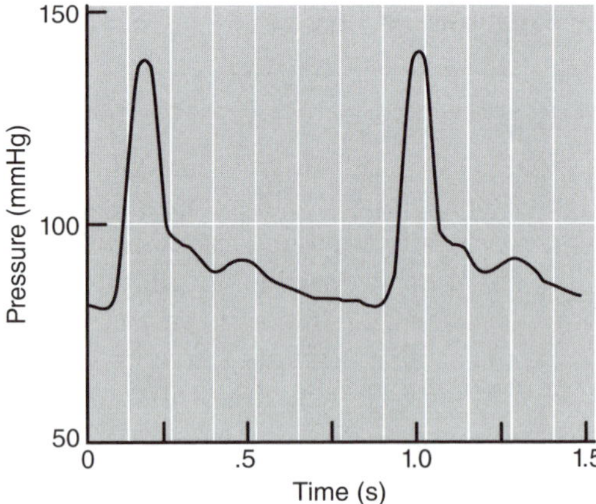

Figure 6.8 Pulse wave morphology. The pressure wave created by the input of blood into the aorta is transmitted throughout the arterial system. The baseline represents diastolic pressure, and the peak of the wave represents systolic pressure.

Several factors contribute to systemic blood pressure. It is common to use a variation of the pressure, flow, and resistance equation to understand systemic blood pressure (BP):

$$\text{MAP} = \text{Cardiac output} \times \text{Total peripheral resistance}$$

From this formula, it is easy to see that an increase in flow ($\dot{Q}$), through an increase in either stroke volume or heart rate, will also increase BP. Similarly, an increase in total peripheral resistance will increase BP. During exercise, $\dot{Q}$ increases and total peripheral resistance decreases, but the increase in cardiac output is greater than the decrease in total peripheral resistance, so MAP increases modestly. The most common cause of hypertension is an increase in total peripheral resistance. However, other factors also influence BP, including arterial stiffness (a stiffer artery will cause higher pressure as discussed earlier) and blood volume. A decrease in blood volume will cause a decrease in BP (or conversely, an increase in BP occurs with an increase in blood volume). This is what happens during hemorrhage, where the loss of blood can decrease the BP to dangerously low levels, limiting blood flow to the brain and other vital organs.

DETERMINING THE CAUSE OF HIGH BLOOD PRESSURE

During a routine preseason physical, a male athlete exhibited high blood pressure. The team physician suspected that the athlete was overtraining and recommended a reduction in training load (overtraining can cause increased BP through overactivation of the sympathetic system). However, follow-up evaluations revealed no change in BP even with severe reductions in training load. The athlete was started on an ACE (angiotensin-converting enzyme) inhibitor, and within four weeks his BP was normal. Thus, his high BP was not caused by overtraining, but by an overactive renin-angiotensin-aldosterone system. He subsequently returned to training and competition without any further complications.

PULSE WAVES AND WAVE REFLECTIONS

The stretch of the aorta accommodating blood from the contracting ventricle creates a pressure wave that is transmitted through the arterial tree. This pressure wave can be measured, and the aortic pressure wave looks much like the wave displayed in figure 6.8. The pressure wave is also transmitted through the arterial tree at a much faster rate than blood. For example, when the pulse is measured at the radial artery (at the wrist), it is the pulse wave pressure that is sensed and counted.

How fast the pressure wave travels throughout the arterial system is directly related to the stiffness of the arteries. The stiffer the arteries, the faster the transmission. Thus, pulse wave velocity is measure of arterial stiffness. However, the pulse wave velocity varies within the arterial tree. The pulse velocity of a central artery, such as the aorta, is much slower (indicating a more compliant or less stiff artery) than the pulse wave velocity in a peripheral artery. Thus, peripheral arteries are stiffer (less compliant) than central arteries, which is a fundamental difference in functional arterial characteristics. This is likely due to the difference in arterial wall composition; the elastic central arteries have relatively less smooth muscle but more elastin, whereas peripheral arteries are characterized by a substantial amount of smooth muscle and less elastin.

The pulse wave changes contour depending on where in the arterial tree it is measured. Figure 6.9 shows how the pulse wave contour changes from the aorta to the upper leg and then down to the ankle. As the pulse wave becomes more distal, the incisura (notch at the end of ventricular ejection) gradually disappears, and the systolic portion becomes more peaked and narrower. Consequently, the SBP is higher at the distal sites compared to the aorta; however, the diastolic pressure decreases, leading to an increase in pulse pressure. Interestingly, the mean pressure changes little; thus the driving pressure for blood flow is similar at the distal and central sites of the arterial tree (for conduit arteries; this is not true for smaller arterioles and the capillary system—see figure 1.4). Also, the pulse contour difference between central and distal arteries is most prominent in young individuals. As people age, the pulse wave contour changes less, which is probably related to the increase in arterial stiffness observed with aging (O'Rourke and Nichols, 2007).

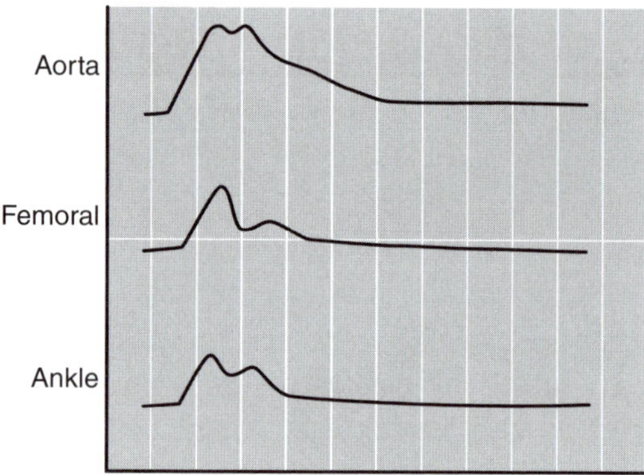

Figure 6.9 Pulse pressure waveforms at different arterial sites.

One of the main reasons the systolic peak increases at distal sites in peripheral arteries is wave reflection. This concept is illustrated in figure 6.10. As the forward pulse wave encounters bifurcations or other changes in arterial anatomy, some portion of that wave is reflected back up the arterial tree. This reflected wave will then merge with the next forward wave, amplifying it and forming a more peaked systolic portion of the pulse wave (augmented pressure). This also leads to an increase in SBP. This concept is illustrated in figure 6.10. Consequently, wave reflection can amplify SBP and increase the load on the heart. Wave reflection is also influenced by arterial stiffness, and a stiffer artery will usually produce greater SBP amplification. In addition, an early return of a reflected wave can decrease the diastolic wave while still increasing the systolic wave, thus increasing pulse pressure as shown in figure 6.10. This may decrease coronary perfusion pressure because diastolic pressure is a

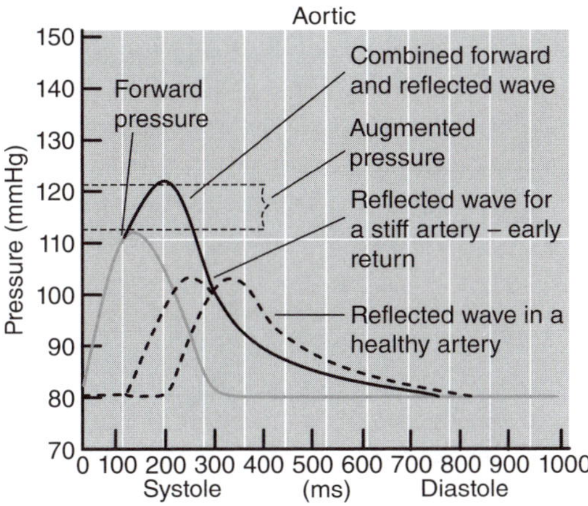

Figure 6.10 Wave reflection and augmentation. When the pressure wave traverses down the arterial tree, it will hit points of bifurcations, and part of the wave will be reflected back up the arterial system. It will then combine with the next pressure wave, increasing systolic pressure and pulse pressure.

major determinant of coronary perfusion, which could have serious implications for persons with coronary artery disease.

BLOOD PRESSURE MEASUREMENT

As mentioned earlier, a means of measuring "true" BP is to insert a pressure transducer into an artery, thereby measuring the pressure directly. However, this measures only the pressure in that artery and is highly invasive, thus not practical for everyday use. Instead, arterial blood pressure is measured via the cuff method; an inflatable cuff is placed around the upper arm.

The cuff method is based on the Korotkoff sounds, and the technique has not changed in over 100 years. The Korotkoff sounds are caused by reestablishment of blood flow after cuff inflation, and are measured through the use of a stethoscope. The stethoscope is placed over the brachial artery, and the cuff is inflated to a pressure above the SBP. This shuts off blood flow through the brachial artery. As the pressure is gradually released, the pressure in the artery becomes greater than the pressure in the cuff, and blood will start to flow through the pressure point again. However, the beginning flow will be turbulent because the size of the artery has been altered by the pressure from the cuff (see figure 6.11 on page 98), and the turbulent flow produces audible vibrations (Korotkoff sounds) that can be heard with the stethoscope. The Korotkoff sounds are classified in five phases. Phase I is the appearance of a clear tapping (pulse) sound, which corresponds to the appearance of a palpable pulse. During phase II, the sounds become longer and softer. This changes again in phase III as the sounds become louder and crisper. Phase IV is characterized by a muffling of the sounds. Finally, phase V corresponds to the complete disappearance of sound (Pickering et al., 2005). The pressure in the cuff when the first sound can be heard (phase I) is the SBP. The air in the cuff is then continually released until the flow is no longer turbulent and no sound can be heard. The pressure where the sound disappears (phase V) is the DBP (Germann and Stanfield, 2002). When the cuff is placed around the upper arm and kept at heart level, these pressures have been assumed to be roughly equal to aortic blood pressure; but this is true only in some people, and aortic pressure can vary considerably from the brachial pressure.

Blood pressure measurements need to be standardized because BP varies with the size of the cuff; body position; room temperature; alcohol, nicotine, and food consumption; exercise; muscle tension; arm position; talking; and background noise. Thus, the American Heart Association has made the following recommendations (Pickering et al., 2005):

1. Blood pressure should be measured in a quiet room, following at least a 5 min rest.
2. The patient should be comfortably seated in a chair, with the back supported and legs uncrossed, and the arm should be supported in a position such that the middle of the cuff is at the level of the right atrium. This level usually corresponds to the midpoint of the sternum.
3. The cuff must be of the correct size. The ideal cuff will have a bladder length that corresponds to 80% of the length of the arm, and a bladder width that corresponds to at least 40% of the arm circumference. A cuff that is too small or too large will provide inaccurate readings; the inaccuracy is greater with a cuff that is too small.
4. To make sure that cuff placement is correct, the brachial artery in the antecubital fossa needs to be palpated, and the cuff should be placed so that the brachial artery corresponds to the midline of the bladder. It is important that the lower end of the cuff be 2 to 3 cm above the antecubital fossa to allow room for the stethoscope.
5. The use of a mercury sphygmomanometer is recommended; however, these devices are being phased out. A calibrated aneroid or oscillometric sphygmomanometer is adequate.
6. A minimum of two measurements should be made, with at least 1 min between readings. The average of the two readings is used if the difference between them is less than 5 mmHg. If the difference is greater than 5 mmHg, additional readings should be obtained and the average of all of the readings should be used.
7. It is important that no talking take place during readings.

Another recent method that also yields arterial pressure waveforms is finger plethysmography. In this method, a small cuff is placed around a finger and inflated to constant pressure. The pressure changes caused by each heartbeat are then measured and SBP, DBP, and MAP can be calculated. The advantage of this method is that beat-to-beat blood pressures can be recorded, but the instrumentation is expensive and the measurement is time-consuming. Therefore, blood pressure is usually measured using the cuff and stethoscope method (standard sphygmomanometry) (Perloff et al., 1993).

Recently developed technology can actually allow estimation of aortic pressure using a combination of brachial blood pressure and pulse contour measurements from the radial artery pulse wave. This methodology also allows measurement of pulse pressure amplification. The technique, called applanation tonometry, involves partially depressing the radial artery with a very sensitive pressure transducer. This produces a pulse waveform, and the wave reflection properties can be measured. Using an algorithm, the aortic pressure is then calculated. This method has been validated against invasively measured BP and has been shown to be more clinically sensitive than standard brachial pressure. However, the technology is expensive, requires a great deal of training, and is not generally available in clinical settings.

CONTROL OF VASOCONSTRICTION AND VASODILATION

Regulation of peripheral blood flow is determined primarily by the vascular tone of small arteries and arterioles. By vasoconstricting or vasodilating, these arteries direct blood flow to where it is needed and direct flow from areas where less flow is needed. Vascular smooth muscle is under the control of both local peripheral factors (intrinsic regulation) and the central nervous system (extrinsic regulation). In vessels supplying skeletal muscle, local control is usually more important, whereas cutaneous vessels are usually more influenced by central factors.

Vasoconstriction and vasodilation of arteries are determined by the degree of contraction of vascular smooth muscle. Contraction of smooth muscle leads to vasoconstriction, while smooth muscle relaxation leads to vasodilation. Smooth muscle is usually in some state of contraction, called vascular tone. Thus, the action of smooth muscle in the small arteries and arterioles provides most of the control of peripheral vascular resistance, which greatly affects both blood flow and blood pressure.

Local Control

There are several local factors that control local blood flow. Perfusion pressure in the arterioles regulates blood flow to the tissues through a process called myogenic autoregulation. If perfusion pressure increases in the arterioles, the smooth muscle will contract in response to increased transmural pressure in the arterioles caused by the increase in perfusion pressure. As a result, blood flow is maintained at very constant levels in the microcirculation and the capillary beds. This concept is shown in figure 6.11. This mechanism also protects the capillaries, maintaining steady pressure and flow through them. Conversely, a decrease in the perfusion pressure leads to arteriolar vasodilation and increased flow until the steady flow is again achieved (figure 6.11). The **myogenic autoregulation** is independent of the endothelium and appears to be directly controlled by smooth muscle, possibly through activation or deactivation of calcium channels that increases or decreases intracellular calcium (Mulvany and Aalkjaer, 1990).

The myogenic autoregulation also provides protection against excessive arterial wall stress. The law of Laplace states that

$$\text{Wall tension} = \text{Transmural pressure} \times \text{Vessel radius}.$$

Thus, at a constant pressure, vasoconstriction actually produces less wall tension. This law applies to thin-walled vessels, especially capillaries. However, when the vessel walls are thicker, wall thickness also needs to be accounted for. The means of doing this is to divide the term transmural pressure × radius by wall thickness:

$$\text{Wall tension} = (\text{Transmural pressure} \times \text{Vessel radius}) / \text{Wall thickness}$$

This formula shows that an increase in vessel wall thickness decreases wall tension. Thus, in the case in which an arteriole responds to increased transmural pressure by

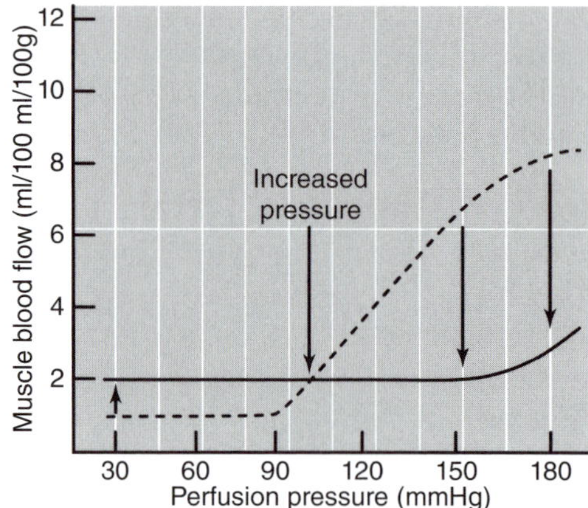

Figure 6.11 Myogenic autoregulation. A change in perfusion pressure causes a temporary change in arteriolar flow, which quickly returns to baseline values (or close to baseline) as a result of myogenic autoregulation.

vasoconstricting through myogenic regulation, this also has the effect of protecting the vessel by decreasing or keeping the vessel wall tension constant. It also protects the capillary, since the transmural pressure in the capillary will then also be kept constant (or decreased); and in turn the wall stress is not increased. Considering that the capillary does not have much wall structure to protect it from rupturing, this is an important mechanism for keeping capillary flow and wall tension within appropriate limits (Berne and Levy, 2001).

The endothelium is also directly involved in controlling the vasoactive response. For instance, shear stress activates an endothelium-mediated pathway leading to vasodilation. The vasodilation is caused by the formation of NO (nitric oxide) as discussed in chapter 7. If the formation of NO is prevented (by infusion of an NO inhibitor), the vasodilation is greatly diminished as shown in figure 6.12. However, NO inhibition

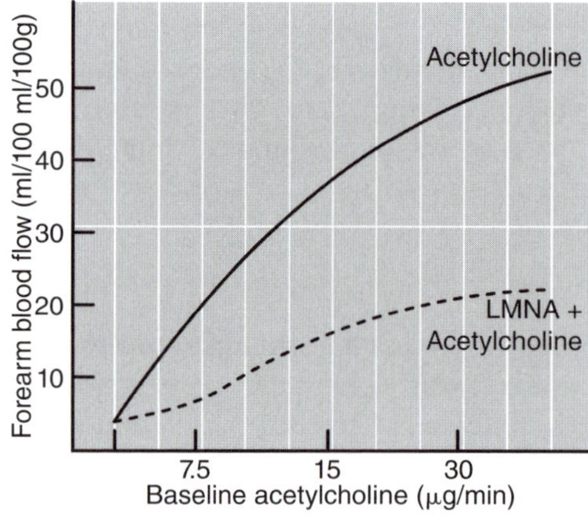

Figure 6.12 Effect of nitric oxide (NO) and NO inhibition on forearm blood flow. Forearm blood flow increases with acetylcholine infusion into the artery due to an increase in NO production, which leads to vasodilation. This response is decreased, but not abolished, when the NO formation is blocked using an NO blocker (LMNA), showing that other vasodilators are also important.

does not abolish vasodilation completely, suggesting that other vasodilators are also involved. These vasodilators are likely also of endothelial origin, since removing the endothelium from a vessel completely abolishes the vasodilatory response.

The endothelium can also release vasoconstrictor substances, such as endothelin-1, which is a powerful vasoconstrictor. Thus, depending on the type of endothelial activation, the endothelium can promote both vasodilation and vasoconstriction. Although the endothelium controls these processes, the actual vasomotion is produced by smooth muscle contraction or relaxation, through several processes initiated by the endothelium (see chapter 7 for more detail) (Williams and Lind, 1979; Koller et al., 1994).

It is also generally accepted that the metabolic activity of the tissue is involved in the regulation of blood flow. This is likely related to the need for oxygen in the tissue, and a decrease in tissue oxygenation is thought to lead to vasodilation. However, this is probably not a direct effect of hypoxia (lack of oxygen) per se, but rather the effect of various vasodilator substances that are released in hypoxic conditions. Although several substances have been proposed, the exact substance has not been identified. It is thought that potassium ions may have an immediate, short-term effect of vasodilation. Adenosine and prostaglandins have also been proposed, together with inorganic phosphates, hydrogen ions, and acetylcholine. There is some support for most of these vasodilators, but the relative importance of each has not been determined. During muscle contraction, it is likely that a combination of these vasodilators interacts to produce the needed increase in blood flow.

If blood flow is inhibited (occluded) to a particular tissue, the blood flow following removal of that inhibition (occlusion) is increased compared to the basal flow. This is called reactive hyperemia. In humans, reactive hyperemia can be induced through placing a blood pressure cuff on a limb, like the upper arm, then inflating the cuff to suprasystolic pressures. This stops arterial inflow to the arm, creating a hypoxic condition in the tissue. After release of the cuff, the blood flow to the arm increases substantially, as illustrated in figure 6.13. **Reactive hyperemia** is greater with longer

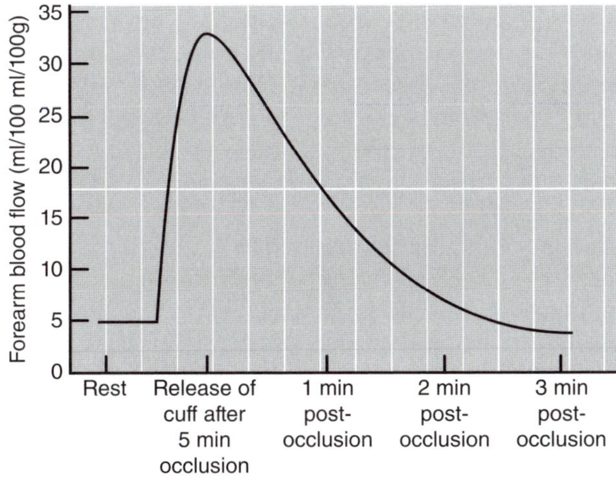

Figure 6.13 Reactive hyperemia in the forearm. Effect of 5 min of blood flow occlusion (using an inflated blood pressure cuff around the upper arm) on forearm blood flow. A reactive hyperemia results in a five times increase in flow after cuff release, and flow gradually returns to baseline within 3 min.

periods of occlusion, but a 5 min occlusion period is commonly used. The reactive hyperemia becomes even greater if the arm is exercised during the occlusion, suggesting that local metabolic factors are responsible for the vasodilatory response.

Several other chemical factors also influence vasomotion. Acetylcholine produces a vasodilatory response through activation of the NO pathway, which may have a role during exercise (acetylcholine is the neurotransmitter at the neuromuscular junction). Other substances such as bradykinin, serotonin, and histamine also have vasodilatory effects; and prostaglandins can have both vasodilatory and vasoconstrictive effects. It is likely that there is a great deal of redundancy in control of vasomotion, suggesting that if a certain pathway is damaged or less functional, another pathway can take over and produce at least a partial effect (Little, 1985).

Extrinsic Control

Extrinsic control of vasomotion in the periphery is achieved by several factors, including the medullar cardiovascular control centers of the brain mediating sympathetic and parasympathetic activity, hormonal factors (mostly involving catecholamines), baroreceptors, chemoreceptors, and the rennin-angiotensin-aldosterone system (RAAS) (related to fluid control and kidney function).

The cardiovascular control center located in the medulla provides primarily sympathetic input to the vasculature. The basal vascular tone is accomplished by tonically active sympathetic stimulation, produced by the release of norepinephrine from the sympathetic nerve endings. Norepinephrine binds to α-adrenergic receptors producing vasoconstriction of the arterioles and small arteries. Inhibition of the sympathetic impulses produces less vasoconstriction and can result in relative vasodilation. In general, sympathetic stimulation results in vasoconstriction, a reduction in local blood flow, and increased peripheral resistance.

Epinephrine and norepinephrine produce the main hormonal vascular effects. Both of these hormones are released from the adrenal medulla and are then transported through the circulation. However, norepinephrine is also a neurotransmitter and is therefore also released from sympathetic nerve endings. As a result, the local concentration of norepinephrine is probably higher than that of epinephrine. Epinephrine has a dual effect on vasomotion. Small concentrations bind to β1-receptors in the coronary vessels and to β2-receptors in blood vessels supplying skeletal muscle. This produces vasoconstriction in the heart vasculature, but vasodilation in skeletal muscle. At higher concentrations, epinephrine also binds to α-receptors, and the net result is vasoconstriction. Norepinephrine always binds to α-receptors and produces vasoconstriction.

Several important hormonal actions related to kidney function have drastic effects on blood pressure and vasomotion (McGill, 2009). Renin is released by the kidney in response to a drop in blood pressure or renal blood flow. As shown in figure 6.14, renin converts angiotensinogen to angiotensin I, which is then converted to angiotensin II. Angiotensin II is a powerful vasoconstrictor, primarily affecting the arterioles. This increases blood pressure and vascular resistance. Angiotensin II also triggers the release of aldosterone from the adrenal cortex. Aldosterone in turn increases fluid retention by conserving renal sodium. This leads to an increase in plasma volume, thus increasing blood pressure. In healthy individuals, these hormones are the primary factors in long-term blood pressure control. However, hypertension and kidney disease are associated with an increase in rennin production resulting in larger than

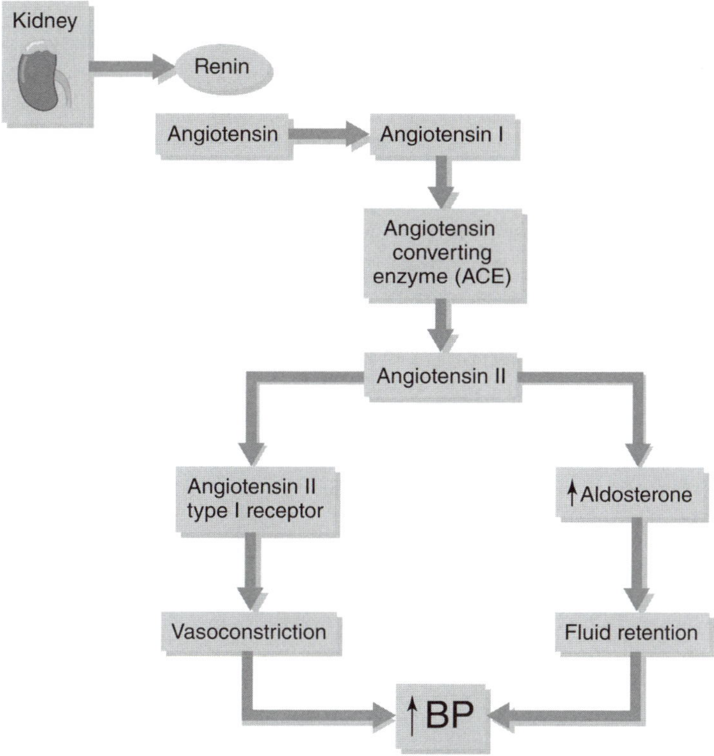

Figure 6.14 The renin-angiotensin-aldosterone system. Renin is released from the kidney and converts angiotensinogen to angiotensin I. Angiotensin-converting enzyme (ACE) then converts angiotensin I to angiotensin II. Angiotensin II is a potent vasoconstrictor and also increases the production of aldosterone, leading to an increase in blood pressure, through two different pathways.

normal increases in angiotensin II, producing substantial vasoconstriction and thus causing an abnormal increase in blood pressure.

REFLEX CONTROL OF BLOOD PRESSURE AND VASOMOTION

Arterial baroreceptors are located in the aortic arch and the carotid sinuses. These receptors are stimulated by a stretch in the artery caused by an increase in BP (or an increase in BP caused by a stretch). A good example of how the baroreceptors work is change in posture. When someone assumes an upright posture from a supine position, a large redistribution of blood volume occurs. Blood leaves the central circulation and pools in the lower extremities. This leads to a decrease in venous return, decreased stroke volume (SV), and decreased cardiac output (Raven et al., 2000; Rowell, 1986). If the BP is substantially lowered, enough blood is not getting to the brain, producing slight dizziness. All these changes lower central blood pressure, which is sensed by the arterial baroreceptors. To correct the situation, BP must be increased. As a result, the medulla oblongata sends sympathetic nerve impulses to the heart and the periphery, resulting in increased HR, peripheral vasoconstriction, and increased venous tone.

These changes will increase SV and increase total peripheral resistance (TPR); and the combination of increased SV and increased HR will also increase cardiac output, resulting in an increase in MAP. Assuming a supine position (from standing) produces the opposite effects. Blood will leave the periphery and pool centrally, increasing central blood pressure. This is sensed by the arterial baroreceptors, resulting in increased parasympathetic stimulation and decreased sympathetic stimulation to the heart and periphery by the medulla oblongata. This will lower HR, decrease TPR and venous tone, and decrease SV, thus ultimately decreasing the cardiac output. These changes will decrease the MAP. The action of the baroreceptors is very quick, occurring within two heartbeats of the change in pressure. Thus, the baroreceptor reflex is one of the main control mechanisms for short-term control of blood pressure.

The changes in HR imposed by the arterial baroreceptors in response to a change in pressure have an effective range within which a relatively small change in pressure will produce a large change in HR, as shown in figure 6.15. The response at either the upper or lower ranges of BP is much smaller, or almost nonexistent. The changes in heart rate are achieved by reciprocal changes in both sympathetic and parasympathetic stimulation.

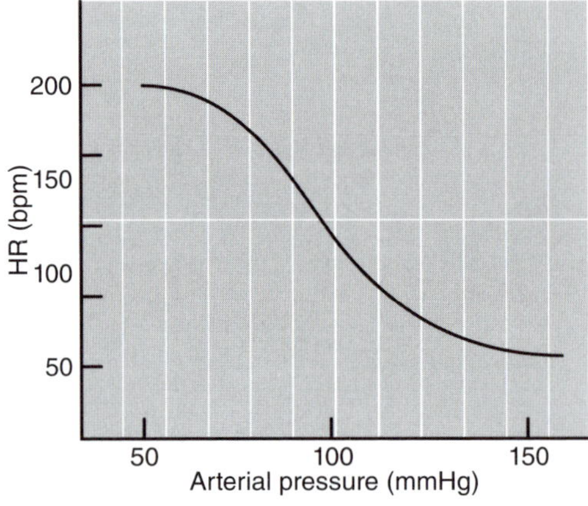

Figure 6.15 The baroreflex arc. The baroreflex has an effective range; at low and high pressures, the HR response to a change in pressure is diminished.

The **Bainbridge reflex** is caused by changes in blood volume, sensed by receptors in the atria and the venous system. When an increase in blood volume causes a stretch of the right atrium, heart rate is accelerated. This results in an increase in cardiac output without a change in SV. A decrease in blood volume, resulting in a decrease in the stretch of the atria, will have the opposite effect: a decrease in HR and cardiac output. At low blood volumes, a forceful contraction of the ventricles may severely affect HR, causing a large, possibly detrimental drop in HR and reducing Q to dangerously low levels (Hainsworth, 1995). However, often the Bainbridge reflex and the baroreceptor reflex oppose each other. For instance, at low blood volumes, HR can increase (HR should decrease if controlled by the Bainbridge reflex), suggesting that the baroreceptor reflex has overridden the Bainbridge reflex (low blood volume would also drop MAP, thus stimulating baroreceptors to increase cardiac output). Although not directly associated with the Bainbridge reflex, a stretch of the atria also causes a stimulation for diuresis in an attempt to reduce blood volume. This is accomplished through the release of hormones, primarily antidiuretic hormone (Berne and Levy, 2001).

Cardiopulmonary baroreceptors are stretch receptors located in the heart and lungs. They are especially sensitive to changes in blood volume. For instance, unloading of these receptors, as would occur during hemorrhage, causes substantial increases in sympathetic neural output, producing substantial vasoconstriction. The cardiopulmonary receptors also have hormonal effects, producing increased release of hormones such as angiotensin, antidiuretic hormone, and aldosterone. The net effect is a reduction in urine output and sodium retention, coupled with vasoconstriction, aimed at preserving fluid. The cardiopulmonary receptors are involved during more common tasks such changes in posture. Transitioning from a supine to an upright position unloads the receptors, causing increased peripheral sympathetic-mediated vasoconstriction in order to protect against a drop in BP. Conversely, loading of the cardiopulmonary receptors causes an increase in parasympathetic output, which may depress sympathetic output. This seems to be the case during mild exercise; here, the exercise-induced increase in central filling pressure loads the cardiopulmonary receptors, causing a vagally mediated reduction in sympathetic output that results in reduced vasoconstriction (Berne and Levy, 2001; Ray and Sito, 2000).

Peripheral chemoreceptors are sensitive to changes in hydrogen ions, and an increase in chemoreceptor stimulation will primarily increase pulmonary ventilation. A mild respiratory stimulus will decrease HR. However, a large respiratory stimulus will inhibit the medullary vagal centers, thus decreasing the parasympathetic influence on the heart, producing an increase in HR through vagal withdrawal (Berne and Levy, 2001; Guyton and Hall, 2000) and thereby increasing BP. The peripheral chemoreceptors also produce increased sympathetic stimulation leading to reflex vasoconstriction and vascular tone, thus increasing BP. A decrease in BP can actually activate the chemoreceptors, because the decrease in pressure decreases blood flow, which causes a decrease in oxygen and a buildup of carbon dioxide. However, in normal situations, the chemoreceptors are not important contributors to control of BP because they are not substantially activated unless arterial pressure drops below 80 mmHg (Guyton and Hall, 2000).

SUMMARY

Blood flow is determined primarily by the pressure differential and the radius of an artery. Blood flow is also influenced by viscosity and by whether the flow is laminar or turbulent. Numerous mechanisms control vasoconstriction/vasodilation and heart rate, thereby altering both total systemic flow and flow distribution throughout the body. Vasodilation/vasoconstriction is controlled by central factors such as sympathetic stimulation, or by local factors such as myogenic autoregulation and the release of local vasoactive dilators like NO, prostaglandins, adenosine, and endothelin-1. Blood pressure is usually measured using the cuff method; but other methods exist, such as intra-arterial pressure transducers, applanation tonometry, and beat-to-beat finger pressures. It is important to standardize BP measurement, as many factors will affect readings. Blood pressure is controlled by several reflex mechanisms, including the baroreceptor, cardiopulmonary receptor, chemoreceptor, and Bainbridge reflexes. These control mechanisms are involved in short-term control of BP in situations such as changes in posture or exercise. Long-term control of posture is achieved through the renin-angiotensin-aldosterone system.

CHAPTER 7

Vascular Structure and Function

The human vascular system is a vast array of blood vessels that serve two primary functions: the distribution of blood throughout the body and the exchange of material between the blood and body tissues. Distribution involves transporting blood to and away from the heart. Physiological mechanisms that control the diameter of arteries and arterioles, causing vasoconstriction or vasodilation, determine blood flow and ensure that organs receive a blood supply that matches metabolic needs of the tissue. Vascular tone also has a large effect on blood pressure, and hence the driving force of blood throughout the vascular system.

The ultimate purpose of the vascular system is to permit exchange of gases, nutrients, and fluids between the vascular space and tissues of the body. Oxygen diffuses into the blood in the pulmonary capillaries and is carried, primarily bound to hemoglobin, to the capillaries where oxygen diffuses into the tissue. The amount of oxygen that diffuses into the tissue is dependent on local metabolic needs of the cells. Nutrients diffuse from the blood into the tissue at the level of the capillaries.

In addition to these basic functions, which are discussed in detail later in the chapter, the vessel wall has several other important functions, including serving as a barrier between blood and body tissues, secreting chemical mediators that help regulate coagulation (discussed fully in chapter 8), initiating angiogenesis, and playing a role in inflammatory defense against pathogens.

The functions of the vascular system are determined largely by the structure of the vessel. Importantly, in the vascular tree, as well as in individual circulations, there are significant structural differences in blood vessels that determine functional differences.

STRUCTURE OF BLOOD VESSELS

Blood vessels vary in size and structure depending on their location in the body and the functions they serve. In general, blood vessels consist of three discrete layers: the adventia, the tunica media, and the intima (figure 7.1). The outermost layer, the adventia, contains nerve innervation and is composed of loosely woven collagen fibrils

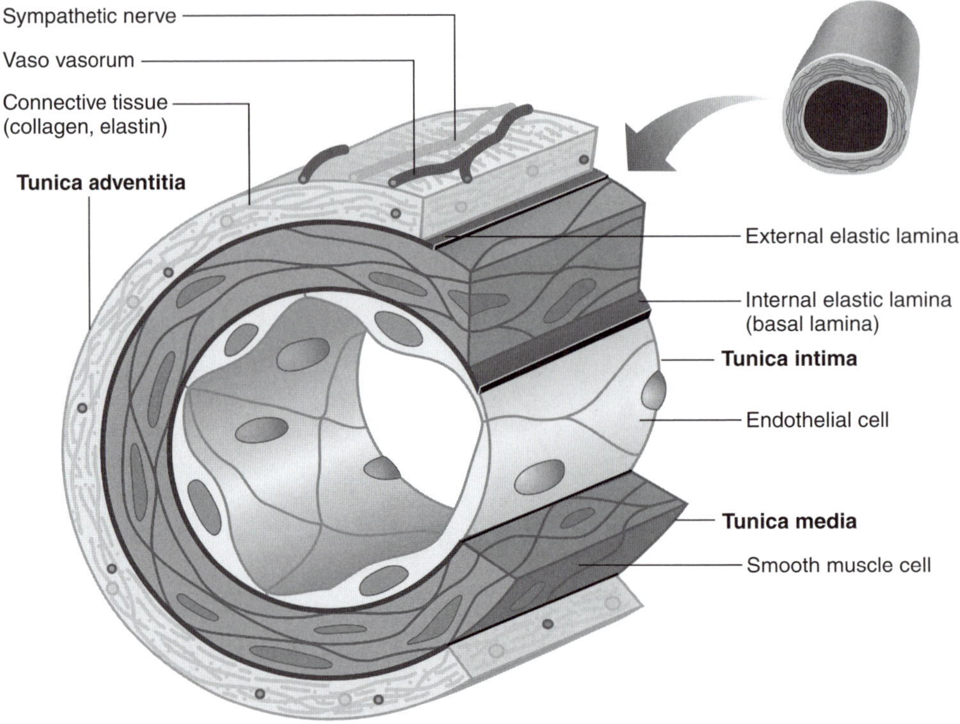

Figure 7.1 Layers of blood vessel. Most blood vessels are composed of three layers: the tunica adventitia, the tunica media, and the tunica intima. Capillaries are the exception; they are composed of the tunica intima only.

that protect and reinforce the vessel. The vaso vasorum provides a blood supply to the vessel wall.

The middle layer, the tunica media, is located between the adventia and the intima. It is separated from the adventia by the external elastic lamina and from the intima by the internal elastic media. The tunica media is composed mostly of smooth muscle cells embedded in a matrix of collagen, elastin, and various glycoproteins. The smooth muscle cells of the tunica media of small arteries and arterioles are largely responsible for determining the diameter of these vessels; hence they play an important role in distributing blood flow and maintaining blood pressure. The ratio of smooth muscle to connective tissue determines the mechanical properties of the vessel. The media of elastic arteries, such as the aorta, contain layers of smooth muscle cells interwoven with a large amount of elastin, enabling the vessel to passively expand as blood is pumped into it. In contrast, smaller arteries and arterioles have a relatively large amount of smooth muscle that is required for these vessels to contract and regulate organ blood flow and arterial blood pressure. Capillaries have no tunica media, only the intimal layer.

The innermost layer, the intima, is often composed primarily of a single layer of endothelial cells and a basement membrane. In the larger blood vessels is also a region of connective tissue between the endothelial cells and the basal lamina. The **endothelium** serves as a barrier between the blood and underlying tissues of the body. Throughout the vascular system, the endothelium acts to minimize friction as blood moves through the lumen of the blood vessel. The endothelium responds to a variety of mechanical forces and chemical signals from the blood and plays a critical role in determining cardiovascular health.

VASCULAR NETWORK

Chapter 1 provided a brief overview of the vascular tree to help orient the reader to the interrelated components of the cardiovascular system. The vascular system consists of a network of vessels that distribute blood and allow for exchange of gases, fluids, and nutrients (figure 7.2). Table 7.1 summarizes the size and primary function of the different vessels. Large arteries branch off the aorta and provide blood flow to specific organs or regions of the body. Once an artery reaches the organ to which it supplies blood, it branches into smaller arteries that distribute blood within the organ. Small arteries continue to branch into smaller vessels. In general, once a vessel reaches a diameter of less than 200 μm, it is termed an arteriole. There is no universal agreement, however, on the exact distinction between a small artery and an arteriole. Arterioles are generally considered *resistance vessels,* although some authors also refer to small arteries and arterioles as resistance vessels. Resistance vessels have a large amount of vascular smooth muscle in the tunica media. The vascular smooth muscle is innervated by autonomic nerves and has membrane receptors that bind circulating hormones or locally produced mediators to affect vasoconstriction or vasodilation.

As arterioles become smaller in diameter, they gradually lose their smooth muscle. Capillaries are composed only of endothelial cells and a **basement membrane**—this single-cell barrier is well suited to the function of exchange. Although capillaries are the smallest vessel in the vascular network, they have, by far, the greatest total cross-sectional area because of their huge number (see figure 1.4, p. 8). Again, this large surface area is a structural element that greatly facilitates their function as exchange vessels.

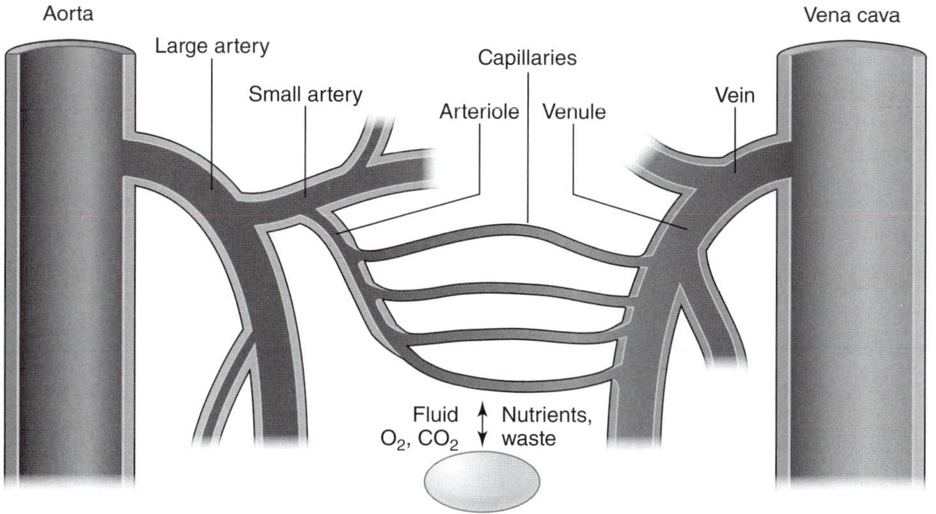

Figure 7.2 The vascular system.

Table 7.1 Size and Primary Function of Blood Vessels

Vessel type	Diameter	Function
Aorta	2.5 cm	Distribution and dampening of pulse
Elastic arteries	1.0-3.0 cm	Distribution
Muscular arteries	6-10 mm	Distribution and resistance
Arterioles	10-30 μm	Resistance (flow regulation)
Capillaries	6-9 μm	Exchange of gases and nutrients
Venules	10-30 μm	Exchange, capacitance, and collection
Veins	2-5 mm	Capacitance and blood return
Vena cava	3.0 cm	Collection and blood return

Capillaries join together to form postcapillary venules, which serve as a site of exchange for fluid and macromolecules. Postcapillary venules merge into venules, and venules merge into veins. Both venules and veins have vessel walls that contain smooth muscle, and these vessels are capable of venoconstriction or venodilation. Venules and veins are known as capacitance vessels because their walls are easily distensible and thus they are the site in the vascular tree that stores most of the blood volume. Venoconstriction can be an important mechanism to increase blood flow return to the heart.

ENDOTHELIUM

The **vascular endothelium** is a single layer of epithelial cells that line all blood vessels. Collectively, the endothelium represents an enormous surface area, estimated to be several thousand square meters (Krogh, 1929). Structurally, it is a semipermeable membrane that forms the biological interface between circulating blood and the tissues of the body. It retains plasma and blood within the circulation while simultaneously allowing rapid nutrient movement between blood and tissue. While the endothelium was once considered a passive barrier between the blood and the underlying tissue, it is now recognized as a dynamic tissue that helps regulate myriad activities, including vascular tone, exchange of gases and nutrients, hemostatic balance, fluid balance, immune function, and angiogenesis. The endothelium has the ability to monitor, integrate, and transduce blood-borne signals and mechanical forces, making it a sensory organ that is able to respond to numerous signals by sending signals both outward to the blood and inward to smooth muscle cells in the vessel wall. Another important feature of the endothelium is its regional specialization (DeCorleto and Fox, 2005). The endothelial lining of blood vessels has specific functions depending upon the location in the vascular tree, and even within specific organs.

Endothelial Structure

The endothelium consists of flattened cells forming a single layer that lines the entire vascular system, providing a barrier between the blood and the underlying tissue. As shown in figure 7.3, the endothelium is separated from the underlying tissue

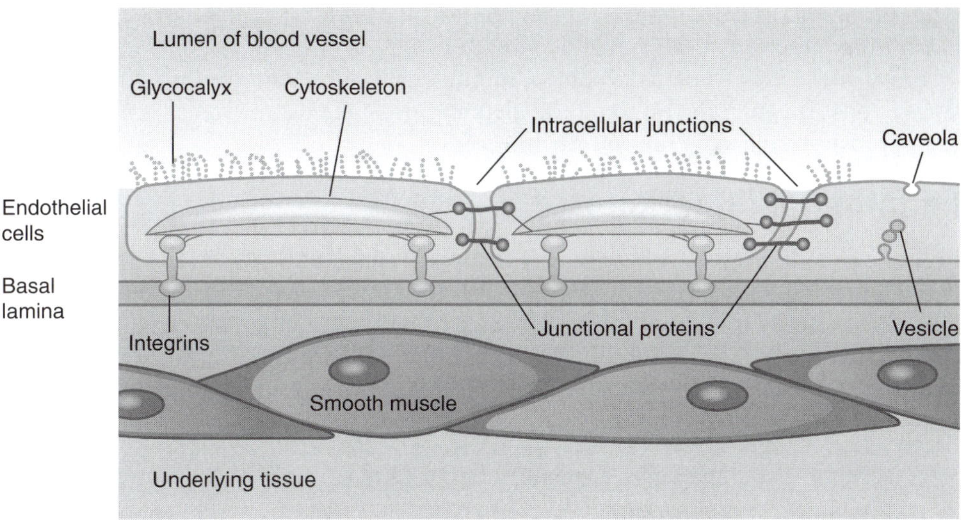

Figure 7.3 Vascular endothelium. The vascular endothelial cells are covered by a glycocalyx. Endothelial cells have an organized cytoskeleton that attaches to the basal lamina through integrins, and the cells are joined together by junctional proteins.

by the basal lamina (basement membrane). In most vessels—capillaries being the exception—the endothelium is positioned above smooth muscle in the tunica media. The enormous surface area of the endothelium facilitates many of its functions, including the exchange of gases and nutrients and the signaling of hemodynamic forces to underlying smooth muscle. The basal lamina is approximately 50 to 100 nm thick and is composed of an extracellular matrix, largely type IV collagen fibers. The endothelial cells are attached to the basal lamina through integrins that are also attached to components of the cytoskeleton on the interior of the endothelial cell.

The luminal surface of endothelium is lined by a negatively charged layer called the glycocalyx. The glycocalyx allows water and small solutes to exit the circulation but functions to keep negatively charged proteins, such as lipoproteins and albumin, in the bloodstream.

Endothelial cells have a highly structured cytoskeleton that provides the scaffolding for the cell and helps attach endothelial cells to one another and to the basal lamina. The cytoskeleton is composed largely of actin and myosin filaments arranged in different formations. The space between adjacent endothelial cells is termed the intracellular cleft. Adjacent endothelial cells are attached to one another via junctional proteins that extend into the intracellular cleft. The degree and functioning of the junctional proteins vary considerably throughout the vascular system. Postcapillary venules have a higher permeability than most capillaries because they have fewer junctional proteins and a wider gap between cells. The junctional proteins are dynamic structures that can alter their functioning rapidly, resulting in acute changes in permeability. For example, in response to tissue trauma or infection the venules become more permeable, allowing immune cells, proteins, and fluid to more easily enter the tissue (as evidenced by swelling). In contrast, brain endothelium has extremely low permeability because it has numerous and complex junctional proteins. This low permeability protects the brain from infections that are carried in the blood.

Large molecules, such as albumin, immunoglobulins, and lipoproteins, are too large to pass through the intracellular cleft and thus rely on a caveola–vesicle system to be transported across the endothelial cell.

Endothelial Function

Properly functioning endothelium is critical to the health and function of the cardiovascular system. This section briefly describes the major functions of the endothelium. Specific functions are addressed in more depth later in this chapter and in subsequent chapters. As shown in figure 7.4, the vascular endothelium has multiple functions, and these functions are often specific to the site in the vascular tree. The primary functions of the endothelium include

1. serving as a selectively permeable barrier to regulate blood–tissue exchange,
2. regulating vascular tone,
3. releasing anticlotting and proclotting factors,
4. participating in inflammatory defense against pathogens, and
5. initiating new blood vessel formation (angiogenesis).

The exchange of fluids, nutrients, and gases between the blood and tissues occurs primarily within the capillaries (figure 7.4, #1). The primary role of the endothelium in regulating this exchange is to form a semipermeable membrane that contains blood within the vessel and allows nutrients and gases to move into the tissue. These functions are due largely to the glycocalyx and the intracellular junctions. While lipid-soluble substances diffuse through the endothelial cells, small, water-soluble substances dif-

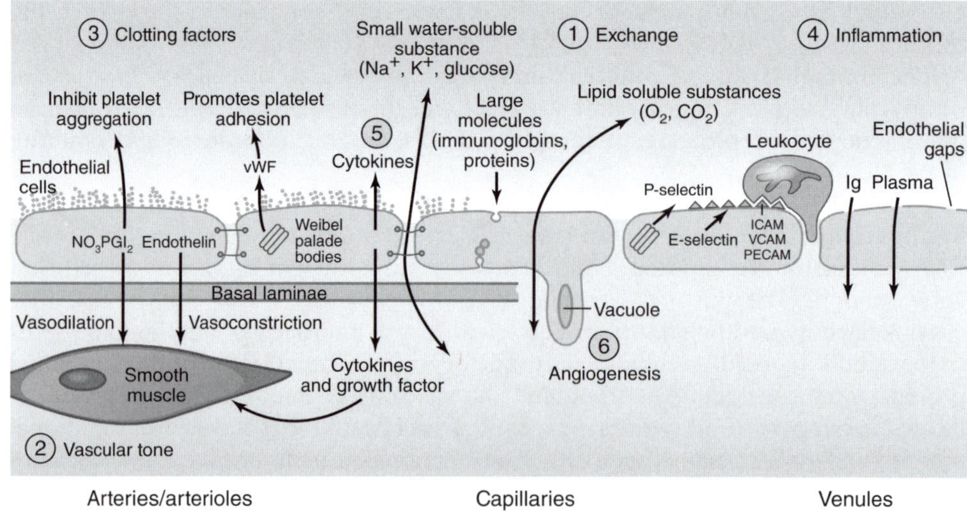

Figure 7.4 Functions of endothelium. The vascular endothelium is involved in many processes, which vary by location within the vascular system. See text for details.

fuse through the intracellular junctions. Large molecules, such as immunoglobulins and protein-bound hormones, pass from the circulation to the underlying tissue via the caveola–vesicle system. These flask-shaped structures are actually invaginations into the cell membrane that permit the endothelial cells to take up substances from the blood by the process of endocytosis. The movement of fluid and nutrients across the endothelium occurs primarily at the capillary level, due to the thinness of the vessel wall.

Vascular tone is determined by the degree of smooth muscle contraction in the tunica media (figure 7.4, #2). Endothelium that lines the arteries and arterioles plays a central role in regulating smooth muscle contraction. Endothelial cells release chemical mediators that lead to vasodilation (nitric oxide, NO; prostacyclin, PGI_2) and vasoconstriction (endothelin). The balance of these mediators plays an important role in determining blood flow distribution to various organs.

Factors released from the endothelium have a potent effect on blood clotting potential (figure 7.4, #3). In fact, some of the same vasodilatory factors, notably NO and PGI_2, that cause relaxation of underlying smooth muscle are also released into the bloodstream where they have an inhibitory effect on platelet aggregation, making a clot formation less likely. Thus, the release of these factors has the simultaneous and mutually reinforcing effect of increasing blood flow and maintaining blood fluidity. In fact, the endothelium is one of the only surfaces, either natural or synthetic, that can maintain blood in its fluid state during prolonged contact. This ability is due to the presence of heparin sulfate proteoglycan molecules on the surface of endothelial cells (Libby, 2005). Under certain circumstances, namely vascular injury, endothelium also secretes von Willebrand factor (vWF), which is produced in specialized organelles called Weibel-Palade bodies. Von Willebrand factor promotes platelet adhesion and plays a role in the coagulatory cascade. Thus, the endothelium plays a critical role in maintaining blood fluidity and fostering clot formation. The role of the endothelium in maintaining homeostatic balance is discussed fully in chapter 8.

The endothelium is also essential in the inflammatory defense against pathogens. In response to injury or invasion, venular endothelium produces adhesion molecules that cause circulating leukocytes to be attached to the endothelium. Several adhesion molecules are involved in causing a leukocyte to adhere to the endothelium. P-selectin, which is released from Weibel-Palade bodies, and E-selectin cause a loose bond between the leukocyte and the endothelium. Intracellular adhesion molecules (ICAMs) and vascular adhesion molecules (VCAMs) then cause a tighter bonding of the leukocyte and endothelium. Finally, the leukocyte begins to move into the intracellular junction; it then moves into the tissue through the process of diapedesis. The final step is dependent on platelet–endothelial cell adhesion molecules (PECAM). Venular endothelium can also produce large endothelial gaps that allow for increased delivery of immunoglobulins (antibodies) to the tissue to provide an immune response in response to infection. Plasma also passes through these endothelial gaps, leading to the swelling that characterizes inflammation.

The endothelium plays a critical role in the development of new vessel formation, termed **angiogenesis.** Capillary sprouting initiates all new blood vessels. Endothelial cells can be stimulated to divide rapidly when there is a need for new vessel formation (for growth or repair or to support new tissue). New vessel formation begins with the breakdown of the basal lamina and the sprouting of the endothelium from the side of a capillary or venule. The cell extensions put out by the endothelium, called pseudopodia, grow toward the stimulus for new blood supply. These pseudopodia are enlarged by cytoplasmic growth until they divide into daughter cells. Vacuoles then digest material within the new daughter cells. Eventually the vacuoles of the daughter cells fuse, resulting in a new lumen. The entire process continues until the new sprout encounters another capillary to connect to.

Endothelial Dysfunction

Pathological changes in the structure and functioning of the endothelium can have devastating consequences for human health. Alterations in endothelial function can lead to increased permeability to plasma lipoproteins, increased adhesiveness to leukocytes, and imbalances in the release of factors that regulate vascular tone and hemostasis. **Endothelial dysfunction** is the term that typically refers to these manifestations of endothelial pathology. It is now accepted that endothelial dysfunction is an early manifestation of atherosclerosis and is associated with increased incidence of cardiac events (Celermajer et al., 1994; Neunteufl et al., 2000). On a positive note, there is growing evidence that endothelial dysfunction can be improved by exercise training. Thus, endothelial dysfunction plays an important role in the initiation, progression, and clinical complications of coronary artery disease and is a target for medical intervention, including exercise training.

ENDOTHELIUM REGULATION OF VASCULAR TONE

As discussed in chapter 6, vascular tone is the primary mechanism for controlling blood flow to tissue at the organ level, and is essential for maintaining total peripheral resistance and hence blood pressure and blood volume at the systemic level. Blood flow through any organ is determined primarily by the arteriole smooth muscle tone. Smooth muscle tone is determined by the balance of intrinsic and extrinsic factors that cause vasoconstriction and vasodilation. Arterioles have a basal tone, meaning that even when all external influences are removed, they remain partially constricted. Factors that lead to smooth muscle contraction enhance vascular tone, whereas factors that lead to smooth muscle cell relaxation decrease vascular tone. Chapter 6 noted that vascular tone is achieved by a combination of extrinsic (neurohormonal) and local factors. Here we discuss one of the important local regulators of vascular tone—the endothelium.

The role of the endothelium in regulating vascular tone began to be appreciated only in the late 1970s with the discovery by Vane and colleagues that the blood vessels secrete a vasodilatory substance, prostacyclin (PGI_2). In 1980, Fruchgott and Zawadzki discovered a potent factor released from the endothelium that also leads to the relaxation of vascular smooth muscle. The substance was originally identified as endothelial-derived relaxing factor (EDRF) but is now recognized as nitric oxide. In 1989, a Japanese group discovered an endothelial substance called endothelin that causes vasoconstriction of the vascular smooth muscle. These substances are produced in the endothelium and secreted as they are produced—that is, they are not stored for later release (Levick, 2003).

The endothelium is essential for maintaining the health of the vessel wall and vasomotor tone in both large arteries and arterioles. These functions are due in large part to the production and release of vasoactive substances (sometimes called autocoids). Nitric oxide is probably the most important and best-characterized vasoactive substance, and its vasodilatory function is often used as an index of endothelial function (Maiorana et al., 2003).

Nitric Oxide

Nitric oxide is produced continually by the endothelium and has multiple roles: It causes vascular smooth muscle (VSM) relaxation, inhibits platelet aggregation, inhibits vascular smooth muscle proliferation, and inhibits the production of adhesion molecules. Despite the importance of these varied roles of nitric oxide, the focus of this section is on its role in vascular smooth muscle relaxation.

Nitric oxide (NO) is the most potent vasodilator secreted by the endothelium. It is synthesized in the endothelium by the enzyme nitric oxide synthase (eNOS), which cleaves NO from the amino acid L-arginine. Nitric oxide then rapidly diffuses from the endothelium into vascular smooth muscle where it leads to relaxation. Nitric oxide synthase, and thus the production of NO, is regulated by shear stress and by several receptor-bound agonists (figure 7.5).

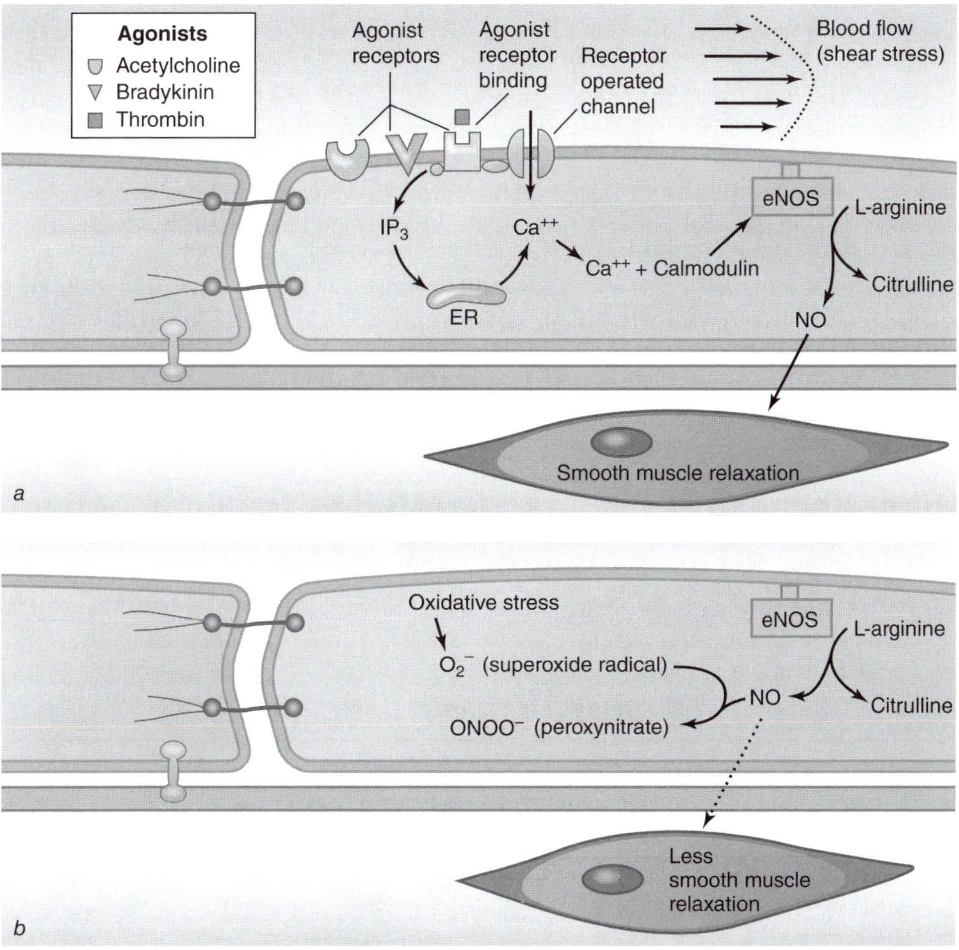

Figure 7.5 Nitric oxide production and degradation. *(a)* Nitric oxide production in the endothelial cell is stimulated by shear stress and agonist receptors. *(b)* Nitric oxide is degraded by superoxide radicals, leading to a decrease in nitric oxide availability and thus a loss of ability for smooth muscle to vasodilate.

The activity of eNOS is continuously stimulated by shear stress in order to maintain basal production of NO. **Shear stress** is the force exerted on the vessel wall, that is, the endothelium, by the sliding action of blood flow. Shear stress is probably the most important stimulus for the production of NO under normal conditions. During exercise, shear stress increases due to increased cardiac output and increased blood flow to the exercising skeletal muscle. The increased shear stress leads to greater production of NO, which causes greater vasodilation—a phenomenon known as flow-mediated dilation. The methods of measuring flow-mediated dilation are discussed later in this chapter. Shear stress appears to be transduced (sensed) by the integrins that anchor the endothelium to the basal lamina. Transduction leads to the phosphorylation of protein kinase B, which increases eNOS activity.

Several agonists can stimulate production of eNOS, including acetylcholine, thrombin, bradykinin, adenosine triphosphate (ATP), and substance P (figure 7.5*a*). Agonists stimulate eNOS activity by raising intracellular calcium concentrations, which promote the formation of calcium–calmodulin complexes. In turn, the calcium–calmodulin complex stimulates eNOS activity and NO production rate. Agonists can increase intracellular concentration by two main mechanisms: activating receptor-operated channels that lead to an influx of calcium or causing the release of calcium from the endoplasmic reticulum. The release of calcium from the endoplasmic reticulum occurs when agonist binding activates a membrane-bound enzyme (phospholipase C), which catalyzes the formation of inositol triphosphate (IP_3). The IP_3 then causes the release of calcium from the endoplasmic reticulum. Agonist-mediated vasodilation leads to NO release primarily in large arteries.

Nitric oxide activity is diminished under conditions of high oxidative stress (figure 7.5*b*). Oxidative stress leads to high levels of superoxide radicals. Nitric oxide reacts with superoxide radicals to produce peroxynitrate, thus reducing the bioavailability of NO to cause smooth muscle relaxation. The decreased bioavailability of NO is one of the primary reasons for impaired endothelial function, as is commonly seen in individuals with atherosclerosis.

Other Vasoactive Factors Secreted by the Endothelium

Endothelium-derived hyperpolarizing factor (EDHF) can cause relaxation of vascular smooth muscle. It appears that EDHF contributes to agonist-induced vasodilation

CALCIUM CHANNEL BLOCKERS

Calcium channel blockers (CCB) are often used clinically to lower blood pressure in hypertensive patients. Calcium channel blockers block voltage-gated channels, leading to a decrease in intracellular calcium and hence a reduction in the force of muscle contraction. In vascular smooth muscle this leads to vasodilation and a decrease in total peripheral resistance. There are two classes of calcium channel blockers—dihydropyridine and non-dihydropyridine. The dihydropyridine calcium channel blockers are used to treat hypertension. You may have heard of some popular calcium channel blockers that are regularly prescribed to treat hypertension—Norvasc (amlodipine), Procardia (nifedipine), and Cleviprex (clevipidine).

in small vessels. Prostacyclin (PGI_2) also contributes to smooth muscle relaxation. Prostacyclin production is initiated by agonist binding to endothelial receptors and is not as sustained as NO production.

Endothelin is a potent vasoconstrictor released from the endothelium, causing a sustained vasoconstriction (2-3 h). There are many isoforms of endothelin in the human body; endothelin-1 (ET-1) is the most prevalent. Under normal conditions, endothelin is synthesized at low basal rates. Endothelin release is stimulated by angiotensin II, antidiuretic hormone, thrombin, cytokines, reactive oxygen species, and shear stress-induced integrin stimulation. Conversely, prostacylin and atrial natriuretic peptide inhibit ET-1 release. Endothelin levels are often elevated in patients with cardiovascular disease (Levick, 2003).

There are complex interactions between NO and endothelin in the regulation of vascular smooth muscle tone. The binding of endothelin to its receptor stimulates eNOS, and it appears that NO modulates the endothelin-mediated contraction of vascular smooth muscle. Furthermore, NO inhibits endothelin activity. Thus, endothelin and NO are continuously acting on vascular smooth muscle cells in a coordinated way to produce finely tuned changes in vessel diameter (figure 7.6).

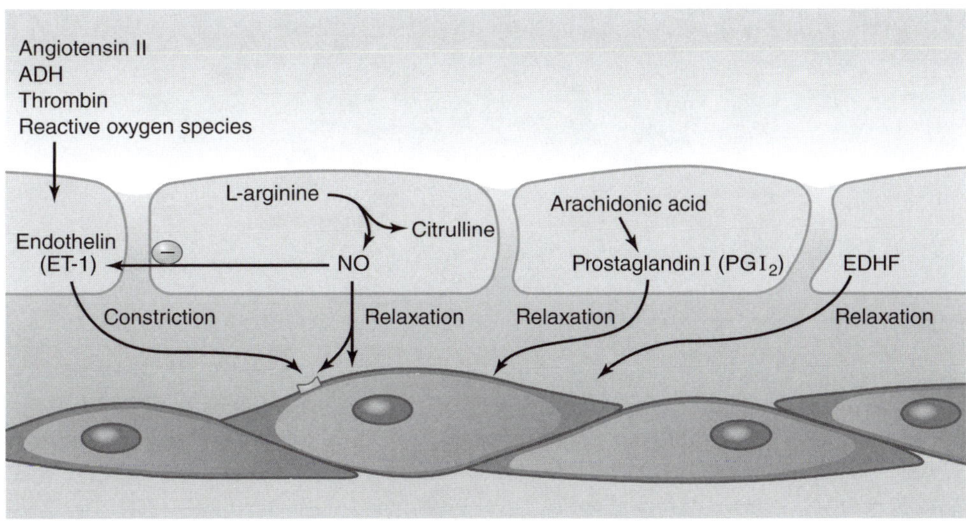

Figure 7.6 Regulation of vascular tone by endothelial factors. Vascular tone represents a balance of influence from vasodilatory and vasoconstrictor mediators released from the endothelium.

VASCULAR SMOOTH MUSCLE

It is the degree of contraction of smooth muscle that determines vessel diameter and blood flow to an organ. Furthermore, the extent of smooth muscle contraction determines total peripheral resistance and has a large influence on blood pressure. (Recall that mean arterial pressure equals cardiac output times total peripheral resistance [$MAP = \dot{Q} \times TPR$]). To provide a better understanding of vascular smooth muscle, a review of the structure of smooth muscle follows.

Vascular Smooth Muscle Structure

Vascular smooth muscle cells are located in the tunica media of arteries, arterioles, and veins. They are small (~5 μm × 50 μm) spindle-shaped cells that are arranged circumferentially in the blood vessel walls (figure 7.7). In many but not all blood vessels, adjacent smooth muscle cells are structurally and functionally connected via gap junctions. The gap junctions are composed primarily of the protein connexin. These specialized junctions allow ion currents to flow between adjacent cells, and hence depolarization can be spread from cell to cell. In small arteries and arterioles, there are also specialized gap junctions between the endothelial cells and myocytes. These gap junctions, termed myoendothelial junctions, permit the transmission of regulatory signals from the endothelium to the vascular smooth muscle.

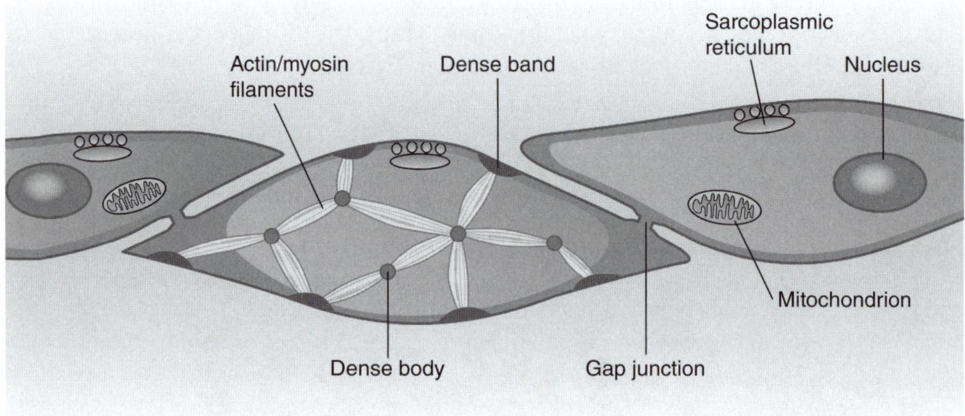

Figure 7.7 Vascular smooth muscle structure.

Vascular smooth muscle cells contain contractile proteins (myosin and actin), but the sarcomere-like contractile units are not as well developed as in skeletal muscle. Thus, smooth muscle lacks the striations seen in skeletal muscle. The actin in smooth muscle is much longer than in skeletal muscle and is attached to the surface of the cell at structures called dense bands. In the interior of the cell, actin filaments are anchored to structures called dense bodies (instead of Z lines in skeletal muscle). These dense bodies are connected to each other and the cell membrane by intermediate filaments that provide the cytoskeleton framework for the smooth muscle cell. Myosin filaments are interwoven between the actin filaments, but in a less structured way than in skeletal muscle. In smooth muscle, the contractile filaments travel obliquely or transversely to the long axis of the cell.

Vascular smooth muscle also contains cellular organelles. Of particular importance here is the fact that the smooth endoplasmic reticulum stores and releases calcium. The smooth endoplasmic reticulum of vascular smooth muscle is poorly developed, compared to that of skeletal muscle, and accounts for only 1% to 4% of the cell by volume (Levick, 2003). Thus, intracellular calcium stores are limited; and vascular smooth muscle, particularly in resistance vessels, is dependent on an influx of extracellular calcium for muscle contraction.

Vascular Smooth Muscle Function

Smooth muscle cells, like skeletal and cardiac muscle cells, produce force due to cross-bridge interaction. However, there are important distinctions between vascular smooth muscle and the other types of muscle in qualities such as

1. speed of contraction,
2. degree of shortening,
3. duration of contraction,
4. energy cost of contraction, and
5. stimulus for contraction.

Vascular smooth muscle contracts very slowly, with a shortening velocity about 1/10th that of skeletal muscle. Because the contractile filaments are relatively long, the degree of shortening can be substantial. Vascular smooth muscle cells may shorten to less than half their original length. Arteries and arterioles maintain a state of partial contraction to maintain vascular tone. Thus, they are in a sustained state of contraction. Despite the extremely long duration of contraction, vascular smooth muscle has a very low energy cost owing to its specialized ability to maintain a long-lasting cross-bridge that either recycles infrequently or not at all and thus consumes very little ATP. This phenomenon is referred to as a **latch state.** Essentially, the contracted smooth muscle is locked into a contracted state. Because of this unique mechanism, vascular smooth muscle can maintain a contraction using just 1/300th of the energy required by skeletal muscle for the same contraction (Levick, 2003). Smooth muscle within walls of blood vessels can be stimulated to contract by action potentials from neurons or changes in their membrane potential caused by circulating agonists or agonists produced locally by the endothelium (the focus of much of this chapter).

Mechanism of Contraction

Vascular smooth muscle contraction is regulated primarily by the concentration of calcium in the cytosol. The level of cytoplasmic calcium is determined by the entry of extracellular calcium (through either voltage-sensitive or receptor-operated channels), the release of stored calcium from the endoplasmic reticulum, and the removal of calcium by the calcium-ATPase pump in the cell membrane (expulsion) and in the sarcoplasmic membrane (sequestration) (figure 7.8). When calcium concentration increases, as with nerve stimulation or agonist binding, tension in the smooth muscle increases. In contrast, when calcium concentration in the cytoplasm falls, tension in the contractile elements falls and the vessel dilates.

Mechanisms of Relaxation

Vascular smooth muscle relaxation may be caused by hyperpolarization of the cell membrane, which occurs when the endothelium releases EDHF, or through a second messenger system. Agonists binding to membrane-bound receptors may activate

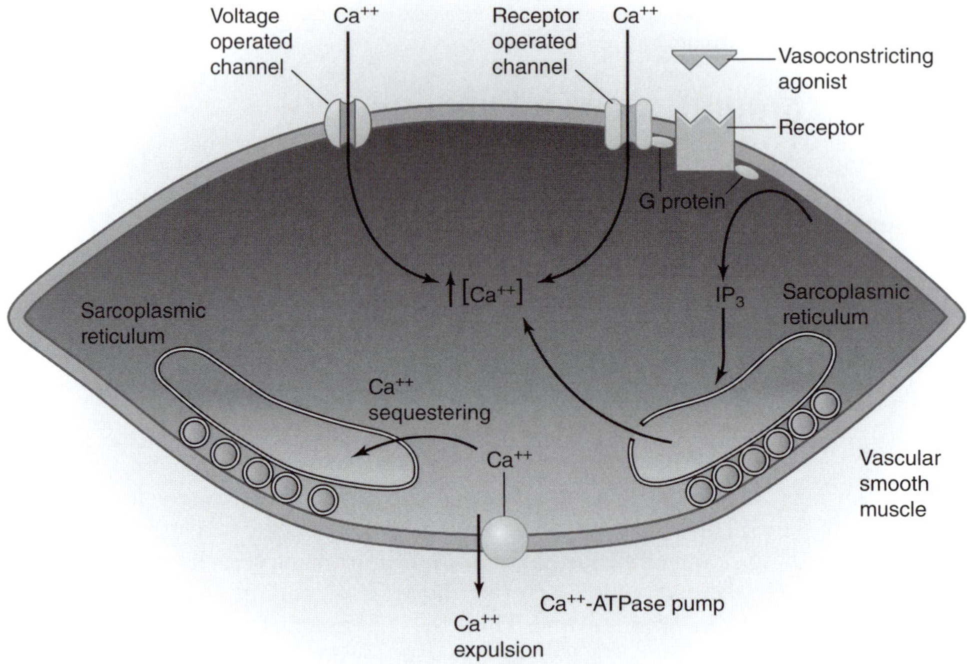

Figure 7.8 Vascular smooth muscle contraction. Vascular smooth muscle tone is primarily determined by intracellular calcium levels.

the second messenger cyclic adenosine monophosphate (cAMP) or cyclic guanosine monophosphate (cGMP). Nitric oxide, the primary signal for vascular smooth muscle relaxation, initiates relaxation via the cGMP pathway. Nitric oxide stimulates the enzyme guanylyl cyclase, which converts guanosine triphosphate (GTP) to cGMP. Cyclic GMP then activates several kinases. These kinases stimulate calcium pump activity, which decreases sarcoplasmic calcium levels, thus decreasing cross-bridge formation and leading to muscle relaxation.

MEASURING ENDOTHELIAL AND VASCULAR FUNCTION

Endothelial dysfunction and impairment in vasodilation is an early indicator of cardiovascular disease and a predictor of cardiovascular events. Thus, assessing vascular function is of great interest to researchers and clinicians. Clinicians are especially motivated by the desire to detect vulnerable individuals before the onset of symptoms. Given that sudden cardiac death or myocardial infarction is the first presentation of cardiovascular disease over 50% of the time, it is critically important to be able to identify those at greatest risk (Zipes, 1998). The traditional approach of measuring the number and extent of traditional risk factors, as is done with the Framingham risk score, and categorizing individuals based on a 10-year risk of myocardial infarction, is popular but problematic. In one study of adults under 65 years who had had a myocardial infarct, only 25% met the criteria for preventive treatment the day before their event (Akosah et al., 2003). Therefore, measures of vascular function have been investigated to help elucidate the role of the endothelium and vascular smooth muscle in regulating cardiovascular physiology and to improve cardiovascular disease detection.

Brachial Artery Vasodilation

Brachial artery ultrasound can be used to measure the ability of the brachial artery to vasodilate in response to increased blood flow following hyperemia (Celermajer et al., 1994). This technique requires that the brachial artery be imaged with a high-resolution ultrasound (figure 7.9). A cuff is placed around the upper arm and inflated above systolic blood pressure to occlude arterial inflow and create ischemia in the forearm. The cuff is rapidly deflated, and blood flows into the distal portion of the arm at a rate that is determined by the relaxant capacity of the microvasculature. The increase in blood flow causes an increase in shear stress on the endothelium, which in turn releases NO and causes arterial dilation. The increase in arterial dilation, called flow-mediated dilation, is typically expressed as a percent change. A greater change in flow-mediated dilation is indicative of healthy endothelium, whereas a muted increase in flow-mediated dilation is indicative of endothelial dysfunction.

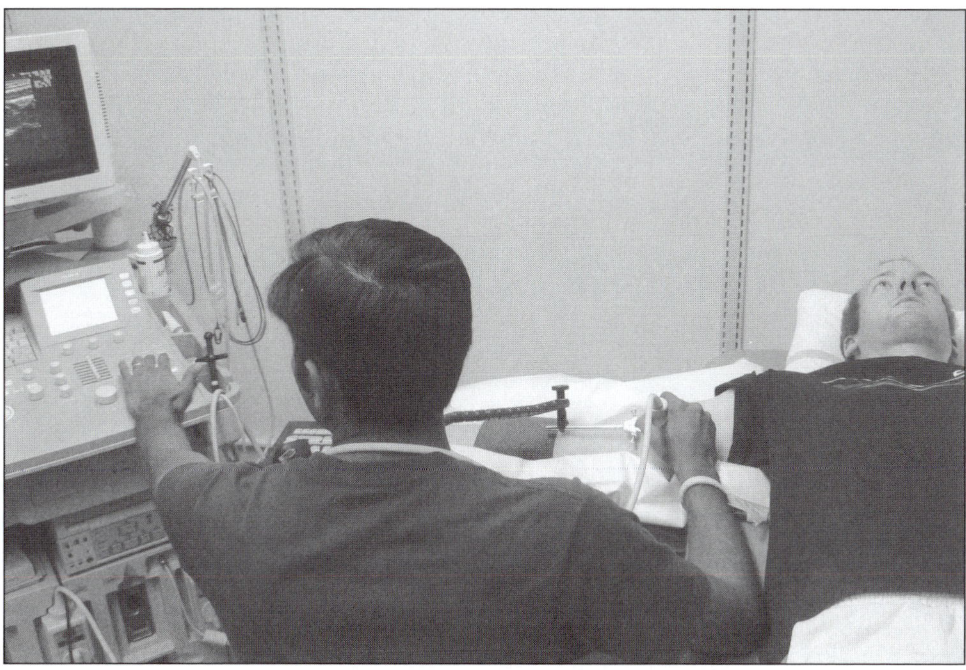

Figure 7.9 Measurement of flow-mediated dilation using brachial artery ultrasound.

Venous Occlusive Plethysmography

Venous occlusive plethysmography, also called strain gauge plethysmography, uses a sensitive mercury-in-silactic gauge wrapped around the forearm to measure small changes in the circumference of the forearm in response to arterial inflow (figure

7.10). In a resting state, the plethysmograph will display a stable baseline with small defections corresponding to arterial inflow with each period of systole. To measure *resting blood flow,* a cuff around the upper arm is inflated to above venous pressure (~50 mmHg) but well below systolic pressure. As arterial inflow continues unimpeded and venous outflow is prevented, the forearm swells. Through measurement of the slope of the line, arterial inflow can be calculated (assuming the arm is a cylinder). To assess *maximal blood flow,* another cuff is placed around the upper arm and inflated to suprasystolic pressure to occlude arterial inflow and create ischemia. The cuff is typically inflated for a period of 2 to 5 min. Following the designated time, the cuff is rapidly deflated, and arterial inflow is measured as already described. A high maximal blood flow in response to the hyperemia indicates good endothelial function, whereas a low maximal blood flow indicates impaired endothelial function.

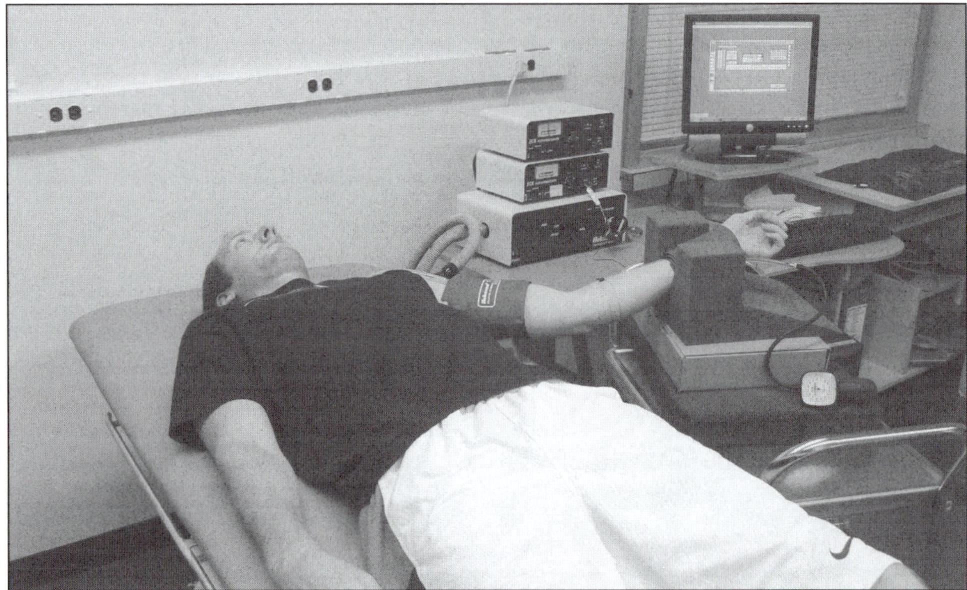

Figure 7.10 Measurement of arterial inflow using venous occlusive plethysmograph.

Applanation Tonography

Applanation tonography employs a high-fidelity transducer to record the pressure of pulse waves. Pulse wave analysis uses waveforms measured at the radial pulse. By means of a generalized transfer function, the radial waveform is used to derive the ascending aortic pulse wave. Analysis of the shape of the derived aortic pressure waveform provides indices of arterial stiffness, the degree of systolic afterload, and central systolic and diastolic blood pressures. Pulse wave velocity can be measured by obtaining a pressure wave at two arterial sites, often the carotid and the femoral artery. The timing of the onset of systole of the pressure wave is compared to the timing of the onset of the R wave on the electrocardiogram. Pulse wave velocity provides an index of large artery stiffness.

SUMMARY

The vascular system is responsible for the distribution of blood throughout the body and for the exchange of gases and nutrient between the blood and body tissue. The structure and function of the vessels are well suited to these tasks. Arterioles are primarily responsible for determining the relative distribution of blood flow; capillaries are the site of gas exchange; venules collect postcapillary blood; and veins return deoxygenated blood back to the right atrium. The innermost layer of the blood vessels, the endothelium has several critical roles, including serving as a barrier between blood and tissue, regulating vascular tone, providing an antithrombotic surface, participating in the inflammatory response, and initiating angiogenesis. Contraction of the vascular smooth muscle in the tunica media is responsible for vessel diameter and largely determines blood flow to various organs of the body.

CHAPTER 8

Hemostasis: Coagulation and Fibrinolysis

Hemostasis is defined as the stoppage of blood. **Hemostatic balance,** however, refers to the dynamic balance between blood clot formation (coagulation) and blood clot dissolution (fibrinolysis). Hemostatic balance is carefully regulated to maintain blood fluidity under normal circumstances and to promote the rapid formation of a blood clot to prevent blood loss when the integrity of the circulation is threatened. The rapid transition from anticoagulatory activity to procoagulatory activity is essential to the maintenance of life and depends on the coordinated activity of the primary components of the hemostatic system, namely

- platelets,
- blood vessel walls, and
- plasma proteins (coagulatory and fibrinolytic factors).

While the process of blood clot formation in response to a cut or traumatic injury is familiar, it also occurs on a regular basis in response to small tears in the endothelial lining of blood vessels. The ability to seal a tear in a blood vessel is essential to prevent loss of blood. On the other hand, excessive clot formation or the inability to dissolve a clot once a wound is repaired can also present life-threatening challenges.

A blood clot that forms in the cardiovascular system is called a thrombus. In most cases, a myocardial infarction occurs when the normally antithrombotic endothelium is disrupted and a thrombus (clot) forms inside a coronary vessel. In most cases, a thrombus is triggered by plaque rupture that exposes the blood platelets and clotting factors to highly thrombotic material inside the vessel wall. This leads to the rapid formation of a platelet plug that is reinforced with fibrin to become a **blood clot (thrombus)**. The thrombus leads to a decrease or complete arrest (occlusion) of blood flow, and myocardial tissue dies. Thus, the delicate balance of hemostasis has important implications for wound healing, normal repair of endothelial damage, and atherothrombotic heart disease. The balance of prothrombotic and antithrombotic activity is affected by many factors, including exercise training. In fact, the cardioprotective effects of exercise can be partially attributed to a decreased propensity to form clots and an enhanced fibrinolytic potential following exercise training.

In the most basic sense, a blood clot is simply a clump of platelets held together by a mesh of fibrin. Thus, the two major processes that form a clot are the formation of a platelet plug and the generation of fibrin. Clots are formed to prevent blood loss from a damaged blood vessel and to provide an environment for tissue repair. Once a clot is no longer needed, it is broken down by the process of fibrinolysis. Fibrinolysis degrades the fibrin that holds the clot together, allowing the clot to dissolve. In short, hemostatic balance is achieved by myriad factors that inhibit or stimulate platelet function, coagulation, and fibrinolysis (figure 8.1).

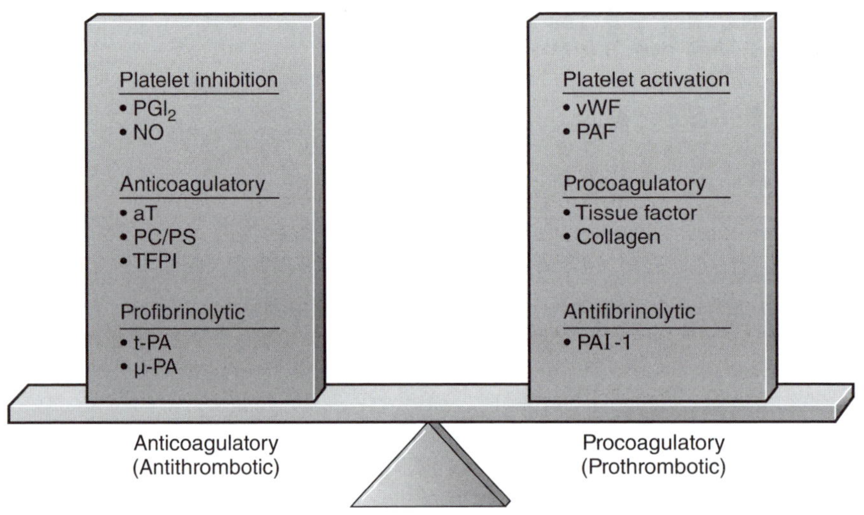

Figure 8.1 Hemostatic balance.

Hemostasis relies on a vast array of highly coordinated interactions among platelets, the endothelium of blood vessels, and coagulatory and fibrinolytic proteins in the blood. Under normal conditions, blood components and the endothelium are highly antithrombotic; that is, they function to maintain blood fluidity and prevent coagulation. However, in response to endothelial damage or injury, prothrombotic mechanisms overwhelm the anticoagulatory mechanisms and lead to the rapid and localized formation of a blood clot (figure 8.2).

The formation of a blood clot is a normal and healthy response to vascular damage. The formation of excessive clots, however, can be fatal. A hypercoagulable state can lead to excessive clotting in response to minor injury and can result in a thrombus that occludes an artery, leading to myocardial infarction, stroke, or peripheral artery disease. In fact, one of the major challenges with atherosclerotic disease is the increased potential for clot formation and the impaired process of fibrinolysis. Many cardiac

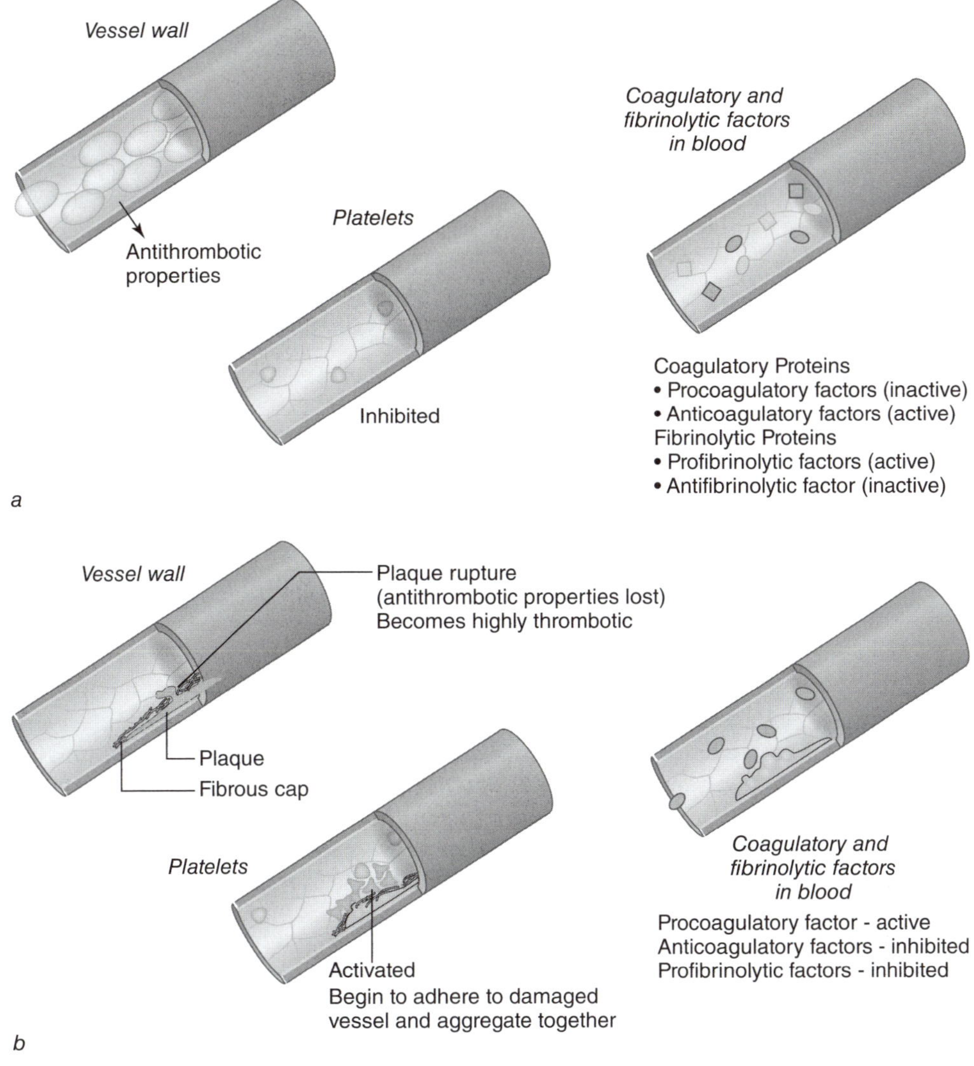

Figure 8.2 Interaction of the major components of the hemostatic system *(a)* at rest and *(b)* in response to vascular damage.

medications function by altering platelet function, coagulation, or fibrinolysis to prevent excessive clotting. Conversely, individuals with hemostatic imbalances that represent a hypocoagulable state are at risk for excessive (or even fatal) bleeding from a minor cut.

VASCULAR INJURY

Vascular injury can occur due to a cut or major trauma. However, vascular injury occurs more often than is apparent in injuries that result in external bleeding. Changes in hemodynamic stress can cause minor tears in the endothelial layer that trigger coagulation and result in tissue repair. In healthy individuals, this process occurs constantly: Small tears occur in the endothelium, and a clot is formed and then dissolved. In individuals with hemostatic imbalances that lead to a hypercoagulable state, however, even small tears in the vessel wall can lead to excessive clotting and to thrombotic events.

The first step in preventing blood loss is the powerful constriction of the injured vessel, sometimes called **vascular spasm.** The constriction of the injured blood vessel occurs almost instantly, causing a reduction in blood flow and thus limiting blood loss. Vasoconstriction is initiated by damage to the blood vessel and is mediated through nervous reflexes, local myogenic spasms, and humoral factors released from the damaged tissue and activated platelets. Depending on the extent of vascular damage, vasoconstriction can last for several minutes or hours to help prevent blood loss.

PLATELETS

A blood clot forms when platelets aggregate and then become enmeshed in fibrin scaffolding that is produced as a result of the coagulatory cascade. Thus, the formation of the platelet plug is the first major process in the formation of a blood clot, and occurs in three specific steps: platelet adhesion, platelet activation, and platelet aggregation. A platelet plug begins to form within seconds of vessel injury and relies on special structural and chemical characteristics of platelets.

Platelet Structure and Function

Platelets play several critical roles in hemostasis. They interact with endothelial cells, leukocytes, and coagulatory factors to ensure the stoppage of blood loss following vascular injury; and they participate in hemostatic processes necessary to maintain the fluidity of blood under normal conditions. Platelets are responsible for the initiation of a platelet plug following vascular injury, provide the physical site for many coagulatory steps, and ultimately contribute to the restoration of the damaged endothelial wall.

Platelets, also called thrombocytes, are tiny cell fragments that result from the breakup of large megakaryocytes in the bone marrow. These (1-4 μm) cytoplasmic fragments are highly structured (figure 8.3). They contain a cell membrane with numerous receptors, a cytoskeleton and canalicular system that permits platelets to undergo dramatic changes in shape upon activation, and specialized organelles in their cytoplasm. These organelles contain granules that store a variety of chemicals that regulate platelet aggregation and participate in coagulation.

Platelet membranes have multiple receptors. Specific glycoprotein receptors (Gp Ib/V/IX) are responsible for the adhesion of platelets to the vessel wall, whereas other receptors (Gp IIb/IIIa) facilitate platelet-to-platelet binding during aggregation. The platelet membrane also has receptors that bind to agonists to help regulate platelet function, including adenosine diphosphate (ADP), thrombin, von Willebrand factor, collagen, fibrinogen, fibrin, epinephrine, platelet activating factor, thromboxane A_2, and prostacyclin receptors (Konkle and Schafer, 2005).

Platelets contain a circumferential band of microtubules that help support platelet shape. An open canalicular system and contractile proteins attached to the cytoskeleton provide platelets with the functional capacity to swell, change shape, and contract. The open canalicular system also provides a way for substances to enter the cell and for the contents of the platelets to be secreted during activation (Rao, 1999).

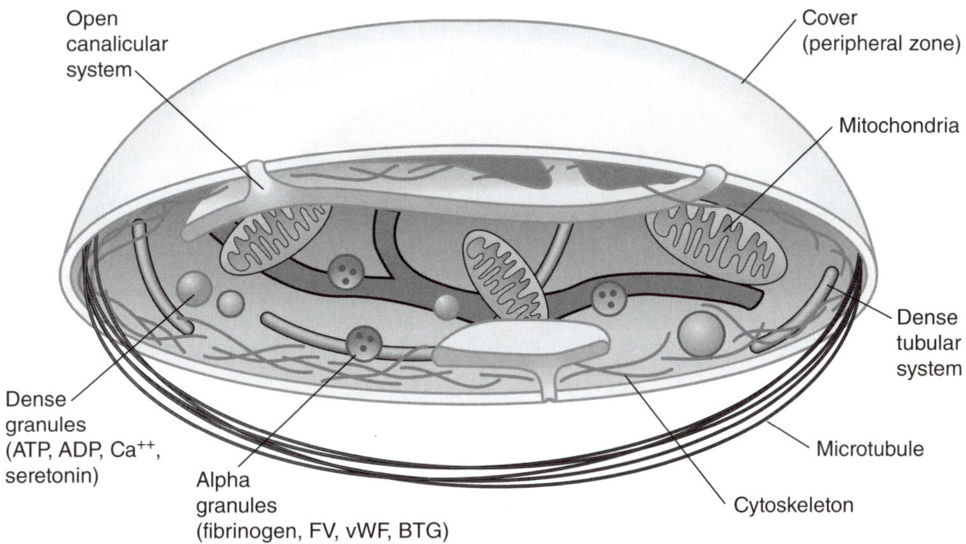

Figure 8.3 Cross section of platelet (thrombocyte).

Reprinted, by permission, from Copyright Platelet Research Lab. Available: www.platelet-research.org.

A dense tubular system (analogous to smooth endoplasmic reticulum) stores calcium and is the site for the synthesis of regulatory mediators (thromboxane A_2, prostacyclin). Other cellular organelles include mitochondria, lysomosmes, and granules. Alpha granules (α-granules) store large molecules such as fibrinogen, thromboplastin, Factor V, and won Willebrand factor. Dense granules store smaller molecules including ADP, serotonin, and calcium.

Mechanisms of Platelet Plug Formation

In the absence of injury, platelets circulate in the bloodstream and are prevented from adhering to the endothelial wall by antiplatelet factors, including nitric oxide (NO) and prostacyclin (PGI_2) released from the endothelium (Coleman et al., 2001). These platelet inhibitors are continually released from the healthy endothelium and serve to inhibit platelet adherence to the vascular lining. Furthermore, the platelet membrane itself is nonadhesive.

When a vessel is damaged, platelets are exposed to subendothelial collagen in the vessel wall. This initiates platelet adherence to the exposed collagen, a change in platelet shape (from discs to spheres with long pseudopods), platelet activation, and secretion of granules. The activation of platelets (and the subsequent secretion of granular contents) helps attract additional platelets to the site of the injury and causes platelets to aggregate, which in turn leads to the development of a platelet plug.

Primary hemostasis is achieved when the platelet plug completely occludes the vessel and stops blood flow. **Secondary hemostasis** is achieved when the activated platelet expresses chemical factors on its surface that help initiate the coagulatory cascade (discussed in a later section) and cause fibrin to form between the platelets. The fibrin mesh strengthens the platelet plug and makes it impermeable to blood cells and plasma (Ryningen and Holmsen, 1999).

Endothelial injury leads to the rapid, local formation of a platelet plug to seal the break in the vessel. Platelet plug formation involves three overlapping processes: platelet adhesion, platelet activation, and platelet aggregation. Figure 8.4 provides an overview of these three processes.

Platelet Adhesion

Platelets adhere to subendothelial tissue to stop the loss of blood from a damaged vessel. Following endothelial damage, von Willebrand factor (vWF) is released from the endothelium, serving as the "glue" that promotes platelet adherence to the injured vessel (Coleman, 2001). Platelet adhesion to the damaged vascular wall is mediated primarily through the interaction of vWF and platelet receptors, specifically Gp Ib/IX/V receptors.

Platelet Activation

Platelet activation occurs when platelets adhere to the damaged vascular wall and in response to humoral factors (epinephrine, ADP, thrombin) bind to platelet receptors. Platelet activation causes platelets to change shape, form pseudopodia, and release potent regulatory factors. Platelet activation causes the release of β-granule contents (fibrinogen, thrombospondin, β-thromboglobulin, platelet factor 4), and initiates a feedforward mechanism in which more platelets are recruited to the site of injury.

The products released from platelet granules during activation, along with thromboxane A_2 synthesized in the dense tubular system, are essential to the final process of aggregation because they recruit additional platelets to the site of injury. During the activation phase, platelets also express specific receptors (Gp IIb/IIIa receptors) on their surface. These receptors serve as an important site for binding fibrinogen, which allows platelets to bind together to form the platelet plug.

Platelet Aggregation

Platelet aggregation, the process by which activated platelets bind to one another, is the final step in platelet plug formation. Mediators released from activated platelets

LOW-DOSE ASPIRIN IN THE PRIMARY PREVENTION OF CARDIOVASCULAR EVENTS

Several organizations (including the American Heart Association and the American College of Chest Physicians) recommend low-dose aspirin (acetylsalicylic acid) for individuals with moderate risk of coronary events (defined as a 10-year risk greater than 10%). Many physicians now regularly recommend that their male patients over 45 years and females over the age of 55 years take a "baby" aspirin a day (81 mg). Aspirin has a powerful antiplatelet effect; it inhibits the formation of thromboxane A_2 (which promotes aggregation), thereby decreasing platelet aggregation. Aspirin also has other antithrombotic effects, and in addition there is mounting experimental evidence to suggest that aspirin has beneficial effects on the arterial wall.

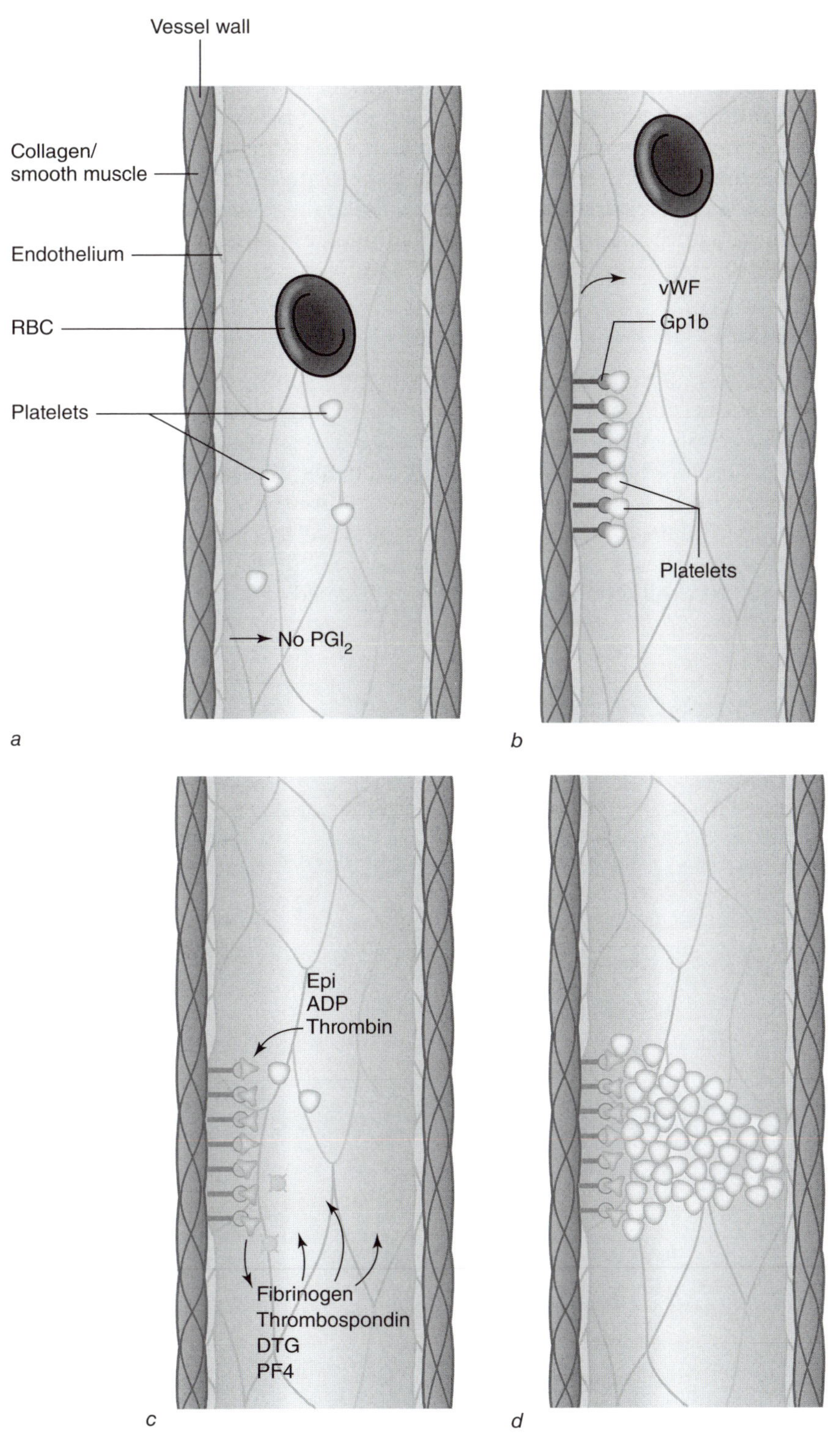

Figure 8.4 Overview of platelet adhesion *(a-b)*, platelet activation *(c)*, and platelet aggregation *(d)*.

cause platelets to change their shape and express receptors (Gp IIb/IIIa) that facilitate platelet-to-platelet binding. Platelets do not bind directly to one another; rather, each platelet binds to fibrinogen (through the Gp IIb/IIIa receptor). Since each platelet has tens of thousands of fibrinogen receptors, bonds to fibrinogen allow many platelets to be bound together rapidly to form a platelet plug (Konkle and Shafer, 2005). This positive feedback loop can lead to an occlusive platelet thrombus within minutes. Platelet aggregation is also intimately connected to the process of coagulation. Platelets provide the surface upon which coagulation takes place, and they release several factors that stimulate and regulate coagulation.

COAGULATION

Coagulation is the series of highly regulated steps resulting in the formation of fibrin that provides the framework for the loose platelet plug. Plasma coagulatory proteins (clotting factors) normally circulate in the blood in their inactive form. Table 8.1 lists the clotting factors in the blood and their synonyms. Activation of the coagulatory system results in the formation of a blood clot composed of platelets and fibrin.

Blood clot formation begins shortly after the vessel has been injured, starting 15 to 20 s after the damage has occurred. The sequence of highly regulated reactions that culminate in the production of fibrin is often referred to as the coagulatory cascade (figure 8.5).

Table 8.1 Blood Clotting Factors

Blood clotting factors (procoagulants)		
Factor number	**Factor name**	**Synonym**
I	Fibrinogen	
II	Prothrombin	
III	Tissue factor (TF)	Tissue thromboplastin
IV	Calcium ions (Ca^{++})	
V	Proaccelerin	
VI*		
VII	Proconvertin	Stable factor
VIII	Antihemophilic factor (AHF)	
IX	Plasma thromboplastin component (PTC)	Christmas factor
X	Stuart factor	Stuart-Prower factor
XI	Plasma thromboplastin antecedent (PTA)	
XII	Hageman factor	
XIII	Fibrin stabilizing factor	
	Prekallikrein	Fletcher factor
	High-molecular-weight kininogen (HMWK)	Fitzgerald factor

*Number no longer used (may be the same as factor V).

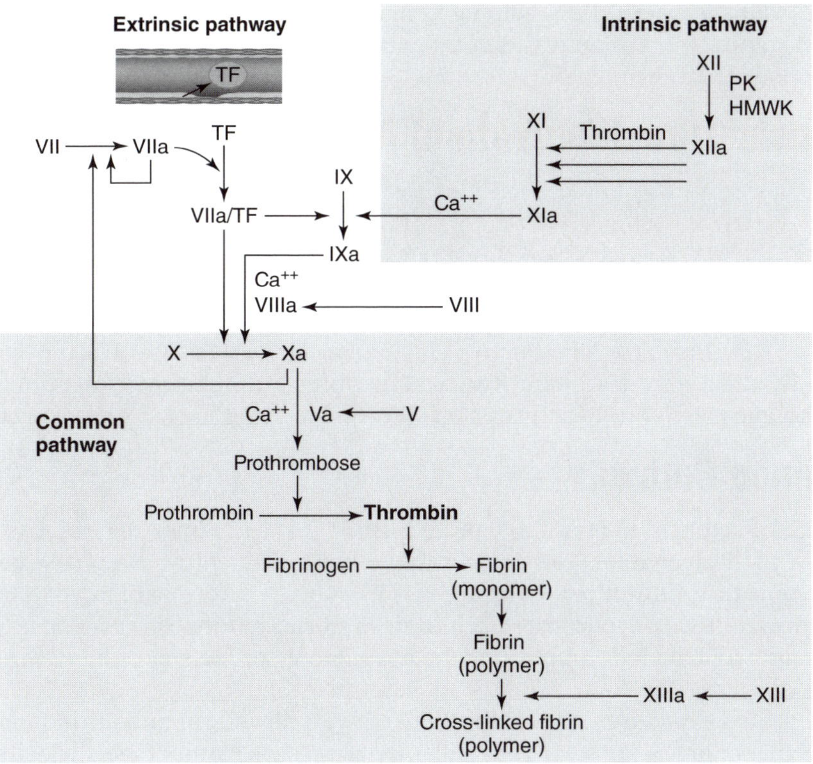

Figure 8.5 The coagulatory cascade.

The cascading series of events that lead to the generation of fibrin can be initiated by trauma to the vessel wall and surrounding tissue, trauma to the blood, or contact between the blood and procoagulatory factors in the vessel wall. Two pathways have generally been recognized, the extrinsic tissue factor pathway and the intrinsic direct activation pathway, although these two pathways interact extensively. These two pathways converge to form a common pathway that leads to the formation of the critical enzyme thrombin. Thrombin catalyzes the conversion of fibrinogen to fibrin and also plays an important positive feedback role at several important sites of the cascade (see figure 8.5).

Extrinsic (Tissue Factor) Pathway

Coagulation in vivo is probably initiated through the extrinsic pathway (also called the tissue factor pathway). Tissue damage causes the expression of tissue factor (TF) on injured endothelial cells, leukocytes, or cells in the subendothelial layers, particularly smooth muscle and fibroblasts (Konkle and Schafer, 2005). Tissue factor binds to Factor VIIa, which normally circulates in small amounts in the blood. The formation of TF–Factor VIIa complex causes the generation of more Factor VIIa from its inactive form (Factor VII), thereby amplifying the initial response. Factors generated later in the cascade, namely Factor Xa and thrombin, also feed back and induce Factor VII activation, helping to magnify the response. The final step in the extrinsic pathway is the activation of Factor X to Factor Xa, which is catalyzed by the TF–Factor VIIa complex.

Intrinsic (Direct Activation) Pathway

The intrinsic pathway (also called the direct activation pathway) is triggered by the autoactivation of Factor XII to Factor XIIa in the presence of high-molecular-weight kininogen (HMWK) and prekallikrein (PK). This pathway is important in the in vitro activation of coagulation, but there is considerable communication between the two pathways in vivo (Konkle and Schafer, 2005). Factor XIIa acts enzymatically to convert Factor XI to Factor XIa. Factor XIa, in turn, converts Factor IX to Factor IXa. Factor IXa then converts Factor X to its active form (Xa). This reaction occurs on the phospholipid membrane of activated platelets and requires Factor VIIIa as a cofactor.

Common Pathway

Activated Factor X (Factor Xa) can be formed through either the extrinsic or the intrinsic pathway. Factor Xa becomes the enzyme of the prothrombinase complex, which converts prothrombin to thrombin. Thrombin is the critical coagulatory factor that catalyzes the conversion of fibrinogen to fibrin; but it also has a host of other functions, which include amplifying the coagulatory process by stimulating several of the coagulatory factors (VII, V, VIII, XI).

The primary function of thrombin is to convert fibrinogen into fibrin monomers. The fibrin monomers automatically polymerize with each other to form fibrin, which forms the meshwork that holds the newly formed clot together. In the presence of fibrin-stabilizing factor (Factor XIII), the loosely formed fibrin strands become cross-linked with covalent bonds (Coleman et al., 2001).

Clot Retraction and Tissue Repair

Within minutes after a clot is formed, the clot begins to contract due to the action of the contractile proteins (including actin and myosin) in the platelets. As the platelets contract, they pull on the fibrin attached to their surface and compress the fibrin meshwork. Clot retraction pulls the edges of the blood vessel closer together and thus facilitates repair of the initial tissue injury.

Anticoagulatory Mechanisms

Under normal circumstances, several anticoagulatory mechanisms function to limit blood clotting and maintain blood fluidity. These mechanisms are also important in limiting the extent of clotting and ensuring that clotting is localized to the site of injury. As depicted in figure 8.6, anticoagulatory mechanisms function at various sites along the coagulatory pathway to limit fibrin accumulation, most notably inhibiting the formation of thrombin.

Antithrombin

Antithrombin inhibits thrombin formation and several other factors in the intrinsic and common pathway. The rate of antithrombin formation increases by a factor of several thousand in the presence of heparin (Konkle and Shafer, 2005). Heparin is

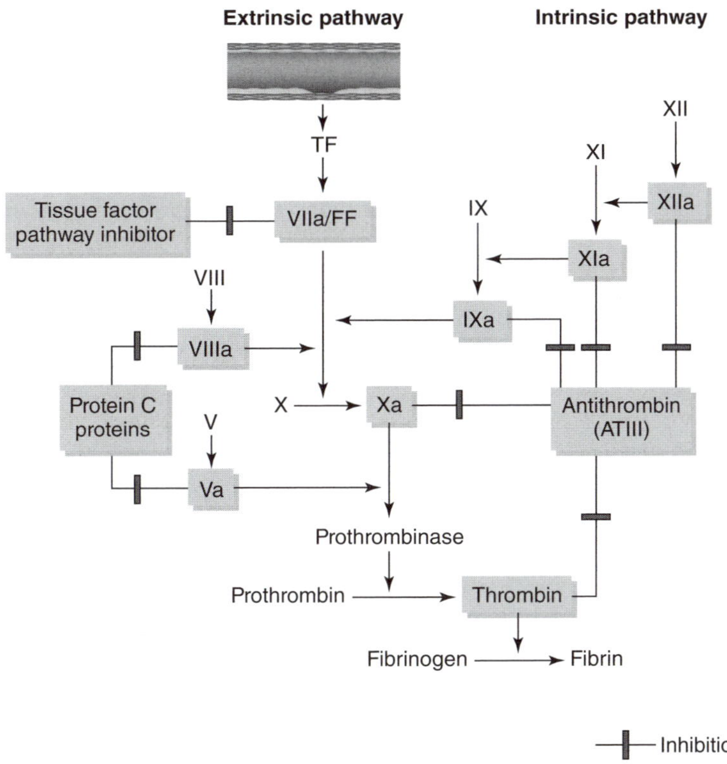

Figure 8.6 Major anticoagulatory pathways.

present in vessel walls, suggesting that the inactivation of thrombin by antithrombin probably occurs on vascular surfaces.

Protein C and Protein S

Activated Protein C is an anticoagulant that destroys Factors Va and VIIIa. Protein S serves as a cofactor for these reactions. Activation of Protein C occurs on thrombomodulin (a binding site for thrombin) on endothelial cell membranes.

Tissue Factor Pathway Inhibitor

The extrinsic pathway (tissue factor pathway) is inhibited by the tissue factor pathway inhibitor. Tissue factor pathway inhibitor has an inhibitory action on both tissue factor and Factor VIIa. Tissue factor pathway inhibitor can be released from endothelial cells (under the influence of heparin) or from platelets. Low levels of tissue factor pathway inhibitor are associated with an increased risk of deep vein thrombosis.

FIBRINOLYSIS—CLOT DISSOLUTION

It is essential to the preservation of life that clots can be formed to seal tears in the vascular system. Once the injury has been repaired, however, it is equally important that clots be dissolved and the fluidity of the blood be maintained. **Fibrinolysis** is the portion of the hemostatic system that is responsible for the dissolution of a blood clot.

It has become increasingly clear that both hypercoagulability and hypofibrinolysis are major factors in atherothrombotic disease.

As seen in figure 8.7, the fibrinolytic system degrades fibrin and thus is responsible for dissolving clots. The primary enzyme of the fibrinolytic system is plasmin, an enzyme that digests fibrin to fibrin degradation products. Plasminogen, the inactive form of plasmin, is produced in the liver and circulates in the blood. The primary physiological plasminogen activators are tissue-type plasminogen activators (t-PA) and urokinase-type tissue plasminogen activator (u-PA). Both t-PA and u-PA are produced by endothelial cells and are released by a wide variety of stimuli, including hormones, cytokines, and hemodynamic forces.

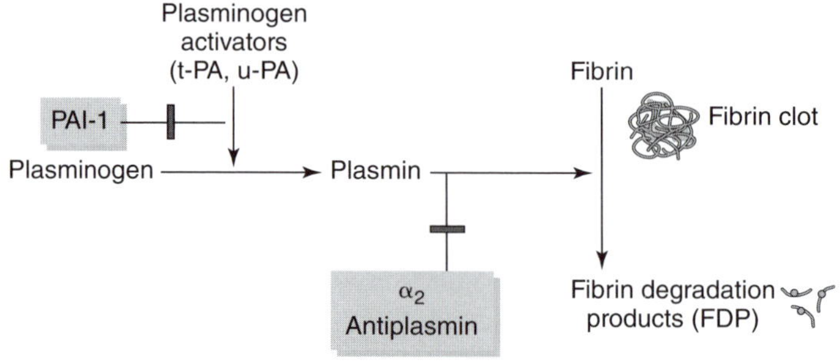

Figure 8.7 Fibrinolytic pathways.

The binding of the plasminogen activators (t-PA and u-PA) and plasminogen occurs on fibrin, thus promoting the localized interaction between the plasminogen activators and plasminogen and accelerating the rate of plasmin formation at the site where it is needed. The initial degradation of fibrin provides additional sites for plasminogen activators and plasminogen binding and therefore accelerates the process of fibrin degradation (Konkle and Schafer, 2005).

There are two inhibitors of the fibrinolytic system: plasminogen activator inhibitors (PAI-1 and PAI-2) and alpha$_2$ antiplasmin. PAI-1 is the primary plasminogen activator inhibitor. It is produced by endothelial cells and functions by inhibiting the plasminogen activators (t-PA, u-PA). Alpha$_2$ antiplasmin is produced by the liver and inhibits the action of plasmin.

ASSESSING HEMOSTASIS

Testing the hemostatic system is necessary for individuals with primary hemostatic diseases (such as hemophilia). However, it is increasingly becoming clear that platelets, coagulation, and fibrinolysis are all intimately involved with the development of atherosclerosis. Today millions of Americans take a daily dose of aspirin because of its antiplatelet properties. When medical treatment is directed at changing hemostatic balance, for instance when anticoagulation medication is prescribed for a patient who is vulnerable to excessive clot formation, frequent blood testing is necessary to ensure that the hemostatic balance is not tipped too far in the opposite direction.

The hemostatic system can be assessed by a variety of blood tests. Bleeding time is a crude test of overall hemostasis. To perform the traditional bleeding time test (Ivy method), a blood pressure cuff is inflated around the upper arm and a lancet or scalpel is used to make a stab wound in the underside of the forearm. The time from when the stab wound was made until all bleeding has stopped is called the bleeding time.

Platelet Function

Platelet function is often tested in a clinical setting using platelet aggregometry methodology. Platelet aggregometry involves a series of tests performed on platelet-rich plasma (or whole blood) exposed to known platelet activators (such as thrombin, ADP, epinephrine, collagen). The platelet activators are added to the plasma, and a dynamic measure of platelet aggregation is acquired using optical density measures.

Platelet function can also be assessed easily using a commercially available device– the platelet function analyzer (PFA-100, Siemens, Eschborn, Germany). The PFA is a small unit that uses a small sample of whole blood to measure the time required for a platelet plug to occlude a small aperture. Samples of blood are aspirated under high shear rates through an aperture (150 μm) cut into a membrane coated with collagen and ADP or collagen and epinephrine. The time to occlude the aperture closure time is reported in seconds.

Coagulatory Tests

Clinical tests of coagulation often include prothrombin time (PT) and activated partial thromboplastin time (aPTT). Prothrombin time is the time required for a clot to form after tissue factor (thromboplastin) and calcium are added to a blood sample. Prothrombin times are used to evaluate the extrinsic coagulatory pathway and are routinely measured in individuals who are taking anticoagulants to prevent excessive blood clotting. Prothrombin time is expressed as an International Normalized Ratio to allow for easy comparison. A value of 1.0 is considered normal, while values of 1.3 to 1.5 are common for individuals taking anticoagulation medication.

Activated partial thromboplastin time is the time required for a clot to form after a phospholipid reagent and calcium are added to the blood sample. Activated partial thromboplastin times are used to evaluate the intrinsic coagulatory pathway.

Coagulatory and Fibrinolytic Factors

Blood samples can be tested for many of the coagulatory and fibrinolytic factors found in the blood. Commonly reported coagulatory factors include fibrinogen (Factor I), Factor V, and Factor VIII. Commonly reported fibrinolytic factors include tissue plasminogen activator (t-PA) and plasminogen activator inhibitor (PAI-1). The availability of new bioassay techniques is making it increasingly possible to measure coagulatory and fibrinolytic factors as well as other mediators of coagulation (such as adhesion molecules, glycoproteins).

SUMMARY

Hemostasis is carefully regulated to maintain blood in the fluid state under normal circumstances and yet promote the rapid formation of blood clots to prevent blood loss when necessary. Disruptions in hemostasis can be fatal—resulting in either lethal blood loss or a coronary thrombus that occludes a coronary artery and leads to a myocardial infarction. Exercise scientists have focused a great deal of energy on the prothrombotic environment that accompanies cardiovascular (and cardiometabolic) disease states, as well as the relationship between exercise and various components of the hemostatic system. Hemostasis is an intricate and overlapping series of pathways involving platelet adhesion, activation, and aggregation; coagulation; and fibrinolysis. Within the coagulatory pathway are procoagulatory and anticoagulatory factors. Similarly, within the fibrinolytic pathway there are profibrinolytic factors and antifibrinolytic pathways. A prothrombotic environment is created by enhanced platelet activation or aggregation, procoagulatory stimuli, or antifibrinolytic factors. In contrast, an antithrombotic environment is created by platelet inhibition, anticoagulatory factors, or profibrinolytic factors.

SECTION II

Exercise Physiology

The second section of the book describes how all the components of the cardiovascular system respond in an integrated fashion to aerobic and resistance exercise and to training programs. It contains a detailed discussion of the acute and chronic effects of aerobic and resistance exercise on cardiac function, vascular function, and hemostatic variables.

Chapter 9 discusses the cardiovascular responses to aerobic exercise and the mechanisms responsible for these exercise responses. Chapter 10 discusses the training adaptations of aerobic exercise and the mechanisms responsible for them. Chapter 11 describes the acute responses of the cardiovascular system to resistance exercise. And chapter 12 is on the chronic adaptations of the cardiovascular system to a regimented resistance training program.

CHAPTER 9

Cardiovascular Responses to Acute Aerobic Exercise

By Danielle Wigmore, Bo Fernhall, and Denise Smith

Energy requirements during aerobic exercise are greater than those at rest. This necessitates an increase in oxygen and nutrient delivery to active tissues. The cardiovascular system responds to the increase in energy requirements during exercise with a highly integrated response designed to increase blood flow to the working muscles while at the same time maintaining arterial pressure within homeostatic limits. This chapter discusses the cardiovascular responses to aerobic exercise and the mechanisms responsible for these exercise responses.

CARDIAC RESPONSES

During exercise, whole body oxygen consumption ($\dot{V}O_2$) increases in proportion to exercise intensity, and is a function of increases in cardiac output ($\dot{Q}$) and oxygen extraction. The relationship between these variables can be described by the Fick equation:

$$\dot{V}O_2 = \dot{Q} \times \text{a-v } O_2 \text{ diff}$$

where a-v O_2 diff is the difference in oxygen content between arterial and venous blood, or oxygen extraction. During an acute bout of aerobic exercise, $\dot{Q}$ increases to accommodate the increased blood and oxygen demand of working tissues. The cardiac response to exercise involves both increased heart rate (HR) and stroke volume (SV), resulting primarily from neural and hormonal mechanisms as well as alterations in venous return and ventricular preload. These factors will be discussed in this section.

Cardiac Output

During aerobic exercise in moderately trained men, $\dot{Q}$ increases from ~5 L/min at rest to ~25 L/min during maximal exercise, a fivefold change (Kanstrup and Ekblom, 1978). The increase in $\dot{Q}$ is even greater in highly trained athletes. During submaximal steady state exercise, cardiac output increases rapidly and may plateau (remaining relatively unchanged) within the first few minutes of exercise. The level of cardiac output during steady state exercise depends on exercise intensity, which determines the oxygen demand of the exercise. Steady state cardiac output is maintained as a result of increases in HR and SV (figure 9.1). During steady state exercise, both HR and SV increase rapidly in proportion to exercise intensity and then plateau within the first several minutes of exercise. During incremental exercise to maximal effort, $\dot{Q}$ initially increases linearly with exercise intensity. As workload continues to increase to maximal, $\dot{Q}$ either experiences an attenuated rise or reaches a plateau (Mortensen et al., 2005) (see figure 9.2).

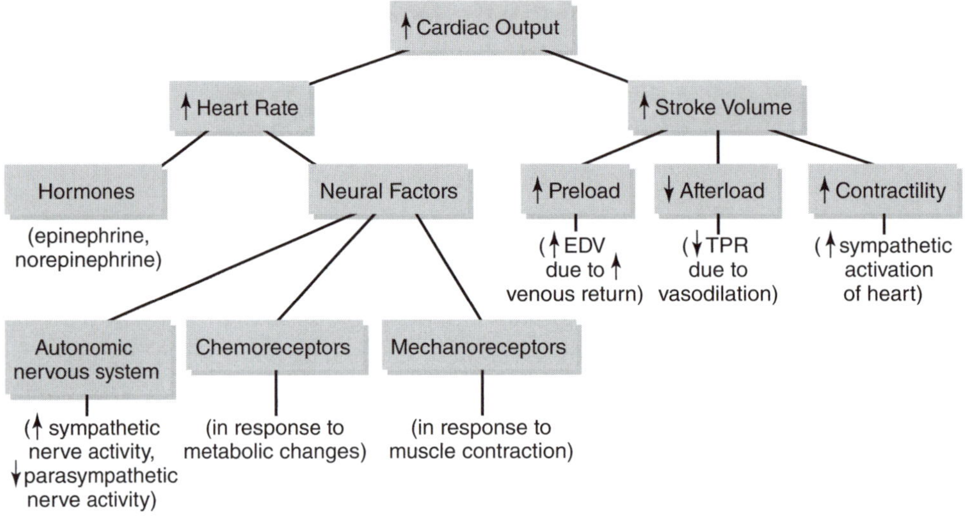

Figure 9.1 Factors affecting $\dot{Q}$ during exercise. During aerobic exercise, increases in heart rate (HR) and stroke volume (SV) contribute to the increase in $\dot{Q}$. In turn, HR is determined by changes in neural and hormonal input. The increase in SV is the net result of changes in preload, afterload, and contractility.

Heart Rate

Heart rate increases or decreases as a result of neural and hormonal influences. During exercise, the increase in HR results from a combination of decreased parasympathetic activity and increased sympathetic activity. When exercise is initiated, there is a parallel activation of the motor cortex and the cardiovascular control centers of the medulla. Thus, at the same time muscles are being stimulated to contract, a coordinated autonomic response is initiated in the medulla, a phenomenon referred to as **central command.** The autonomic response is composed of an immediate withdrawal of parasympathetic outflow to the heart followed by increased sympathetic activation

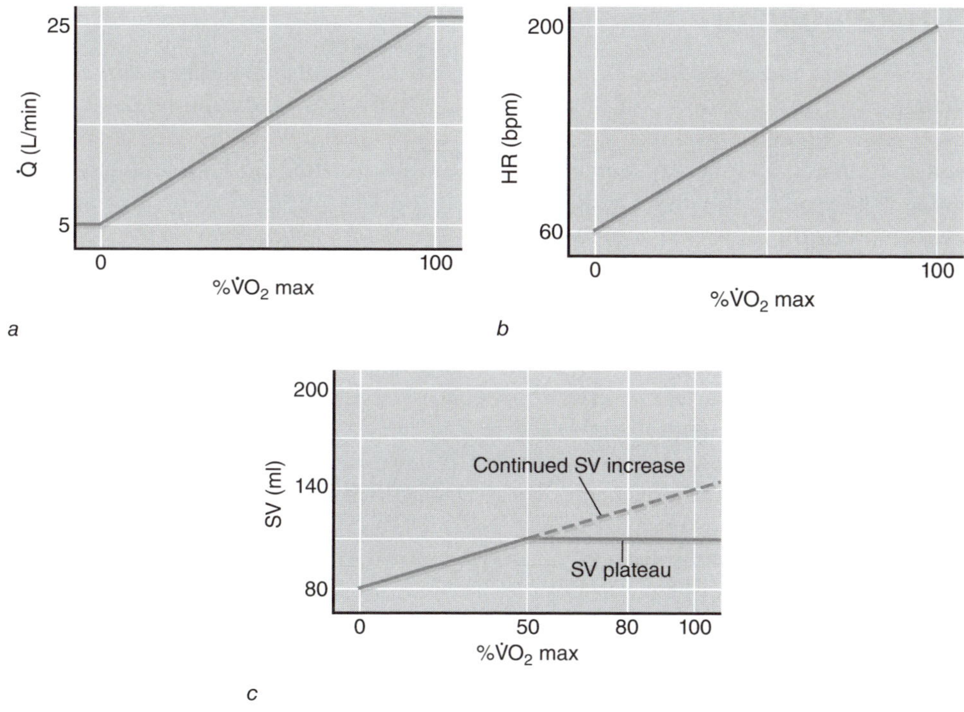

Figure 9.2 Changes in cardiac output (Q̇), heart rate (HR), and stroke volume (SV) during incremental exercise to maximum. For SV, the solid line represents the traditionally reported SV response, while the dotted line shows the continued increase in SV with exercise intensity observed in some studies.

of the heart and blood vessels (Maciel et al., 1986). The net result is an immediate rise in HR, which contributes to the rise in cardiac output.

Mechanoreceptors and chemoreceptors that are stimulated by muscle contraction contribute to the continued rise in HR (Alam and Smirk, 1937; McCloskey and Mitchell, 1972; Rotto and Kaufman, 1988). Once muscle contraction has begun, mechanoreceptors located within the muscle sense changes in muscle length and tension and relay this information to the cardiovascular control centers, resulting in a further increase in sympathetic outflow. Chemoreceptors relay messages in a similar manner; they sense changes in the muscle's chemical environment that result from the accumulation of metabolic by-products as exercise progresses and relay this information to the cardiovascular control centers (Alam and Smirk, 1937; Rotto and Kaufman, 1988).

Reinforcing the direct effect of sympathetic nerves on HR is the indirect effect of catecholamines released from the adrenal medulla. Release of the catecholamines epinephrine and norepinephrine is elevated during moderate- to high-intensity exercise. These catecholamines have a cardioacceleratory effect. Thus, the rise in HR associated with exercise is the result of an integrated neurohormonal response.

Heart rate increases linearly in proportion to exercise intensity (figure 9.2) and may plateau as maximal exercise is approached. The continued rise in HR plays an important role in enabling cardiac output to increase with incrementing exercise intensity, but it also has the potential to limit cardiac output via its effect on SV. The relationship between HR and SV at high exercise intensities is discussed in the following section.

Stroke Volume

Stroke volume increases during exercise and is a function of changes in preload, afterload, and intrinsic cardiac contractility. Greater return of venous blood to the heart increases ventricular preload, subsequently resulting in a more forceful ventricular contraction and a higher SV, a phenomenon known as the Frank-Starling law of the heart. The level of intrinsic cardiac contractility (as compared to greater force of contraction with increased stretch), on the other hand, will depend on the degree of sympathetic stimulation of the heart. Afterload decreases as a result of a declining total peripheral resistance, which results in a greater SV.

The relative contribution of these factors depends on exercise intensity. There appears to be a greater contribution of the Frank-Starling mechanism at low-intensity exercise, which wanes as exercise approaches maximal intensity. Stroke volume increases ~30% at low-intensity exercise and changes little as exercise intensity increases (Plotnick et al., 1986). This is due, in part, to a small yet significant rise in end-diastolic volume (Rowland and Roti, 2004), which in turn results from a greater venous return.

This initial rise in end-diastolic volume is related to mobilization of blood from the legs as exercise begins (the rise is not seen in situations where gravity is not a factor, such as in supine exercise or exercise in weightlessness). During exercise, the skeletal muscle pump, the abdominothoracic pump, and venoconstriction facilitate a greater venous return to the heart. Also at the onset of exercise, rapid vasodilation results in a drop in systemic vascular resistance, and hence a drop in afterload. Thus, the lower resistance to ventricular outflow likely plays an important role in the rapid rise in SV observed early in exercise. At moderate-high intensities, venous return becomes limited by the continued and dramatic drop in vascular resistance, and end-diastolic volume plateaus and may even begin to decline (Rowland and Roti, 2004). During this time, however, SV is maintained by an increase in cardiac contractility, reflected by a decrease in end-systolic volume (Higginbotham et al., 1986; Plotnick et al., 1986).

During exercise, ventricular filling is enhanced by an increase in the pressure gradient between the atria and ventricles. This gradient, in turn, results primarily from increased **ventricular suction.** Ventricular suction refers to the suction-like movement of blood from the left atrium into the left ventricle. Recent data suggest that ventricular suction is augmented during progressive aerobic exercise as a result of both enhanced ventricular contractility and relaxation (Rowland et al., 2006). Left atrial pressure appears to change little during progressive exercise, implying that the higher atrial-ventricular pressure gradient during exercise is primarily the result of ventricular suction due to rapid ventricular relaxation. This effect, however, appears to maintain, rather than increase, ventricular filling and SV at high exercise intensities (Rowland et al., 2006).

Historically, SV was thought to plateau or decrease slightly at workloads beyond ~50% $\dot{V}O_2$max, due to the effect of HR on SV (Higginbotham et al., 1986; Spina et al., 1992) (see figure 9.2). As HR increases, the duration of diastole decreases, allowing less time for ventricular filling. Thus, SV, and ultimately cardiac output, is not able to increase further at high HR.

The plateau in SV that is often reported at approximately 30% to 50% of $\dot{V}O_2$max is not a universal observation. Some investigators report an increase in SV up to, and including, maximal exercise in highly trained athletes (Ferguson et al., 2001;

Gledhill et al., 1994; Krip et al., 1997; Zhou et al., 2001) (see dotted line, figure 9.2). Using the acetylene rebreathing technique, Gledhill and associates (1994) found that SV increased up to maximal workload in highly trained cyclists. Zhou and colleagues (2001) had similar findings in elite runners, using a similar technique. In both studies, the continued increase in SV was limited to highly trained individuals, as untrained subjects experienced a plateau (Gledhill et al., 1994) or a slight decrease (Zhou et al., 2001) in SV. Further, Zhou and colleagues (2001) observed this phenomenon only in elite runners, as trained cross country runners experienced the typical plateau in SV beyond a HR of ~110 beats/min. Data from two additional studies demonstrate a higher SV at maximal compared to submaximal workloads in both trained and untrained subjects, resulting from either a shallow continued rise in SV beyond light workloads (Krip et al., 1997) or a secondary rise in SV at high workloads (Ferguson et al., 2001).

Although the precise mechanism responsible for the continued rise in SV with increasing exercise intensity observed in some studies is unknown, it has been suggested that augmentations in ventricular filling (greater preload) and emptying (greater contractility) may play a role. Supporting these suggestions, higher rates of ventricular filling and emptying have been reported in trained compared to untrained subjects (Ferguson et al., 2001; Gledhill et al., 1994; Krip et al., 1997). Adaptations to endurance training include plasma volume expansion, which would enhance ventricular preload, and ventricular hypertrophy, which would enhance cardiac contractility. These adaptations, however, are not limited to elite athletes and are also observed in regularly trained individuals (see chapter 10 for a detailed discussion of aerobic training adaptations). Further, some investigators report the traditional SV response both before and after endurance training (Spina et al., 1992; Wolfe and Cunningham, 1982) or in trained and untrained individuals (Rowland and Roti, 2004). The SV response to incremental exercise to maximum continues to be an area of controversy; but the generally accepted response in normal, healthy people is a plateau in SV that occurs between 30% and 50% of maximum. For a more detailed discussion of this debate, see Gonzalez-Alonso, 2008.

Cardiovascular Drift

During prolonged aerobic exercise, especially at high intensities, central cardiovascular variables that had previously been at steady state levels may begin to drift, a phenomenon referred to as **cardiovascular drift.** The traditional theory suggests that as exercise progresses, body temperature rises, necessitating increased skin blood flow to aid in the release of body heat. Additionally, plasma volume is reduced as a result of increased perspiration and increased respiration. The combination of a reduced blood volume and the distribution of a massive proportion of cardiac output to the skin and exercising skeletal muscle reduces venous return and, ultimately, end-diastolic volume. This, in turn, causes a downward drift in SV. Heart rate, however, compensates by drifting upward to maintain cardiac output in most situations.

This theory of cardiovascular drift has been challenged in recent years. Data suggest that the increase in HR with prolonged, moderate-heavy steady state exercise effectively lowers ventricular filling time, thus reducing ventricular preload and SV. Support for this theory comes from studies using pharmacological blockade to prevent the typical rise in HR observed with prolonged exercise (Fritzsche et al., 1999). Under these conditions, preventing the rise in HR also prevented the decline in SV (Fritzsche et al., 1999).

Additionally, under control conditions, SV declined progressively from 15 to 55 min of exercise despite a plateau in skin blood flow during that time period (Fritzsche et al., 1999). Thus, an increase in skin blood flow was not necessary to induce the typical changes in HR and SV that characterize cardiovascular drift. Instead, elevations in core temperature and sympathetic nervous system activity may be responsible for the progressive rise in HR with prolonged exercise (Coyle and Gonzalez-Alonso, 2001).

VASCULAR RESPONSE

During exercise, there is a redistribution of $\dot{Q}$ ($\dot{Q}$ = cardiac output), with a greater proportion of $\dot{Q}$ going to the skin and skeletal muscles compared to at rest. The proportion of cardiac output supplying the heart remains the same; but because this is the same proportion of a higher cardiac output, the total amount of blood delivered to the heart is greater during exercise than at rest. Thus, during exercise, blood flow increases to the tissues that are highly active and decreases to the tissues that are less active (e.g., the viscera). In this way, the body prioritizes blood flow to its tissues based on need. This section deals with the factors involved in the redistribution of cardiac output and changes in blood flow during exercise.

As stated earlier, whole body oxygen consumption ($\dot{V}O_2$) increases with increasing exercise intensity. The increase in whole body $\dot{V}O_2$ is largely the result of higher cardiac and skeletal muscle $\dot{V}O_2$. The increased metabolic demands of the heart and exercising skeletal muscles require greater delivery of oxygen and nutrients to these tissues, as well as sufficient removal of metabolic by-products. Thus, blood flow must increase to accommodate these needs. During vigorous aerobic exercise, cardiac blood flow and muscle blood flow increase approximately 4- and 20-fold, respectively (Rowell, 1993). Further, skin blood flow increases ~8- to 18-fold, depending on exercise intensity and the rise in body temperature. The rise in skin blood flow is needed to release heat and aid in temperature regulation during exercise. Blood flow is directly proportional to perfusion pressure and inversely proportional to vascular resistance. Thus, exercise blood flow increases as a result of both an increased arterial pressure and a decreased vascular resistance.

Mean Arterial Pressure

During aerobic exercise, mean arterial pressure (MAP) increases as a function of a rising systolic blood pressure (SBP). The rise in SBP is mediated through the increase in cardiac output already discussed. Additionally, sympathetic vasoconstriction of blood vessels in nonexercising tissues aids in the maintenance of arterial pressure, a concept that will be discussed in greater detail later. Meanwhile, local release of vasodilators causes dilation of blood vessels in active muscle beds and a decrease in total peripheral resistance (TPR).

The net effect of opposing vasoactive signals (i.e., systemic vasoconstriction and local vasodilation) is an unchanging diastolic blood pressure (DBP) during exercise. The result of these changes in SBP and DBP is a modest rise in MAP that is proportional to exercise intensity at submaximal workloads. Note that SBP begins to plateau as exercise approaches maximum. This phenomenon is the result of the limits in cardiac output described earlier. As a result of the changes in SBP and HR, **rate–pressure product (RPP)** also increases with increasing workloads and reaches a plateau as maximal exercise is approached. Remember that RPP is the product of SBP and HR and reflects myocardial oxygen consumption.

Stroke work will also increase during exercise, as a result of increases in both SV and ventricular pressure. Remember from chapter 2 that stroke work is the work performed by the ventricles during ejection and equals the area inside the ventricular pressure–volume relationship. During exercise, the enhanced end-diastolic volume and cardiac contractility described earlier in this chapter contribute to a higher stroke work during exercise compared to rest. Figure 2.5 demonstrates the effect of exercise on the ventricular pressure–volume relationship and stroke work. See chapter 2 for further discussion on stroke work.

Vascular Resistance

Total peripheral resistance decreases during aerobic exercise, the result of vasodilation in active tissues, primarily the exercising muscles. Vasodilation is elicited by chemical substances released from the exercising muscles and surrounding vessels (for a review of mechanisms of vasodilation, see chapter 6). The release of vasodilators is dependent on activity of skeletal muscle; at higher exercise intensities, there is a greater vasodilatory stimulus and thus a greater drop in resistance. The reduction in vascular resistance works to increase blood flow to exercising muscles while preventing arterial pressure from rising too high.

Importantly, different local circulations experience different changes in vascular resistance during exercise. This in turn governs the distribution of flow to the various circulations. For example, increased sympathetic outflow from the cardiovascular control centers of the brain results in systemic vasoconstriction and increased vascular resistance. This, in turn, causes a reduction in blood flow to many body tissues. It is well known, however, that skeletal muscle blood flow increases, rather than decreases, during exercise. Exercise blunts vasoconstriction mediated by alpha1- and alpha2-adrenergic receptors (Rosenmeier et al., 2003), thus reducing sympathetically mediated vasoconstriction in active muscle beds. The net effect is vasodilation and lower vascular resistance in exercising muscle. The ability of exercise to overcome sympathetic vasoconstriction in active muscle beds is termed **functional sympatholysis.**

Sympathetic vasoconstriction and functional sympatholysis are responsible for the proper distribution of blood flow during exercise. Vasoconstriction causes increased resistance and a reduction in blood flow to a number of organ systems, such as the kidneys, liver, and digestive tract. These tissues are less active during exercise and therefore require less blood flow. Without this vasoconstriction, vascular resistance would fall excessively, resulting in a MAP too low to maintain sufficient perfusion pressure. Meanwhile, sympatholysis ensures that tissues that are highly active during exercise (i.e., skeletal muscle) receive the increased blood flow that they need. It is important to note, however, that there is an upper limit to $\dot{Q}$. Thus, when the competition for blood flow is very high and $\dot{Q}$ cannot meet the demand, vascular beds in active muscle begin to vasoconstrict to redistribute blood flow to other areas in need of increased flow. This phenomenon is known as the steal effect. Such a competition has been observed between respiratory muscles and active skeletal muscles, where loading the respiratory muscles results in vasoconstriction and reduced blood flow in leg muscles (Harms et al., 1997).

Mechanisms of Exercise Hyperemia

The rise in muscle blood flow observed during exercise is commonly referred to as **exercise hyperemia.** A number of factors contribute to the onset and maintenance of exercise hyperemia, which can be grouped into two main categories: vasodilation and mechanical actions (figure 9.3*a*). Note that during active muscle contraction, vasodilatory and mechanical actions have opposite effects on blood flow (figure 9.3*b*). During

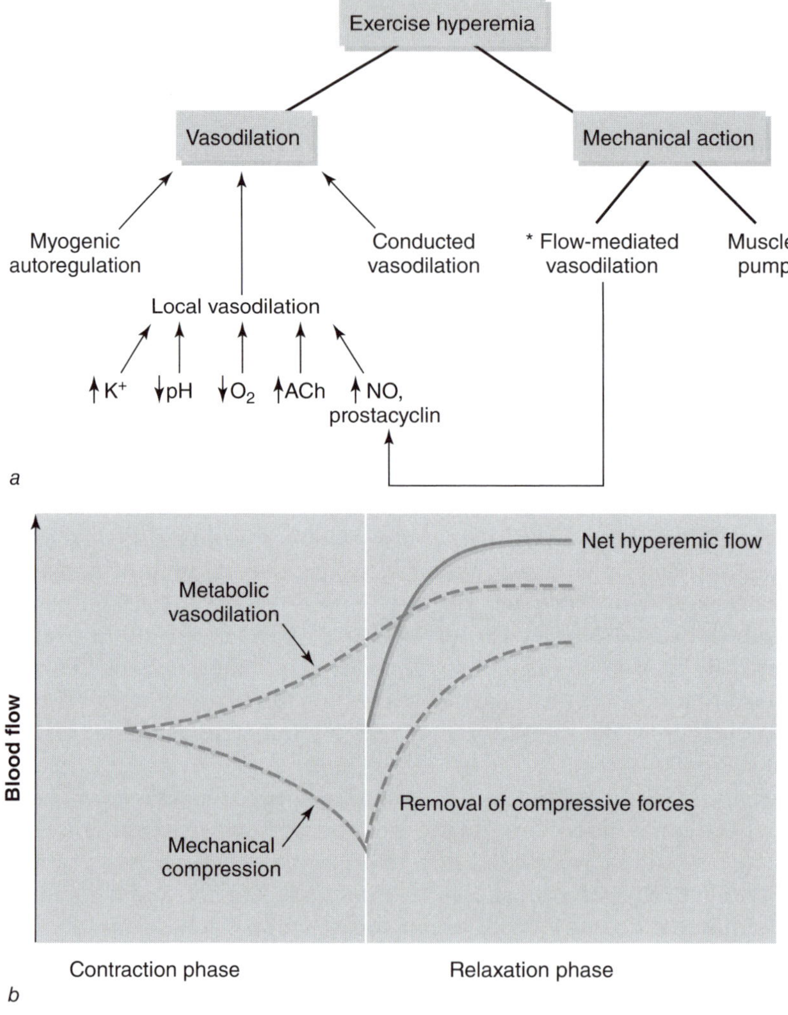

Figure 9.3 Mechanisms of exercise hyperemia. *(a)* Both vasodilatory and mechanical factors contribute to exercise hyperemia. Vasodilation results from a combination of myogenic autoregulation and local and conducted vasodilation. Mechanical factors include both flow-mediated dilation and the skeletal muscle pump. *Although flow-mediated dilation requires a mechanical stimulus to induce vasodilation, it does so by triggering the release of nitric oxide and prostacyclin. *(b)* During a brief contraction, vasodilation and mechanical forces oppose one another. However, they work together immediately postcontraction to increase blood flow, resulting in a net hyperemic response.

Figure 9.3b is based on D.M. Wigmore, K. Propert, and J.A. Kent-Braun, 2005, "Blood flow does not limit skeletal muscle force production during incremental isometric contractions," *European Journal of Applied Physiology* 96: 370-378.

relaxation between contractions, both factors work in concert to enhance blood flow. The mechanisms responsible for these actions are discussed next.

Vasodilation

A number of vasoactive substances have been linked to exercise hyperemia, including hydrogen ions, oxygen, potassium, adenosine triphosphate, acetylcholine, nitric oxide (NO), and prostacyclin. Many of these substances are local tissue factors released from skeletal muscle as a result of increased metabolic activity during exercise. For example, the rise in interstitial potassium concentration during muscle contraction leads to smooth muscle cell hyperpolarization and vasodilation. Similarly, adenosine, which is a potent vasodilator, accumulates in the interstitium during contractions and has long been believed to contribute to exercise hyperemia. However, the importance of adenosine to in vivo vasodilation during exercise has been called into question (Delp, 1999; Hester, Guyton, and Barber, 1982), as adenosine may not be present in high enough amounts to induce any significant vasodilation (Phair and Sparks, 1979). Decreased pH, which results from increased metabolic activity, has also been proposed to cause vasodilation during exercise. It is believed, however, that pH may cause vasodilation indirectly, via its effect on other dilatory mechanisms (Clifford and Hellsten, 2004). Low muscle oxygen levels also trigger vasodilation. Intramuscular oxygen levels may fall as a result of reduced oxygen delivery (e.g., hypoxia) or increased oxygen consumption during exercise. There is evidence that acetylcholine released from the neuromuscular junction during muscle contraction may also induce vasodilation. In one study, stimulation of muscle resulted in arteriolar vasodilation despite the fact that muscle fiber contraction was pharmacologically inhibited with tubocurarine (Welsh and Segal, 1997). Further, infusion of a cholinesterase inhibitor (i.e., a substance that prevents acetylcholine breakdown) enhanced vasodilation, while infusion of a muscarinic receptor antagonist (i.e., substance that prevents binding of acetylcholine to receptors on vessels) inhibited vasodilation (Welsh and Segal, 1997). Acetylcholine induces vasodilation through an endothelial cell-mediated pathway, as described next.

Factors released from endothelial cells, primarily NO and prostacyclin, also play a role in exercise hyperemia. Shear stress, resulting from a rapid increase in blood flow (e.g., during exercise), stimulates the release of various substances from the vascular endothelium, which include adenosine triphosphate and substance P (Mo et al., 1991; Niebauer and Cooke, 1996; Ralevic et al., 1990); these in turn cause calcium channels on the endothelial cell membrane to open (Davies, 1995) (see figure 9.4). The concentration of intracellular calcium increases, triggering release of NO and prostacyclin from the endothelial cells. Further, increased sheer stress causes acetylcholine to be released from the endothelial cells (Martin et al., 1996) (see figure 9.4). Acetylcholine, in turn, binds to muscarinic receptors on neighboring endothelial cells, causing NO release. Nitric oxide and prostacyclin both cause the vascular smooth muscle to relax, resulting in dilation of the vessel. The elevation in shear stress not only stimulates NO release but also stimulates NO production, as evidenced by increased endothelial nitric oxide synthase (eNOS) and messenger RNA (mRNA) expression (Noris et al., 1995; Uematsu et al., 1995). Thus, both increased NO availability and increased NO release contribute to vasodilation during exercise. Because it is initiated by an increase in blood flow, this process is termed **flow-mediated dilation.** Due to the time course of action, flow-mediated dilation is not considered to play an important role

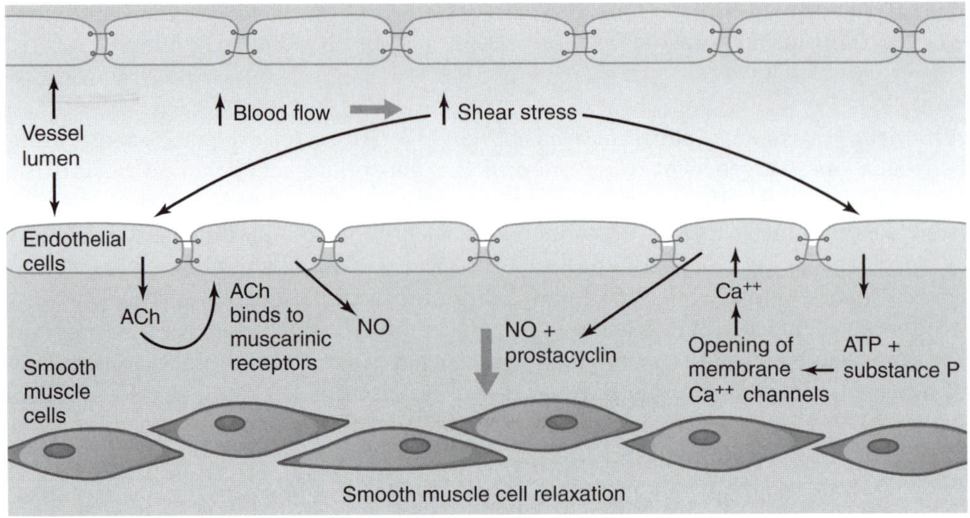

Figure 9.4 Flow-mediated vasodilation. The increase in limb blood flow that accompanies exercise causes shear stress to rise. This in turn triggers a cascade of events, leading to the release of nitric oxide and prostacyclin from the vascular endothelium. Nitric oxide and prostacyclin both elicit smooth muscle cell relaxation and vasodilation.

in increasing blood flow at the onset of exercise, but rather is thought to contribute to the maintenance of exercise hyperemia.

Exercise blood flow is also controlled by myogenic autoregulation. Deviations in blood flow, induced by rapid changes in perfusion pressure, are largely corrected by compensatory changes in vascular tone (Walker et al., 2007). Walker and colleagues (2007) also observed a rapid increase in forearm blood flow between muscle contractions when the arm was lowered from above heart level to below heart level, effectively increasing perfusion pressure by ~30 mmHg. Forearm blood flow was decreased within a matter of seconds, but remained higher than when the arm was above heart level. Similar results, opposite in direction, were observed when the arm was moved from below heart level to above heart level. These data support roles for rapid myogenic autoregulation and changing perfusion pressure in determining muscle blood flow responses during exercise.

Conducted Vasodilation

There is a large body of experimental evidence supporting a phenomenon known as **conducted vasodilation,** in which vasodilation in distal vessels spreads proximally through the vasculature. Conducted vasodilation occurs in the resistance vessels of the microcirculation, where vasodilatory signals are initiated in the distal arterioles and conducted to proximal arterioles and feed arteries, causing vasodilation (figure 9.5*a*). This occurs via cell-to-cell communication between smooth muscle cells and endothelial cells, between adjacent smooth muscle cells, and between adjacent endothelial cells (figure 9.5*b;* Bartlett and Segal, 2000; Emerson and Segal, 2000a). Vasodilation results from hyperpolarization of smooth muscle cells. The coupling between endothelial and smooth muscle cells is such that hyperpolarization in one cell type (endothelial or smooth muscle) is transmitted to the other cell type (Bartlett and Segal, 2000; Emerson and Segal, 2000a). In small arterioles, hyperpolarization is then conducted bidirectionally along both endothelial cells and smooth muscle cells. In larger feed arteries, however, hyperpolarization can be spread only through endothelial cells (Emerson and Segal, 2000b; Segal and Jacobs, 2001). As the signal spreads, each endothelial cell communicates with the adjacent smooth muscle cells, causing hyperpolarization and vasodilation. Evidence supports the existence of gap junctions in both endothelial cell layers and smooth muscle cell layers, allowing for the electrical coupling observed in both cell types (Little, Beyer, and Duling, 1995). Further, myoendothelial gap junctions may be responsible for communication between endothelial and smooth muscle cells (Little, Beyer, and Duling, 1995). The microcirculation is thus able to elicit a highly integrated response to ensure sufficient perfusion of active muscle; increased metabolism results in dilation of distal arterioles via release of a variety of vasoactive substances, increasing capillary perfusion. Subsequent dilation and increased flow in feed arteries ensure sufficient perfusion pressure in the distal vessels.

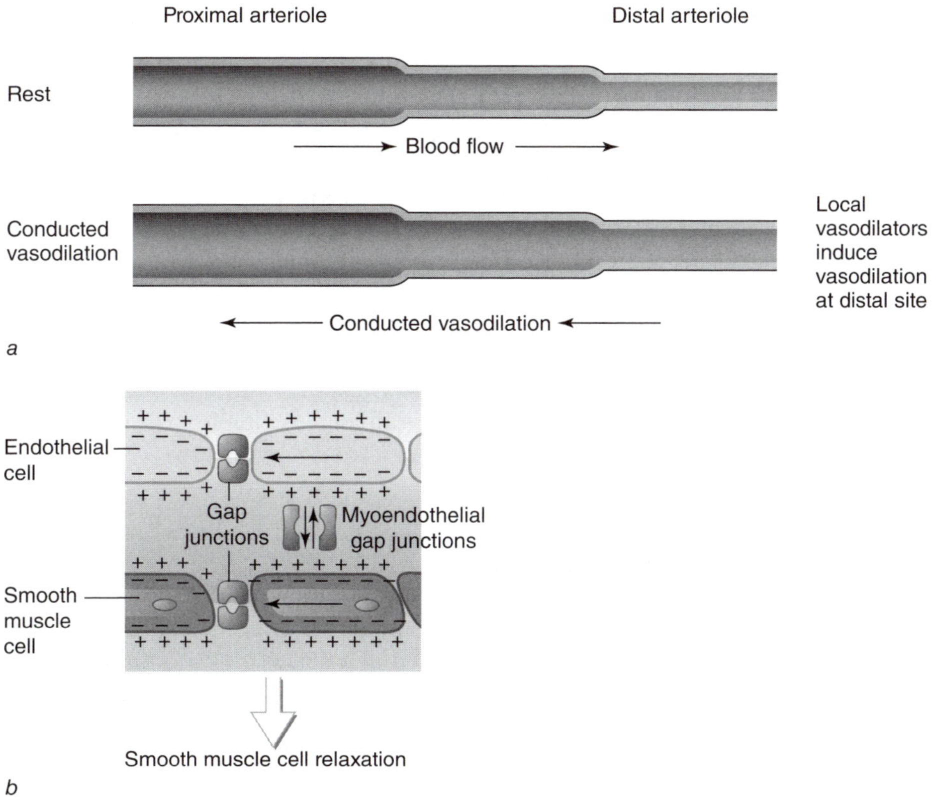

Figure 9.5 Cell-to-cell communication in conducted vasodilation. *(a)* Proximal arterioles branch into smaller distal arterioles. As seen in the lower part of the diagram, when vasodilation is initiated in the distal arteriole by local tissue factors, the signal is transmitted proximally, inducing vasodilation in the larger arterioles. *(b)* The wave of hyperpolarization spreads proximally through both smooth muscle cells and endothelial cells, inducing smooth muscle cell relaxation and vasodilation. The arrows represent the cell-to-cell communication and the direction of the spread of hyperpolarization (between endothelial and smooth muscle cells, between adjacent smooth muscle cells, and between adjacent endothelial cells). Electrical signals are spread from cell to cell through gap junctions.

Mechanical Actions

There are two main mechanical mechanisms that contribute to exercise hyperemia: flow-mediated dilation and the **skeletal muscle pump.** Flow-mediated vasodilation, described earlier, is considered a mechanical mechanism because it is the shearing force associated with increased blood flow that elicits flow-mediated dilation.

Mechanical forces imparted to the vasculature during rhythmic contractions result in a pumping effect, whereby venous return and perfusion of skeletal muscle are enhanced—a phenomenon that is referred to as the skeletal muscle pump. During the contraction phase, forces from the muscles compress the veins, essentially squeezing blood from the veins. Because venous valves allow flow in only one direction (toward the heart), this squeezing effect works to move blood along the venous circulation back toward the heart. In this way, the muscle pump aids venous return and ventricular filling, as mentioned earlier. This squeezing action also empties the veins, such that venous pressure is dramatically reduced upon relaxation. The result is an increase in the pressure gradient across the capillary bed and a large increase in muscle blood flow. It has also been proposed that during the relaxation phase of contraction, the small veins, which are fused to skeletal muscles, are pulled open, potentially causing venous pressure to become negative, thus further enhancing the pressure gradient (Delp, 1999).

Just as veins are compressed by intramuscular forces during contraction, so are arteries. The degree of arterial occlusion depends on the force produced by the muscle and the resulting increase in intramuscular pressure, with stronger contractions producing a greater limitation in arterial blood flow (Degens, Salmons, and Jarvis, 1998; Sjøgaard et al., 1986). This mechanical compression causes a transient reduction in blood flow during the contraction phase, with a subsequent increase in blood flow during the relaxation phase (figure 9.3*b,* p. 146). Due to the rapid nature of the contraction–relaxation cycle during dynamic aerobic exercise, this mechanical compression does not limit the overall blood flow response during exercise. Rather, the net effect of rhythmic muscle contractions is an increase in blood flow, that is, an exercise hyperemia.

Contributions of Vasodilation and the Muscle Pump to Exercise Hyperemia

To date, there is still disagreement about the roles vasodilation and the muscle pump play in initiating and maintaining exercise hyperemia. Proponents of the muscle pump theory claim that the muscle pump is activated at the onset of exercise and is responsible for the initial rise in blood flow (Sheriff, Rowell, and Scher, 1993; Tschakovsky, Shoemaker, and Hughson, 1996). According to this hypothesis, the accumulation of vasodilators associated with increased muscle metabolism is too slow to account for the rapid rise in blood flow that occurs during the first few seconds of exercise; rather, these vasodilators contribute to the maintenance of hyperemia as exercise progresses. Further, increases in muscle blood flow have been shown to be more closely related to treadmill speed than to grade, suggesting that contraction frequency is more important than contraction intensity in determining the blood flow response during exercise (Sheriff and Hakeman, 2001). Similarly, in experiments in which the metabolic demand, and therefore vasodilatory stimulus, of the exercise is maintained by opposing alterations in treadmill speed and grade (increased grade and decreased speed, or increased speed and decreased grade), muscle blood flow changes were coupled to changes in speed (or contraction frequency) rather than changes in grade. These observations are consistent with the muscle pump theory that it is the repetitive contraction–relaxation cycle that aids blood flow.

Other investigators argue against the role of a muscle pump in exercise hyperemia and suggest that rapid vasodilation is responsible for the initial hyperemic response during exercise. Support for this theory comes from studies in which blood flow at the onset of exercise was observed to be related to contraction intensity (Shoemaker, Tschakovsky, and Hughson, 1998), even under conditions in which the contribution of a potential muscle pump was minimized experimentally (Tschakovsky et al., 2004). Also, studies in which maximal vasodilation is induced pharmacologically, prior to exercise, show no additional rise in blood flow upon commencement of exercise (Hamann et al., 2003), which would be expected if a muscle pump contributed to the rise in blood flow. It is possible, however, that under high flow conditions, venous filling is too rapid to take advantage of a muscle pump. Further support of the vasodilatory theory comes from observations in stimulated animal muscle of

1. rapid dilation of terminal arterioles (~2 s within start of exercise [Marshall and Tandon, 1984]) and
2. loss of initial hyperemia when smooth muscle cell hyperpolarization, and therefore vasodilation, is experimentally inhibited (Hamann, Buckwalter, and Clifford, 2004).

Matching Perfusion to Muscle Activity

There is a tight coupling between muscle perfusion and metabolic activity, which causes an increase in perfusion to active muscle that is proportional to exercise intensity. As muscle activity increases, more vasoactive substances are released from the muscle and endothelial cells, resulting in a greater reduction in resistance and increased blood flow. Thus, the muscle facilitates an increase in oxygen delivery to meet the rising metabolic demands of the muscle during exercise. From this scenario, it would appear that individual muscle fibers regulate their own flow; the picture, however, is not that simple. This section details what happens in the microcirculation during exercise and how this affects the coupling of muscle activity and perfusion.

Arterioles branch into capillaries, which are the site for gas exchange between the blood and the muscle fibers. The smallest and most distal of the arterioles are referred to as terminal arterioles, and they control which capillaries are perfused. Each terminal arteriole and the capillaries it supplies is called a microvascular unit. Dilation of a terminal arteriole causes the capillaries in that microvascular unit to become perfused. Capillary perfusion here is defined as the number of capillaries receiving blood flow, rather than the rate of flow through the capillaries.

During exercise, capillary perfusion increases in proportion to the amount of active muscle mass. Vasoactive signals released from active muscle fibers will cause dilation of adjacent terminal arterioles and perfusion of microvascular units. This would appear to be a good system for tightly matching capillary perfusion to individual muscle fiber metabolic demand. The organization of microvascular and motor units in skeletal muscle, however, precludes this arrangement. Microvascular units are aligned such that capillaries in one microvascular unit are all close together (figure

9.6). Thus, when a given microvascular unit is recruited, the perfused capillaries are all in close proximity to one another and supply blood to adjacent muscle fibers. In skeletal muscle, the fibers belonging to a given motor unit are dispersed throughout the entire muscle in a mosaic pattern. Thus, recruitment of one motor unit results in contraction of many muscle fibers throughout the muscle. There is not a one-to-one alignment of microvascular units to muscle fibers, and each microvascular unit is likely to supply multiple muscle fibers, often belonging to different motor units. Because the increase in metabolic activity in a muscle fiber will result in local vasodilation of nearby terminal arterioles, all microvascular units that are adjacent to active muscle fibers will be recruited. Consider figure 9.6. If motor unit (MU) 1 is recruited, this will cause vasodilation of the terminal arterioles of MVU (microvascular unit) 1 and MVU 3, as these MVUs are adjacent to the active muscle fibers of MU 1. Note here that muscle fibers from MU 4 will also be perfused because they are supplied by the same MVUs (1 and 3). However, if MU 4 is recruited, all four MVUs will be recruited because each muscle fiber from MU 4 is adjacent to at least one of the MVUs. This, in turn, causes perfusion of some muscle fibers in each motor unit, even though the only active motor unit is MU 4. There is thus the tendency for some perfusion of inactive muscle fibers.

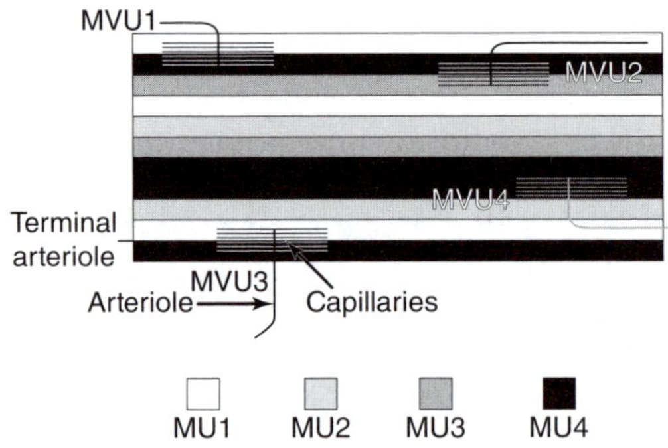

Figure 9.6 Schematic representation of the organization of microvascular units. When a motor unit (MU) becomes active, vasoactive substances are released from the muscle fibers of that MU in proportion to the level of metabolic activity. These vasoactive substances then dilate terminal arterioles of adjacent microvascular units (MVUs), causing the capillaries of those MVUs to be perfused. Microvascular units are spaced throughout the muscle bed in such a way that recruitment of a given MVU will often perfuse muscle fibers from more than one MU.

Experimental data and mathematical models have both suggested that capillary recruitment peaks at low exercise intensities and remains maximal as exercise intensity continues to increase (Fuglevand and Segal, 1997; Honig, Odoroff, and Frierson, 1980). One model simulating physiological recruitment of motor units during exercise showed that nearly all MVUs (90%) were perfused when only ~3% of motor units were active (Fuglevand and Segal, 1997). The rate of blood flow through capillaries, however, increases with increasing exercise intensity, primarily as a result of arteriolar dilation proximal to the terminal arterioles. The degree of vasodilation in microvessels is proportional to the increase in muscle fiber recruitment (see figure 9.7 and VanTeeffelen and Segal, 2000) and contraction frequency (Gorczynski, Klitzman, and Duling, 1978). Thus, the muscle is able to regulate its own blood flow to accommodate increased oxygen demands during exercise.

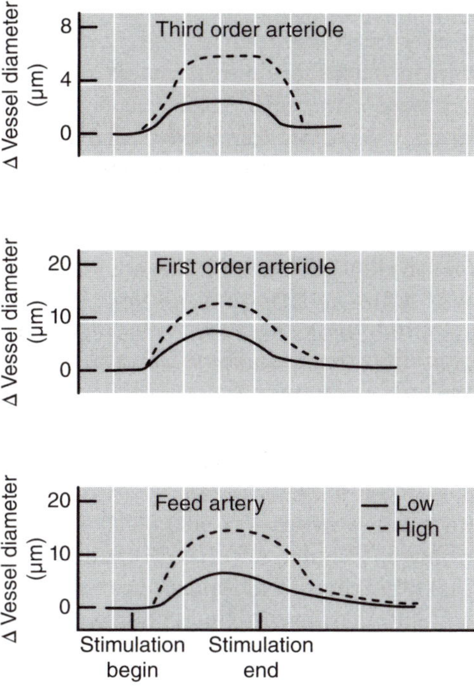

Figure 9.7 Relationship between motor unit recruitment and vasodilation. The degree of vasodilation increases with the level of motor unit recruitment, and this is true for various levels of the arteriolar circulation. Data from VanTeeffelen and Segal, 2000. Note that recruitment was manipulated in such a way that total force production was the same for both recruitment levels.

Based on J.W. VanTeeffelen and S.S. Segal, 2000, "Effect of motor unit recuitment on functional vasodilatation in hamster retractor muscle," *The Journal of Physiology* 524 Pt 1: 267-278.

As discussed earlier, exercise increases metabolic demand and therefore oxygen consumption. The increase in muscle oxygen consumption causes the partial pressure of oxygen (PO_2) in the muscle to drop, thereby increasing the PO_2 gradient between muscle and capillary blood. During exercise, capillary perfusion increases through dilation of the terminal arterioles by the various factors already described. Thus, there is an increase in the capillary surface area, thereby reducing the distance between capillaries and muscle fibers and increasing diffusion of oxygen into active skeletal muscle. In other words, there is an increase in a-v O_2 diff, as more oxygen is extracted from capillary blood. The drop in muscle PO_2 at the onset of exercise aids in the movement of oxygen into the muscle fibers to ultimately meet the increased oxygen demand during exercise.

Skin Blood Flow

During exercise, an increased proportion of $\dot{Q}$ is distributed to the cutaneous circulation to allow heat loss and maintain core temperature within a fairly narrow range. There are several competing mechanisms, however, that determine the net skin blood flow response to exercise. At the onset of exercise, vasoconstriction of cutaneous blood vessels results in decreased skin blood flow (Kellogg, Johnson, and Kosiba, 1991a; Taylor, Johnson, and Kosiba, 1990), contributing to the redistribution of $\dot{Q}$ to exercising skeletal muscle. This, in turn, lowers heat loss, and core body temperature begins to rise as exercise continues. Once a threshold core body temperature is reached, thermoregulatory centers in the hypothalamus mediate cutaneous vasodilation and an increase in skin blood flow, allowing greater heat dissipation (Johnson and Park, 1981). Figure 9.8 illustrates changes in core temperature and cardiovascular variables during exercise in thermoneutral and in hot conditions. The rise in core body temperature is thus the stimulus for increasing skin blood flow. The temperature threshold for initiating this increase in skin blood flow is higher during exercise compared to rest (Johnson and Park, 1981; Kellogg, Johnson, and Kosiba, 1991b), and is thought to result from a delayed activation of cutaneous vasodilation during exercise (Kellogg, Johnson, and Kosiba, 1991b).

Skin blood flow will continue to rise during exercise in proportion to the rise in core body temperature. There is an upper limit, however, to skin blood flow during exercise, with an attenuated rise or plateau being observed beyond a core body temperature of ~38 degrees C (Kenney et al., 1991; Nadel et al., 1979). This plateau in skin blood flow results from an attenuated vasodilatory signal to the cutaneous circulation, rather than activation of cutaneous vasoconstriction (Kenney et al., 1991). Given the limits of cardiac output described earlier in this chapter, the upper limit of skin blood flow likely serves to preserve venous return and cardiac output. However, this comes at the expense of temperature regulation, as core body temperature will rise with continued metabolic heat production.

Coronary Blood Flow

One unique feature of cardiac muscle is that it is never truly at rest; even under "resting" conditions, the heart still beats ~60 to 70 times per minute in order to supply the body with blood. This activity results in a resting cardiac oxygen consumption of 60 ml $\cdot$ min^{-1} $\cdot$ g^{-1} (Tune, Gorman, and Feigl, 2004). During exercise, cardiac oxygen consumption increases up to ~300 ml $\cdot$ min^{-1} $\cdot$ g^{-1} (Tune, Gorman, and Feigl, 2004), in proportion to the increase in HR. As in skeletal muscle, cardiac oxygen consumption is determined by blood flow and oxygen extraction. At rest, oxygen extraction is high; approximately 60% to 70% of available oxygen is extracted from the coronary circulation to meet the oxygen needs of cardiac muscle (Rowell, 1993). Oxygen extraction increases little with further increases in cardiac oxygen consumption as during exercise. Thus, the increased oxygen demand of cardiac muscle during exercise is met almost exclusively by an increase in coronary blood flow, which is proportional to the rise in HR. Similar to what occurs in skeletal muscle, local vasodilation reduces resistance and increases blood flow in the coronary circulation. Also as in skeletal muscle, coronary blood flow can be transiently impeded by mechanical forces. During systole, the increased extravascular pressure resulting from ventricular contraction results in compression of coronary blood vessels, which in turn reduces blood flow. This transient flow limitation results in a much higher coronary perfusion during diastole compared to systole.

Cardiac oxygen consumption is often assessed noninvasively using the rate–pressure product (RPP). The RPP, calculated as the product of HR and SBP, provides an index of cardiac work. This index is particularly useful in assessing the level of stress placed on the heart during exercise in cardiac patients and other clinical populations. As HR and SBP both increase with exercise, so will the RPP. Often, cardiac patients

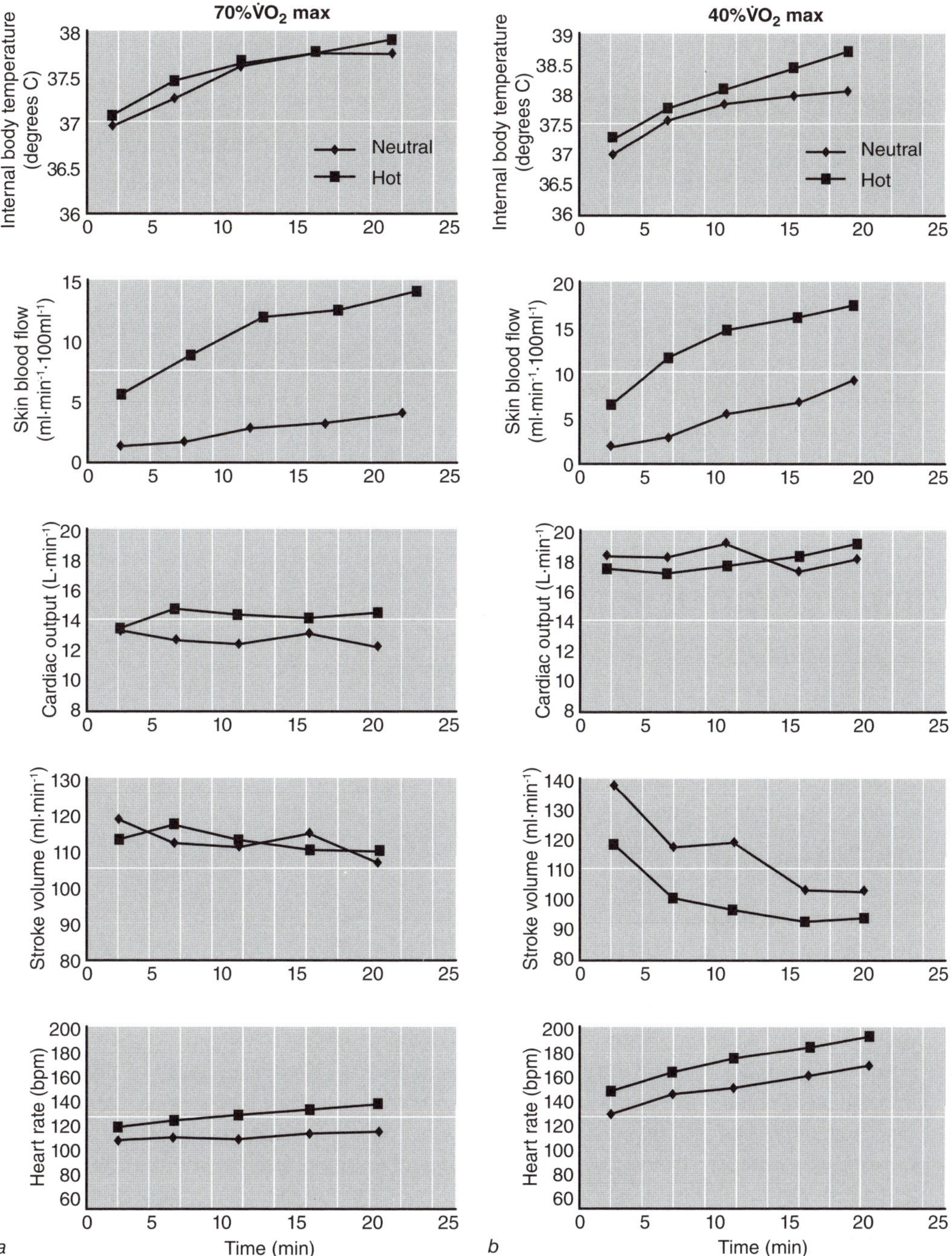

Figure 9.8 Thermal and cardiovascular responses to exercise in thermoneutral and hot environments. Changes in internal body temperature, skin blood flow, cardiac output, stroke volume, and heart rate are shown for *(a)* moderate (40% $\dot{V}O_2$max) and *(b)* heavy (70% $\dot{V}O_2$max) exercise in neutral and hot environments.

(a) is based on data from E.R. Nadel et al., 1979, "Circulatory regulation during exercise in different ambient temperatures," Journal of Applied Physiology 46: 430-437. (b) is based on data from E.R. Nadel et al., 1979, "Circulatory regulation during exercise in different ambient temperatures," *Journal of Applied Physiology* 46: 430-437.

are instructed to exercise at a rate–pressure product below that which elicits clinical symptoms, such as chest pain or abnormal heart rhythms.

Arterial Compliance

Acute aerobic exercise has been shown to increase arterial compliance in healthy young men, as indicated by an increase in whole body arterial compliance 30 min postexercise (Kingwell et al., 1997). Pulse wave velocity, which is inversely related to arterial compliance, is decreased for up to 1 h following acute aerobic exercise, and this is true in both central and peripheral arteries (Kingwell et al., 1997; Naka et al., 2003). Although changes in MAP could influence arterial compliance, data suggest that MAP changes little during the recovery period, particularly beyond ~10 min post-exercise (Kingwell et al., 1997; Naka et al., 2003). Thus, changes in compliance appear to be the result of real changes in properties of the arterial wall rather than an artifact of changes in arterial pressure (Kingwell et al., 1997). Vasodilation and reduced vascular resistance most likely explain the increase in arterial compliance following exercise. In turn, a variety of factors, such as changes in sympathetic tone or circulating hormone levels, and release of tissue metabolites or endothelial factors, may contribute to exercise-induced vasodilation.

HEMOSTATIC RESPONSES

In addition to alterations in cardiac and vascular function, aerobic exercise elicits changes in hemostasis. An acute exercise stimulus alters the functioning of various blood factors involved in coagulation and fibrinolysis, promoting both a procoagulatory and a profibrinolytic state. Further, plasma volume decreases as a result of fluid shifts and fluid loss during exercise. The hemostatic changes that occur during a bout of aerobic exercise are discussed in this section.

Blood Volume

Plasma volume decreases during aerobic exercise, the result of fluid shifts between intra- and extravascular compartments and fluid loss through evaporation. The largest decline in plasma volume is observed during the first ~5 min of steady state exercise (figure 9.9). During this time, a rise in MAP occurs, which in turn creates a pressure gradient that forces fluid from the intravascular space. As steady state exercise progresses, plasma volume may continue to decrease slightly as a result of fluid loss through sweat and evaporation in the lungs. During intense exercise or incremental exercise to maximum, plasma volume will continue to decline beyond the first 5 to 10 min due to greater fluid loss. Exercise in extreme heat exacerbates fluid loss, and hence reductions in plasma volume, as the sweat rate increases to release body heat and maintain body temperature.

Exercising in a hot environment poses a special cardiovascular stress. During exercise, the cardiovascular system has two competing goals: supplying sufficient blood to the working muscles to maintain oxygen consumption and energy production, and supplying sufficient blood to the skin to aid in heat loss and temperature regulation. Under thermoneutral conditions, the cardiovascular system is able to accommodate the increased blood flow requirements of muscle and skin by increasing cardiac output and redistributing that cardiac output to the working tissues, such that

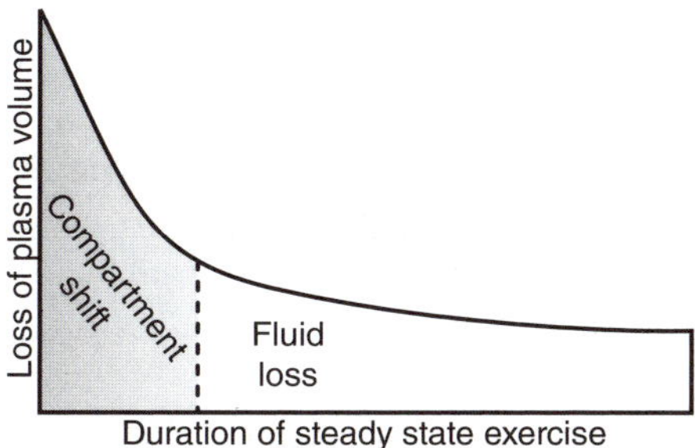

Figure 9.9 Changes in plasma volume during exercise. The largest decline in plasma volume occurs during the first ~5 min of exercise, when fluid moves from the intravascular to the interstitial space (shaded area). Beyond this, little change in plasma occurs under thermoneutral conditions. Plasma volume may decline slightly as exercise continues, particularly if exercise is of high intensity, as a result of fluid lost through sweating.

blood flow increases to both skeletal muscle and skin during aerobic exercise. This is accomplished by vasodilation in the skeletal muscle and cutaneous circulations. In this way, oxygen consumption can increase appropriately with exercise intensity, and sufficient heat is lost to allow thermal balance to be maintained.

During exercise in a hot environment, the competition between skeletal muscle and cutaneous circulations increases. There are several factors that must be considered here. First, it becomes more difficult for the body to lose heat when exercise is performed in a hot environment because when the air is already hot, the temperature gradient between the skin and the air is reduced. When the environment is also humid, the water vapor gradient between the skin and the air will also be reduced. This results in less heat loss through evaporation of sweat. Thus, even though sweating rates increase, less sweat evaporates. As a result, body temperature rises, thus necessitating a further rise in cutaneous blood flow and sweating rate. This creates a greater cutaneous circulatory demand. To meet these demands, skeletal muscle blood flow would have to be reduced, thereby limiting oxygen consumption and exercise performance. Further, exercise performed in the heat leads to excessive fluid loss through sweat, and subsequently a greater decline in plasma volume (Montain and Coyle, 1992). The lower plasma volume, in turn, contributes to a lower venous return and reduced SV. While HR will increase in a compensatory manner, $\dot{Q}$ is reduced if HR cannot fully compensate, as occurs during maximal exercise. Figure 9.8 (p. 155) compares internal body temperature, skin blood flow, $\dot{Q}$, SV, and HR in hot and thermoneutral environments. Fluid ingestion prior to and during prolonged exercise in the heat prevents the decline in plasma volume and attenuates both the drop in SV and the increase in HR (Montain and Coyle, 1992). Further, $\dot{Q}$ is maintained throughout 120 min of exercise in the heat when fluid is replaced (Montain and Coyle, 1992). These observations support the important role of dehydration in the cardiovascular adjustments to exercise in the heat.

The reduction in plasma volume, coupled with the widespread vasodilation in both the skeletal muscle and cutaneous circulations, decreases arterial blood pressure. As discussed earlier in this chapter, MAP must be maintained to supply an adequate driving force to move blood through the circulation and maintain adequate oxygen delivery to the body's tissues. If a person continues to exercise in the heat, not only will arterial pressure be compromised, but core body temperature will also rise to

HOW DOES THE HEAT AFFECT MARATHON PERFORMANCE?

Environmental conditions can have a dramatic effect on exercise performance. Take marathon running, for example. Sustaining moderate- to high-intensity exercise for several hours already presents the body with a strong thermal load in the form of metabolic heat production. However, heat loss mechanisms are activated in response to the increased metabolic heat production and the subsequent rise in core body temperature. In thermoneutral conditions, the body is able to dissipate enough heat through convection, radiation, and evaporation to maintain core temperature within safe limits. Now, consider running that same marathon on a warm day. As the ambient temperature rises, the temperature gradient between the body and the air decreases, thus limiting heat dissipation. While most marathons are not held in extreme weather conditions, there has been the unseasonably warm race day. If race day temperature exceeds or equals core body temperature, then heat dissipation will rely completely on evaporation. While an increase in sweating rate will help the body dissipate heat, it results in greater fluid loss and decreased plasma volume. This in turn decreases SV. When HR cannot compensate, $\dot{Q}$ will decrease too, limiting $\dot{V}O_2$ and exercise performance.

Ely and colleagues (2007) used the wet bulb globe temperature (WBGT) to investigate the effect of weather conditions on marathon finishing times in seven different cities. The WBGT is a weather index that takes solar radiation and humidity into account, along with ambient temperature. When finishing times for elite male competitors were compared to the course record, they were 1.7%, 2.5%, 3.3%, and 4.5% slower at WBGTs of 5 to 10 degrees C, 10.1 to 15 degrees C, 15.1 to 20 degrees C, and 20.1 to 25 degrees C, respectively. Elite female runners demonstrated a similar trend. Interestingly, the impact of temperature on running performance seems to depend on running ability, with more significant performance decrements in slower runners. Runners who placed 25th, 50th, 100th, and 300th, on average, demonstrated a 7.9% decline in performance at 21.1 to 25 degrees C compared to only a 4.5% decrease in the top finishers at the same temperature. Possible reasons are that slower runners had a longer exposure to the environmental conditions or that they experienced less convective and radiative heat loss due to running with a larger group (Montain, Ely, and Cheuvront, 2007).

dangerous levels, as heat production during exercise will exceed the body's ability to release heat. **Heat exhaustion,** a condition whose symptoms include fatigue, weakness, disorientation, and nausea, can result from severe dehydration. A more serious condition is **heatstroke,** which occurs when the body's temperature rises to a dangerous level that leads to multiple system failure. Heatstroke is a life-threatening condition that results in disorientation and often loss of consciousness.

Platelets

As discussed in chapter 8, platelets play an important role in hemostasis by stimulating thrombin generation and thrombus formation. Aerobic exercise has been shown to increase platelet activation and aggregation and to stimulate thrombosis in healthy, sedentary individuals (Cadroy et al., 2002; Ersoz et al., 2002; Wang and Cheng, 1999). Most data suggest that exercise of a high intensity is needed to increase thrombotic potential (Cadroy et al., 2002; Sakita, Kishi, and Numano, 1997; Wang and Cheng, 1999). This may explain why acute exercise, particularly strenuous exercise, frequently precipitates ischemic events.

Exercise-induced platelet activation is thought to be mediated through elevated catecholamine levels during exercise (Hjemdahl, Larsson, and Wallen, 1991; Shattil, Budzynski, and Scrutton, 1989). More specifically, epinephrine stimulates alpha2-adrenergic receptors, ultimately enhancing platelet adhesiveness and aggregation as well as the binding of fibrinogen to platelets (Figures et al., 1986; Wang and Cheng, 1999). While there is debate regarding whether physiological levels of epinephrine are sufficient to stimulate this pathway (Plow and Marguerie, 1980; Shattil, Budzynski, and Scrutton, 1989), epinephrine in the presence of adenosine diphosphate (ADP) has been shown to activate platelets in vivo (Figures et al., 1986; Plow and Marguerie, 1980). Importantly, strenuous exercise results in higher epinephrine levels than moderate-intensity exercise. This is consistent with observations of increased platelet function during high-intensity, but not moderate-intensity, exercise.

Some evidence supports inhibition of platelet activity following acute exercise (Petidis et al., 2008). Petidis and colleagues (2008) recently reported that ADP- and collagen-induced platelet aggregation decreased following 15 min of a modified Bruce treadmill test. The change in platelet aggregation was observed in healthy subjects, hypertensive subjects, and subjects with known coronary artery disease; but the response was more pronounced in healthy individuals. The changes in platelet activity were accompanied by a much greater increase in plasma norepinephrine compared to epinephrine. Norepinephrine has been shown to decrease thrombosis, and this response may be mediated through the release of platelet-inhibiting factors such as prostacyclin and NO (Lin and Young, 1995). Release of NO and prostacyclin from endothelial cells is also stimulated by the increased shear stress that is associated with increased blood flow (Mo et al., 1991; Ralevic et al., 1990). Both NO and prostacyclin have been shown to inhibit platelet aggregation (Jones et al., 1993; Ruschitzka, Noll, and Luscher, 1997). Thus, healthy endothelial cell function may protect against platelet aggregation during acute aerobic exercise. This notion gains merit from observations that patients with coronary artery disease, who exhibit endothelial cell dysfunction, experience less inhibition of platelet aggregation—that is, greater platelet activation—during exercise than healthy individuals (Petidis et al., 2008).

Coagulation

An acute bout of aerobic exercise is associated with a hypercoagulable state, as evidenced by elevated levels of coagulation Factor VIII, thrombin-antithrombin complex, prothrombin fragments 1 and 2, fibrinogen, and fibrinopeptide A (Andrew et al., 1986; Arai et al., 1990; Bartsch et al., 1995; Cadroy et al., 2002; Weiss, Seitel, and Bartsch, 1998), as well as shortened partial thromboplastin and nonactivated partial thomboplastin times (Arai et al., 1990; Weiss, Seitel, and Bartsch, 1998). While there is some discrepancy in the literature regarding the dose of exercise required, most reports suggest that strenuous exercise or very-long-duration moderate-intensity exercise is needed to induce this hypercoagulable state and increase fibrin formation in vivo (see table 9.1) (Andrew et al., 1986; El-Sayed, El-Sayed, and Ahmadizad, 2004; Weiss, Seitel, and Bartsch, 1998).

This exercise-induced increase in coagulation is most likely explained by activation of beta-adrenergic receptors during exercise (Cohen, Epstein, and Cohen, 1968). Further, this response may ultimately be mediated through NO production (Jilma et al., 1997). Data show that both beta-adrenergic blockade and NO blockade are associated with blunted increases in coagulation Factor VIII (Cohen, Epstein, and Cohen, 1968; Jilma et al., 1997).

The hypercoagulable state is also maintained for a period of time following cessation of exercise, with some markers of coagulation remaining altered up to 21 h postexercise (table 9.1) (Bartsch et al., 1995; Hegde, Goldfarb, and Hegde, 2001; Weiss, Seitel, and Bartsch, 1998). The coagulation potential appears to be highest during the first hour following heavy exercise, where thrombin-antithrombin complex, prothrombin fragments 1 and 2, fibrinopeptide A, and coagulation Factor VIII activity all remain elevated (Hegde, Goldfarb, and Hegde, 2001; Weiss, Seitel, and Bartsch, 1998), and prothrombin fragments 1 and 2 may even continue to rise (Bartsch et al., 1995). Further, during this time, activated partial thromboplastin time is still shorter compared to resting values (Weiss, Seitel, and Bartsch, 1998). As time progresses, thrombin-antithrombin complex, prothrombin fragments 1 and 2, and fibrinopeptide A fall toward baseline values, with the greater fall observed during the early part of recovery. While levels of thrombin-antithrombin complex, prothrombin fragments 1 and 2, and fibrinopeptide A return to baseline within 21 h postexercise, activated partial thromboplastin time remains lower at this time (Bartsch et al., 1995). Together, these data suggest that the risk of an acute coronary event may be elevated during the early stages of recovery from exercise as well as during the exercise period itself.

Table 9.1 Comparison of Coagulation Markers Before and Following Exercise

		Rest	Immediately postexercise	1 h postexercise	2 h postexercise	22 h postexercise
PTF 1 and 2 (nmol/L)	Moderate	0.6	0.65	0.6		
	Heavy	0.6	0.8	0.75		
		0.6	**0.8**		**0.9**	**0.6**
TAT (ng/ml)	Moderate	1.5	1.75	1.75		
	Heavy	1.5	3	2.5		
		3	**5.75**		**3.5**	**3.0**
FPA (ng/ml)	Moderate	1.5	1.25	1.6		
	Heavy	1.0	2.25	2.0		
		1.2	**2.2**		**1.2**	**1.4**
aPTT (s)	Moderate	38.5	35.3	35.1		
	Heavy	37.6	32.6	33.3		
		37.7	**28.6**		**30.9**	**35.9**

Prothrombin fragments 1 and 2 (PTF 1 and 2), thrombin-antithrombin complex (TAT), fibrinopeptide A (FPA), and activated partial thromboplastin time (aPTT) are shown at rest and during recovery from moderate- and high-intensity exercise.

Data from P. Bartsch et al., 1995, "Balanced activation of coagulation and fibrinolysis after a 2-h triathlon," *Medicine and Science in Sports and Exercise* 27:1465-1470; C.G. Weiss, G. Seitel, and P. Bartsch, 1998, "Coagulation and fibrinolysis after moderate and very heavy exercise in healthy male subjects," *Medicine and Science in Sports and Exercise"* 30:246-251.

Fibrinolysis

The exercise-induced hypercoagulable state just described is accompanied by an equal or greater potential for fibrinolysis. Tissue plasminogen activator antigen and plasmin-antiplasmin complexes are elevated during, and following, prolonged moderate- and high-intensity exercise, indicating enhanced plasmin formation (Weiss, Seitel, and Bartsch, 1998). This precipitates fibrin degradation, as indicated by increased levels of fibrin degradation products (Weiss, Seitel, and Bartsch, 1998). As with coagulation, activation of fibrinolysis appears to increase with exercise intensity; this is indicated by

1. higher levels of plasminogen activator following high-intensity, compared to moderate-intensity, exercise (Szymanski and Pate, 1994; Szymanski, Pate, and Durstine, 1994) and
2. increased levels of fibrin degradation products during high-intensity, but not moderate-intensity, exercise (Weiss, Seitel, and Bartsch, 1998).

It is possible, however, that lack of elevated fibrin degradation products during moderate-intensity exercise reflects the lack of a significant increase in coagulation rather than a true limitation in fibrinolysis (Weiss, Seitel, and Bartsch, 1998). In other words, if moderate-intensity exercise is not a significant stimulus for coagulation and little fibrin is formed with this type of exercise, then less fibrin would be available for breakdown.

The enhancement of fibrinolysis appears to be greatest immediately after heavy exercise, although fibrinolytic markers such as tissue plasminogen activator and plasmin-antiplasmin complexes remain elevated during the first several hours post-exercise (table 9.2) (Bartsch et al., 1995; Hegde, Goldfarb, and Hegde, 2001; Weiss, Seitel, and Bartsch, 1998).

Table 9.2 Comparison of Fibrinolytic Markers at Rest and Following Exercise

		Rest	Immediately postexercise	1 h postexercise	2 h postexercise	22 h postexercise
t-PA (ng/ml)	Moderate	4	14	5		
	Heavy	4	25	6		
		2	**14**		**4**	**2**
PAP (nmol/L)	Moderate	2	4	4.5		
	Heavy	2	7	7		
		5	**45**		**44**	**5**
FbDP (ng/ml)	Moderate	100	150	150		
	Heavy	100	115	115		
		110	**200**		**160**	**120**

Tissue plasminogen activator (t-PA), plasmin–antiplasmin complex (PAP), and fibrin degradation products (FbDP) are shown at rest and during recovery from moderate- and high-intensity exercise.

Data from P. Bartsch et al., 1995, "Balanced activation of coagulation and fibrinolysis after a 2-h triathlon," *Medicine and Science in Sports and Exercise* 27:1465-1470; C.G. Weiss, G. Seitel, and P. Bartsch, 1998, "Coagulation and fibrinolysis after moderate and very heavy exercise in healthy male subjects," *Medicine and Science in Sports and Exercise"* 30:246-251.

Taken together, the evidence suggests that acute aerobic exercise increases both coagulatory and fibrinolytic potential. However, when normal limits of coagulatory and fibrinolytic markers are considered, it appears that exercise stimulates fibrinolysis to a greater degree than it stimulates coagulation in healthy individuals. This may not be the case in patients with cardiac disease, who already have a higher coagulatory, and a lower fibrinolytic, potential at rest compared to healthy age-matched individuals (Acil et al., 2007). Evidence suggests that these patients may also experience a greater exercise-induced increase in coagulation compared to healthy subjects (Lee et al., 2005). Further, fibrinolysis appears to recover more quickly than coagulation following acute exercise, which may explain the increased vulnerability to acute cardiac events during this time period (Hegde, Goldfarb, and Hegde, 2001). Recently, Acil and colleagues (2007) observed delayed recovery of coagulation following exercise in cardiac patients compared to healthy individuals. It is clear that hemostatic changes following acute aerobic exercise have important implications for cardiovascular health, particularly among cardiac patient populations.

SUMMARY

Aerobic exercise causes responses in both the heart and the vasculature that ultimately satisfy the increased blood and oxygen needs of the body while maintaining adequate MAP. The increases in HR, venous return, and cardiac contractility, along with the decrease in ventricular afterload, allow for a larger $\dot{Q}$ and that larger $\dot{Q}$ is redistributed to the active tissue circulations. This redistribution of blood flow is accomplished by vasoconstriction of inactive tissue circulations and concomitant vasodilation in active muscle beds. Blood flow during rhythmic aerobic exercise is further enhanced by the skeletal muscle pump, which creates a greater pressure gradient across the capillary bed, thereby increasing the driving force for blood flow. These vascular changes that occur during aerobic exercise also result in a drop in TPR, elevations in SBP and MAP, and little change in DBP. Finally, acute aerobic exercise induces changes in coagulation and fibrinolysis that may foster a greater coagulatory potential, particularly in cardiac patients.

CHAPTER 10

Cardiovascular Adaptations to Aerobic Training

By Danielle Wigmore, Denise Smith, and Bo Fernhall

Aerobic exercise training improves cardiorespiratory fitness, as demonstrated by higher maximal oxygen consumption ($\dot{V}O_2max$) and higher attainable peak workloads in trained compared to untrained individuals (Clausen et al., 1973). Such improvements in aerobic fitness are achieved through a complex set of central and peripheral adaptations. Training induces increases in type I muscle fiber area as well as mitochondrial number and function, increasing the capacity to use oxygen. Endurance training also enhances peak cardiac output ($\dot{Q}$) and skeletal muscle blood flow to increase oxygen delivery to exercising skeletal muscles to support their increase in oxygen utilization. Meanwhile, greater capillarity ensures that oxygen diffusion will be adequate to meet the demands of the muscle tissue and sustain higher oxygen utilization. This chapter discusses these training adaptations and the mechanisms responsible for them.

CARDIAC ADAPTATIONS

Chronic aerobic exercise training induces changes in the heart that contribute to the greater peak $\dot{Q}$ and aerobic capacity. Changes in cardiac dimensions and blood volume play the primary role in enhancing stroke volume (SV), and hence $\dot{Q}$, following a period of training. In addition, cardiac adaptations to training allow the heart to maintain a given submaximal work level at a lower heart rate (HR), and therefore with less strain on the heart. The cardiac adjustments that are responsible for these changes are described in this section.

Cardiac Dimensions

Aerobic exercise training produces adaptations in cardiac dimensions, primarily **left ventricular hypertrophy,** increased left ventricular end-diastolic diameter, and increased ventricular wall thickness. While most studies support the changes in left ventricular mass and diameter (Cohen and Segal, 1985; Pluim et al., 2000), the data regarding changes in ventricular wall thickness are not as consistent. Aerobic exercise produces a "volume load" on the heart, due to elevated venous return and the high cardiac output sustained during aerobic exercise. This volume load is thought to be responsible for the high left ventricular end-diastolic volume observed in trained endurance athletes (Cohen and Segal, 1985; Pluim et al., 2000). Additionally, the heart sustains repeated forceful contractions throughout endurance training, which provides a stimulus for hypertrophy of the ventricular wall and hence increased ventricular wall thickness (Pluim et al., 2000; Schmidt-Trucksass et al., 2003; Vinereanu et al., 2001). Endurance training typically results in proportional increases in diastolic diameter and ventricular wall thickness, such that the relative wall thickness (represented by the ratio of ventricular wall thickness to internal diameter) is maintained with training (Cohen and Segal, 1985; Schmidt-Trucksass et al., 2003). However, a careful meta-analysis including data from 59 studies demonstrated a higher relative wall thickness in endurance athletes compared to nonathletes (Pluim et al., 2000), suggesting a greater increase in ventricular wall thickness than in end-diastolic diameter.

Resistance training also provides a stimulus for cardiac hypertrophy, although the nature of the hypertrophy is different than that observed with endurance training. The transient elevations in arterial pressure and ventricular afterload associated with resistance training provide a "pressure load" on the heart. Typically, the increase in ventricular wall thickness exceeds the increase in end-diastolic ventricular diameter; thus the relative wall thickness is higher in strength-trained compared to untrained individuals (Cohen and Segal, 1985; Pluim et al., 2000). Cardiac adaptations to resistance training are discussed in more detail in chapter 12. Table 10.1 compares cardiac dimensions in untrained, endurance-trained, and resistance-trained individuals.

The adaptations in cardiac dimensions following aerobic training contribute to enhanced cardiac function, as will be discussed in the following section. This is in contrast to ventricular hypertrophy associated with pathological conditions, such as heart failure. Consider the following two examples. Prolonged exposure to a high pressure load, which would occur with chronic hypertension, can cause the ventricle

Table 10.1 Average Values for Cardiac Dimensions From Healthy Control Subjects, Athletes, and Patients With Hypertrophic Cardiomyopathy

	Control	Endurance trained	Strength trained	Hypertrophic cardiomyopathy
End-diastolic diameter (mm)	49.6	53.7	52.1	45.0
Posterior wall thickness (mm)	8.8	10.3	11	14.0
Relative wall thickness	0.36	0.39	0.44	0.72

Data from B.M. Pluim et al., 2000, "The athlete's heart. A meta-analysis of cardiac structure and function," *Circulation* 101:336-344; D.N. Vinereanu et al., 2001, "Differentiation between pathologic and physiologic left ventricular hypertrophy by tissue doppler assessment of long-axis function in patients with hypertrophic cartiomyopathy or systemic hypertension and in athletes," *American Journal of Cardiology* 88:53.

walls to thicken without a concomitant change in ventricular diameter, resulting in a stiffer ventricle. This, in turn, may impede ventricular filling and reduce SV. Alternatively, when the heart becomes weak and loses its elasticity, the ventricles will dilate with no change or a decrease in wall thickness. In this condition, the ventricle is not strong enough to sustain forceful contractions, despite an increase in chamber size. This results in reduced ejection of blood, increased left ventricular pressure, and buildup of fluid in the pulmonary circulation. Cardiac dimensions for patients with pathological ventricular hypertrophy are compared to those of healthy trained and untrained individuals in table 10.1.

Cardiac Output

Resting $\dot{Q}$ remains unchanged following a period of endurance exercise training. The same $\dot{Q}$, however, is maintained via a higher resting SV and lower resting HR (Rowland and Roti, 2004). Following aerobic training, $\dot{Q}$ is slightly lower or unchanged at a given submaximal workload yet higher at a given proportion of maximal intensity (figure 10.1; Clausen et al., 1973; Hammond and Froelicher, 1985; Wolfe and Cunningham, 1982). Furthermore, trained individuals are able to achieve a higher $\dot{Q}$ during maximal exercise (Hammond and Froelicher, 1985; Rowland and Roti, 2004), which

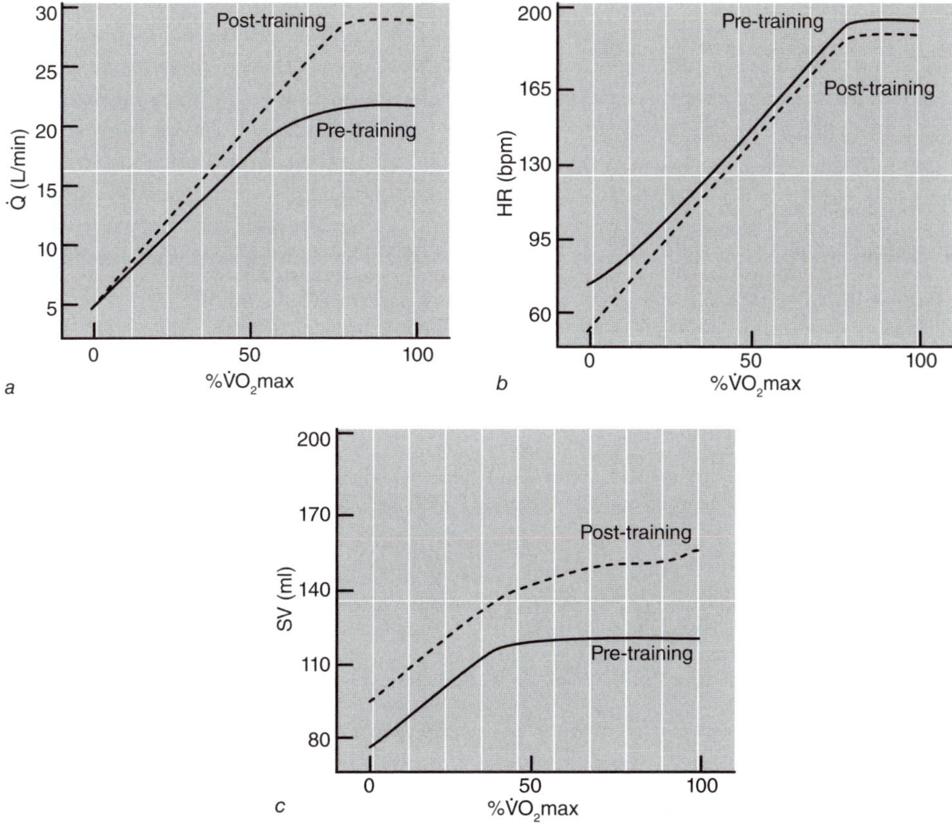

Figure 10.1 The effect of aerobic training on major cardiovascular variables during incremental exercise to maximal.

plays a role in the increase in $\dot{V}O_2max$ with aerobic training. The higher $\dot{Q}$ max is a function of a higher maximal SV following training, rather than a change in maximal HR (Rowland and Roti, 2004).

Stroke Volume

Aerobic training increases SV, both at rest and during exercise. Higher SV results from the changes in cardiac dimensions described earlier, as well as increases in blood volume and venous return. Blood volume expansion, a common finding in trained individuals, is discussed in more detail later. During submaximal and maximal exercise, this higher blood volume leads to a greater venous return, thereby enhancing diastolic ventricular filling (Krip et al., 1997; Gledhill, Cox, and Jamnik, 1994) and end-diastolic volume (Warburton et al., 2004). This, in turn, causes a greater SV via enhanced preload. More specifically, the ventricular filling rate during submaximal and maximal exercise is higher in trained compared to untrained individuals (Ferguson et al., 2001; Gledhill, Cox, and Jamnik, 1994). While ventricular emptying is also augmented in trained subjects, as evidenced by a greater ventricular emptying rate, the enhancement in ventricular filling is thought to play a more important role in the higher SVs observed following training (Ferguson et al., 2001; Gledhill, Cox, and Jamnik, 1994). The higher rate of ventricular emptying is also thought to be related to the Frank-Starling mechanism, as indices of cardiac contractility are unchanged with exercise training (Pluim et al., 2000; Warburton et al., 2004).

As discussed in chapter 9, there is debate regarding the effect of endurance training on the nature of the SV response to incremental exercise to maximum. Some researchers report no change in the pattern of SV response to incremental exercise, with both trained and untrained subjects demonstrating an initial rise in SV with low-intensity exercise followed by a plateau in SV with further increases in intensity (Rowland and Roti, 2004; Wolfe and Cunningham, 1982). Others report that SV continues to rise with exercise intensity up to maximal workloads in highly trained individuals (Gledhill, Cox, and Jamnik, 1994; Zhou et al., 2001). The SV response to training is shown in figure 10.1. Possible mechanisms for these responses were discussed in more detail in chapter 9.

Blood Volume

Aerobically trained individuals have a 20% to 25% larger blood volume compared to untrained individuals—an adaptation that is independent of age and gender (Convertino, 1991). This expansion of blood volume occurs primarily via increased plasma volume, and secondarily from increased red blood cell volume. During the first week of training, the expansion of plasma volume is responsible for the rise in blood volume, with little change in red blood cell volume (Convertino et al., 1980). Plasma volume increases within four days of initiating an endurance training program and reaches a plateau after approximately one week of training (Convertino, 1991). Possible mechanisms for the training-induced elevation of plasma volume include an increase in total protein content and subsequent binding of water to protein or greater sodium and water retention following training (Convertino, 1991; Convertino et al., 1980). Despite a significant plasma volume expansion after training, resting plasma aldosterone and antidiuretic hormone (ADH) concentrations do not change with training (Convertino, Keil, and Greenleaf, 1983; Shoemaker et al., 1997).

Aerobic training also affects the plasma volume response to an acute bout of exercise. Remember from chapter 9 that plasma volume decreases during exercise. This response is blunted following as few as six days of aerobic training (Convertino, Keil, and Greenleaf, 1983; Shoemaker et al., 1997). However, plasma volume remains higher throughout exercise following training due to the training-induced plasma volume expansion. The smaller percent decline in plasma volume after training is accompanied by a smaller increase in aldosterone during exercise posttraining compared to pretraining (Shoemaker et al., 1997). This is also true for plasma ADH concentration at high exercise intensities (Convertino, Keil, and Greenleaf, 1983; Shoemaker et al., 1997). One explanation is that the smaller perturbation to plasma volume results in a smaller stimulus for ADH and aldosterone release (Shoemaker et al., 1997).

Increases in red blood cell volume lag behind those of plasma volume, with one study showing no change in red blood cell volume after four months of endurance training (Oscai, Williams, and Hertig, 1968). Regardless of the time course of these changes, endurance athletes have been reported to have a ~25% higher total hemoglobin mass and red blood cell volume compared to their untrained counterparts—an effect that is magnified when values are scaled to the athlete's body mass, which is typically 35% to 40% higher (Heinicke et al., 2001; Schmidt et al., 2002). This reflects an important enhancement in the oxygen-carrying capacity of the blood of aerobically trained individuals. Despite an increase in the total number of red blood cells, trained individuals often have a lower hematocrit or hemoglobin concentration (Convertino et al., 1980; Weight et al., 1992), a condition that is sometimes referred to as **sports anemia.** This is particularly apparent during the initial stages of training when the expansion of plasma volume far outweighs any increase in red blood cell production. The result of these adaptations is a dilution effect on hemoglobin and red blood cell concentrations. However, because the red blood cell numbers are not actually reduced, the condition is not a true anemia and does not result in any impairment in oxygen-carrying capacity of the blood. In fact, this situation may enhance oxygen delivery by making the blood less viscous and reducing the resistance to blood flow (Plowman and Smith, 2008).

Heart Rate

One of the most well-known adaptations to chronic aerobic training is a reduction in resting HR. While HR increases with increasing intensity in trained individuals as it does in untrained individuals (figure 10.1), HR at a given absolute submaximal intensity is lower after aerobic exercise training. Maximal HR, however, is unchanged or only slightly decreased with training.

The precise mechanism of the changes in HR remains unclear. The most well-accepted explanation is that cardiac vagal tone is enhanced following aerobic training and results in a lower cardiac rate. Support for this hypothesis comes from studies employing a variety of study designs:

1. Pharmacological blockade of vagal activity (Shi et al., 1995; Smith et al., 1989)
2. Respiratory sinus arrhythmia, a technique assessing changes in HR associated with respiration, which are mediated through vagal activity (Kenney, 1985)
3. Changes in exercise HR during discrete periods when the HR response is regulated by changes in either parasympathetic or sympathetic activity (Gallo et al., 1989)
4. The rate of HR recovery after exercise, which reflects vagal reactivation (Imai et al., 1994)

5. Spectral analysis of HR variability, which can be used to indicate the level of vagal tone (Sakuragi and Sugiyama, 2006; Shin et al., 1997)

This last technique involves spectral analysis of fluctuations in the R-R interval of the cardiac cycle and subsequent decomposition of those signals into high-frequency and low-frequency components. The high-frequency component is an indicator of parasympathetic tone, while the low-frequency component reflects a combination of sympathetic and parasympathetic tone. Using this technique, Sakuragi and Sugiyama (2006) found that four weeks of brisk walking lowered resting HR, and that changes in HR were related to changes in the high-frequency signal component. In other words, enhanced parasympathetic activity was responsible, at least in part, for the lower resting HR. Further, Imai and colleagues (1994) reported a more rapid recovery of HR following submaximal and maximal exercise in endurance athletes compared to untrained subjects, reflecting an enhanced postexercise reactivation of parasympathetic tone following aerobic training. A number of studies, however, oppose this view and report that indices of vagal tone are similar in trained and untrained individuals or following a period of training in previously sedentary individuals (Katona et al., 1982; Maciel et al., 1985; Scott et al., 2004). An alternative explanation is that aerobic training induces a decrease in intrinsic HR (Katona et al., 1982; Smith et al., 1989). In other words, the rate at which the heart naturally beats without any nervous innervation would be lower following endurance training. Thus, regardless of changes in autonomic tone, the heart would be starting at a lower intrinsic rate. Using combined sympathetic and parasympathetic blockade, Shi and colleagues (1995) observed no change in intrinsic HR following eight months of aerobic training, supporting the view of a training-induced enhancement of vagal tone.

VASCULAR ADAPTATIONS

Improvements in aerobic capacity are due, in part, to exercise-induced vascular adaptations. Vascular remodeling and enhanced vasodilatory capacity result in reduced vascular resistance. This, in turn, enables greater blood flow during high-intensity exercise. Furthermore, regular aerobic exercise provides a cardioprotective effect, resulting in improved coronary blood flow, reduced platelet activation and aggregation, and increased fibrinolysis. The following sections outline the various vascular changes that occur with regular aerobic exercise.

Blood Pressure

Aerobic exercise training produces small, yet significant, decreases in resting systolic and diastolic blood pressure (SBP and DBP) in normotensive individuals (Kelley and Tran, 1995). Decreases of ~4 mmHg have been reported following training and, although small, are considered to reduce risk for future hypertension and cardiovascular disease (Kelley and Tran, 1995; Pescatello et al., 2004). Aerobic exercise has a more profound blood pressure–lowering effect in hypertensive individuals and has long been considered an important part of treatment for hypertension (Pescatello et al., 2004).

Aerobic training also alters exercise blood pressures (figure 10.2). At a given absolute workload, SBP, DBP, and mean arterial pressure (MAP) are all lower posttraining (Van Hoof et al., 1989; Wilmore et al., 2001). At the same relative workload, SBP is unchanged while DBP and MAP are either unchanged or slightly lowered after training (Van Hoof et al., 1989; Wilmore et al., 2001). Aerobic training results in a slight

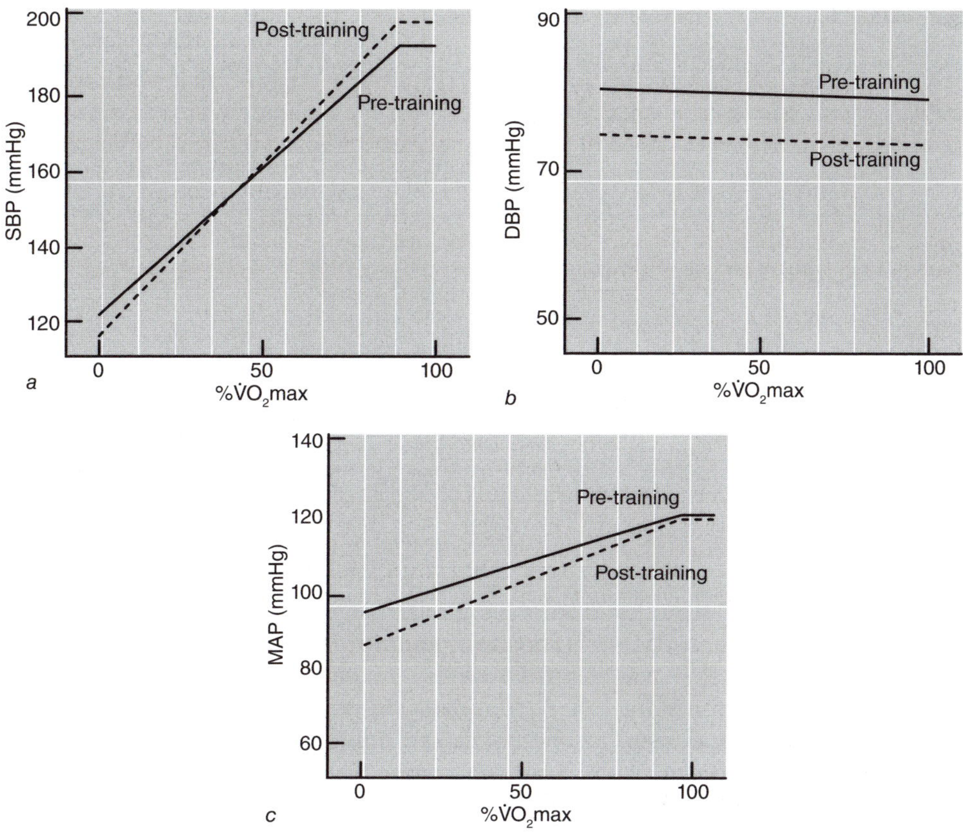

Figure 10.2 The effect of aerobic training on systolic blood pressure, diastolic blood pressure, and mean arterial pressure during incremental exercise to maximum.

increase in SBP and a slight decrease in DBP at maximal exercise, such that MAP is unchanged (Wilmore et al., 2001).

Mean arterial pressure is the product of cardiac output and vascular resistance (MAP = $\dot{Q}$ × TPR [total peripheral resistance]). Vascular remodeling, enhanced vasodilatory capacity, and changes in autonomic tone with training contribute to a greater ability to reduce vascular resistance at rest and during exercise. The mechanisms by which vascular tone is altered with training are discussed in the next section. The lower vascular resistance and unchanged $\dot{Q}$ at rest explain the slight decrease in resting arterial pressure following endurance training. The greater $\dot{Q}$max following training contributes to the higher SBP at peak exercise, while the lower DBP at peak exercise is likely the result of the lower TPR associated with training.

Muscle Blood Flow

The change in muscle blood flow following a period of endurance exercise training varies by condition. Resting muscle blood flow is unaffected by training status (Dinenno et al., 2001). During moderate exercise (<75% $\dot{V}O_2$max), muscle blood flow at a given absolute submaximal intensity is either unchanged or reduced following training (Delp, 1998; Proctor et al., 2001; Putman et al., 1998). Further, animal data support a redistribution of flow within exercising skeletal muscle, such that perfusion to high-oxidative fibers is higher and perfusion to low-oxidative fibers is lower following training (Armstrong and Laughlin, 1984). In this way, the body prioritizes the delivery of oxygen to the muscle fiber types that are most likely to use it in the process of energy production. Confirmation from human studies, however, is lacking, and a reduced heterogeneity of muscle perfusion has been reported in several muscle groups in endurance-trained individuals (Kalliokoski et al., 2001). Finally, at higher exercise intensities (e.g., between 75% and 100% $\dot{V}O_2$max), muscle blood flow is greater after high-intensity aerobic training compared to before (Putman et al., 1998; Roca et al., 1992).

The training-induced changes in muscle blood flow must result from changes in the arterial pressure gradient or vascular resistance, according to the following equation: Flow = $\Delta P / R$, where ΔP is the arterial pressure gradient and R is the vascular resistance (see chapter 6 for a detailed discussion of this relationship). The modest blood pressure changes with training described earlier in this chapter are unlikely to elicit dramatic changes in the pressure gradient posttraining, and therefore are unlikely to play a large role in the blood flow adaptations associated with exercise training. Vascular resistance, however, may be greatly altered.

Effect of Aerobic Training on Vascular Tone

As discussed in chapter 6, the primary factor determining vascular resistance is arterial radius. Thus, factors affecting vasomotor tone will ultimately alter blood flow. One possible explanation for lower blood flow during low- to moderate-intensity exercise after training is a smaller metabolic vasodilator signal at the same absolute exercise intensity. This could result from a reduction in the metabolic stress associated with the exercise, subsequent to changes in muscle metabolic properties, increased mitochondrial oxidative capacity, or increased muscle efficiency following training (Proctor et al., 2001; Starritt, Angus, and Hargreaves, 1999). Consistent with this idea, endurance training increases the number of mitochondria as well as the activity of enzymes involved in oxidative metabolism (Chesley, Heigenhauser, and Spriet, 1996; Hoppeler et al., 1985; Spina et al., 1996).

The heightened blood flow response to high-intensity exercise following a training program could be explained by a reduced sympathetic vasoconstriction, enhanced vasodilation, or both. There is evidence to suggest that muscle sympathetic nerve activity during high-intensity exercise may be reduced by exercise training (Delp, 1998). This would result in decreased sympathetic vasoconstriction, and ultimately greater vasodilation.

Vasodilation is largely a function of endothelium-dependent mechanisms. Thus, one would expect changes in endothelial function with training to play a role in vascular responses. A number of studies have shown that moderate aerobic training increases endothelium-dependent vasodilation in human and animal blood vessels (Clarkson et al., 1999; Goto, 2003; Higashi et al., 1999; Woodman et al., 2005, 2003). For example, Clarkson and coworkers (1999) reported improvements in brachial artery flow-mediated dilation following a 10-week combined aerobic and anaerobic training program in healthy, young military recruits. Woodman and colleagues (2005, 2003) found 16 weeks of treadmill running to improve endothelium-dependent vasodilation in the brachial arteries of hyperlipidemic and hypercholesterolemic pigs. Training has also been shown to improve endothelial cell-mediated vasodilation in skeletal muscle arterioles of rats (Spier et al., 2004; Sun et al., 1994).

It is important here to understand the various methodologies used to study endothelium-dependent vasodilation. Two methods are commonly employed for in vivo studies: flow-mediated vasodilation and receptor-mediated vasodilation. The first technique requires a brief period of vascular occlusion (5-10 min) to induce a high-flow (hence high-shear-stress) condition upon reintroduction of blood flow. The elevated shear stress induces vasodilation as described in chapter 9, and the dilation of the artery is observed and quantified using Doppler ultrasound imaging. Alternatively, arterial infusions of endothelium-dependent vasodilators, such as acetylcholine, are used in conjunction with venous occlusion plethysmography, a technique used to noninvasively measure limb blood flow. Typically, various doses of the vasodilator are sequentially infused into the artery, and blood flow responses to each dose are measured. Most commonly, forearm blood flow is measured in response to the infusion of vasoactive substances into the brachial artery. Administration of vasodilators to harvested arterial ring segments is a commonly used method for assessing vasoreactivity in vitro. Additionally, most studies, in vivo and in vitro, assess responses to a direct smooth muscle cell vasodilator to differentiate endothelial-dependent and -independent responses. Importantly, all of the studies described in this section used this control to determine that smooth muscle cell function was unaffected by training. Thus, improvements in flow-mediated and receptor-mediated vasodilation with training can be attributed to improvements in endothelial cell function, as shown in figure 10.3.

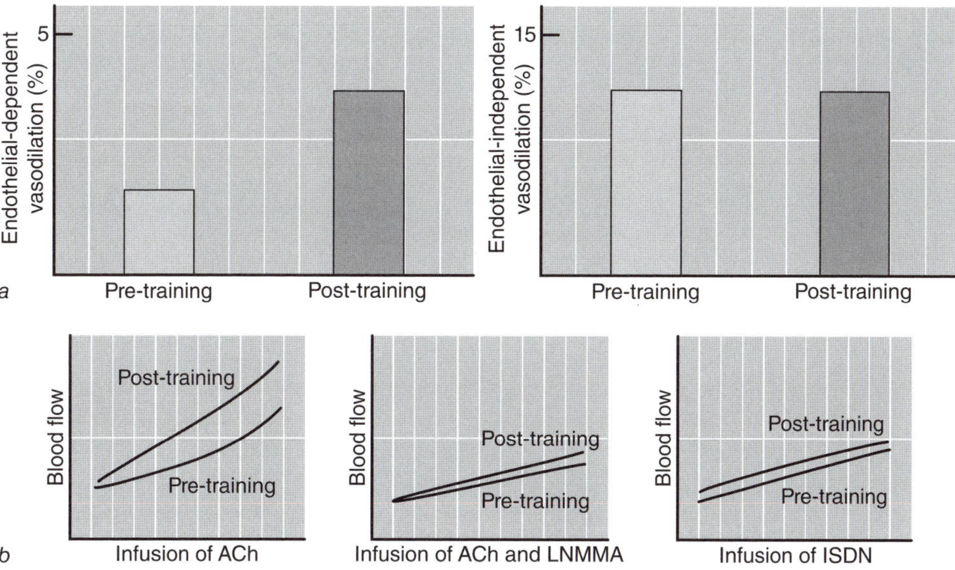

Figure 10.3 Changes in endothelium-dependent and -independent vasodilation with endurance training. *(a)* Aerobic training improves flow-mediated, endothelial-dependent vasodilation (first panel), but direct activation of smooth muscle cells shows that smooth muscle cell function is unchanged following training (second panel). *(b)* The increase in forearm blood flow in response to acetylcholine infusion is higher following aerobic training (first panel); and this effect is mediated through the nitric oxide (NO) pathway, as inhibition of NO abolishes the improvements associated with training (second panel). Again, smooth muscle cell function is unchanged with training (third panel), supporting the idea that improvements in vasodilatory capacity with training are the result of improved endothelial, rather than smooth muscle cell, function.

Figure 10.3a is based on P. Clarkson, 1999, "Exercise training enhances endothelial function in young men," *Journal of the American College of Cardiology* 33: 1379-1385. Figure 10.3b is based on C.Y. Goto et al., 2003, "Effect of different intensities of exercise on endothelium-dependent mitric oxide and oxidative stress," *Circulation* 108: 530-535.

Some data suggest that improvements in endothelium-dependent vasodilation are dependent on nitric oxide (NO), as training-induced increases in endothelium-dependent vasodilation disappear in the presence of a NO inhibitor (Goto, 2003; Higashi et al., 1999; Muller, Myers, and Laughlin, 1994; Spier et al., 2004; Sun et al., 1994; Wang, Wolin, and Hintze, 1993). Other studies, however, suggest that exercise may improve endothelium-dependent vasodilation via increased production of prostacyclin or endothelial hyperpolarizing factor (Woodman et al., 2005).

Although the mechanisms by which training enhances the production of prostacyclin and endothelial hyperpolarizing factor are less well studied, training-induced changes in the NO pathway have received much attention (see figure 10.4). Endurance training is associated with increased NO production, which results from the increased amount and activity of endothelial nitric oxide synthase (eNOS), the enzyme responsible for catalyzing NO synthesis (Hambrecht et al., 2003). Accordingly, eNOS gene expression and protein content have been shown to increase with aerobic exercise training (Hambrecht et al., 2003; Sessa et al., 1994). Further, phosphorylation of eNOS is enhanced with training, which increases the enzyme's activity and the subsequent production of NO (Hambrecht et al., 2003). Endurance training is also associated with a reduction in oxidative stress (Edwards et al., 2004). Because free radicals inactivate NO, a reduction in oxidative stress will increase NO bioavailability.

The intensity of training also appears to be an important factor determining improvements in endothelial cell function. Aerobic training of moderate (50% $\dot{V}O_2max$), but not low (25% $\dot{V}O_2max$) or high (75% $\dot{V}O_2max$), intensity improves endothelium-dependent vasodilation through enhanced NO production (Goto, 2003).

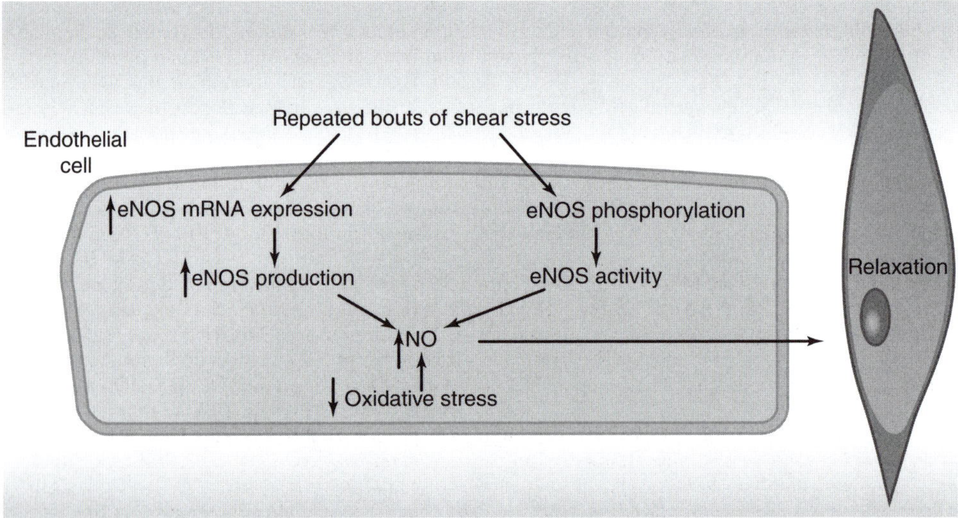

Figure 10.4 Enhancement of nitric oxide (NO)-mediated vasodilation with aerobic training. Improvements in endothelial cell function with aerobic training are largely attributed to repeated exposure to periods of high shear stress. This, in turn, activates endothelial NO synthase (eNOS) gene expression, resulting in increased eNOS production. At the same time, shear stress activates the eNOS phosphorylation, enhancing its activity. Combined, these factors increase NO production. Further, moderate aerobic training is associated with a reduction in oxidative stress, which prevents inactivation of NO by free radicals. Ultimately, training results in a greater concentration of NO available for release from endothelial cells. This NO subsequently binds to smooth muscle cells, causing relaxation.

Low-intensity exercise may not provide a sufficient stimulus for improvements in endothelial cell function. On the other hand, high-intensity training may upregulate eNOS production and subsequently NO production, but it also increases oxidative stress. Thus, the newly formed NO is rendered inactive by oxygen free radicals, resulting in a similar NO bioavailability and a similar degree of endothelium-dependent vasodilation compared to pretraining levels (Goto, 2003).

Increased NO availability and endothelium-dependent vasodilation following endurance training are likely the result of accumulated changes resulting from acute exercise bouts. As described in chapter 9, blood flow through the vessels increases during exercise, producing a larger shear stress on the vessel walls. The increased shear stress, in turn, stimulates eNOS mRNA expression (Uematsu et al., 1995), eNOS phosphorylation (Fisslthaler et al., 2000), NO production (Nakano et al., 2000) and NO release (Mo, 1991).

Vascular Remodeling and Aerobic Training

Aerobic exercise training is also known to stimulate vascular remodeling. Endurance-trained individuals have a larger arterial diameter than untrained individuals, and arterial diameter increases following a period of aerobic exercise training (Dinenno et al., 2001; Schmidt-Trucksass et al., 2003). These vascular adaptations appear to be specific to the trained limb, as evidenced by increased arterial diameter in femoral, but not brachial, arteries with lower body aerobic training (e.g., jogging or cycling; Dinenno et al., 2001; Huonker et al., 2003) and increased arterial diameter in the subclavian, but not femoral, arteries in trained arms of tennis players and in paraplegic athletes (Huonker, 2003). There is also evidence of growth of collateral blood vessels following exercise training. Repeated exposure to increased blood flow and shear stress is believed to stimulate growth of collateral blood vessels, including an increased size and number of vessels (Lloyd et al., 2005). Thus, dilation of newly formed arterioles may produce a greater reduction of vascular resistance across the muscle bed, resulting in higher blood flow to active muscles (Delp, 1998).

Angiogenesis, the growth of new capillaries, also occurs in response to exercise training. Capillarity and capillary-to-muscle fiber ratio increase in active muscles following endurance training (Charifi et al., 2004; Hepple et al., 1997). This occurs primarily as a result of the upregulation of VEGF (vascular endothelial growth factor), which stimulates sprouting of new capillaries from existing ones (Bloor, 2005; Prior et al., 2003). An increased number of capillaries will provide a greater surface area for gas exchange and a shorter diffusion distance from capillary to muscle fiber, therefore increasing the oxygen-diffusing capacity in skeletal muscle (Bloor, 2005). Consequently, endurance-trained individuals exhibit greater oxygen extraction during exercise compared to sedentary individuals (Hammond and Froelicher, 1985; Kalliokoski et al., 2001), which may explain why trained individuals can maintain a similar $\dot{V}O_2$ at a given submaximal workload despite having a lower rate of blood flow.

Another possibility is that capillaries become more tortuous with training. That is, existing capillaries form new sprouts that loop around rather than simply running parallel to muscle fibers (Charifi et al., 2004). Thus a single capillary could be counted more than once on the same cross-sectional muscle sample. Nonetheless, an increase in tortuosity would translate to greater capillary-to-muscle fiber interface and enhanced oxygen extraction. Studies regarding the effect of aerobic training on capillary tortuosity, however, are sparse and inconsistent. Data from animal models suggest that training does not alter capillary tortuosity in skeletal muscle (Poole, Mathieu-Costello, and West, 1989), but Charifi and colleagues (2004) observed increased tortuosity in the vastus lateralis muscles of elderly men following 14 weeks of cycle training. Further, these authors found the degree of tortuosity to be related to the level of aerobic enzyme activity, suggesting a link between capillary tortuosity and muscle oxidative capacity.

It should be noted here that a longer blood transit time following training likely complements the increased surface area for gas exchange (Kalliokoski et al., 2001). In trained subjects, the longer blood transit time results from the slower rate of blood flow during submaximal exercise that was discussed earlier. Slower movement of the blood allows a longer time during which gas exchange can take place, and ultimately a greater oxygen extraction.

Coronary Blood Flow

Aerobic exercise training induces adaptations in the coronary circulation that result in enhanced blood flow capacity and oxygen delivery to cardiac muscle. This occurs as a result of vascular remodeling and enhanced vasodilatory capacity in the coronary circulation. Aerobic training results in larger diameters of conduit arteries and arterioles (Brown, 2003; Currens and White, 1961; Kozakova et al., 2000) and growth of new arterioles (Breisch et al., 1986). Interestingly, evidence supports a progressive remodeling in the cardiac microvasculature with aerobic training. The early stages of training stimulate an increase in capillarity in cardiac muscle, but this adaptation is followed by a reversal of the capillary changes and concomitant increase in the number of large arterioles as training progresses (White et al., 1998). It is believed that the newly formed capillaries grow into larger arterioles, and that this increase in number and size of arterioles plays an important role in increasing coronary blood flow (White et al., 1998). It has also been suggested that the vascular changes in the coronary circulation reflect the changes in cardiac dimensions and are necessary to maintain adequate perfusion of a larger heart (Brown, 2003).

Perhaps more important is the change in the vasodilatory capacity of coronary blood vessels. It is important to note, however, that not all branches of the coronary circulation respond to aerobic training in the same way. Most studies demonstrate that following aerobic training, endothelium-dependent vasodilation is unchanged in conduit arteries (Oltman, Parker, and Laughlin, 1995; Oltman et al., 1992), but greater in smaller resistance arteries and arterioles (Muller, Myers, and Laughlin, 1994; Parker et al., 1994). Changes in endothelium-dependent vasodilation along the coronary vascular tree are related to training-induced adaptations in NO production. Recent studies demonstrate a nonuniform pattern of eNOS expression following training (Laughlin et al., 2001). From these studies, it appears that eNOS protein content is unchanged in conduit arteries and intermediate-sized arterioles but higher in smaller coronary arteries and small and large arterioles following aerobic training (Laughlin et al., 2001). Prostacyclin-induced vasodilation may be enhanced in coronary resistance arteries following exercise training (Muller, Myers, and Laughlin, 1994). There is also evidence to support enhanced sensitivity of smooth muscle cells to vasodilators, another factor contributing to an enhanced capacity to reduce vascular resistance in the coronary circulation (Oltman et al., 1992).

HEMOSTATIC ADAPTATIONS

Aerobic exercise training has a variety of hemostatic effects that together provide a cardioprotective effect. Platelet activation and aggregation are reduced with training, while fibrinolysis is enhanced in trained compared to untrained individuals. Although training is thought to have little effect on coagulation in healthy individuals, it may play an important protective role in cardiac patients.

Platelets

As discussed in chapter 9, acute aerobic exercise, particularly of high intensity, enhances platelet adhesion and aggregation. However, chronic exercise training serves a protective function and has been shown to reduce platelet adhesion and aggregation, both at rest and in response to an acute bout of exercise (Wang, Jen, and Chen, 1997, 1995). This has important implications for cardiovascular health, as acute bouts of exercise have been associated with myocardial infarction and sudden cardiac death (Bartsch, 1999). Thus, sedentary individuals are more susceptible to exercise-related cardiac events than trained individuals.

There are several ways in which aerobic exercise training may improve platelet function. Training increases levels of prostacyclin and NO, which inhibit platelet aggregation. Further, training decreases levels of oxidized low-density lipoproteins (LDL), which enhances platelet activation via inactivation of NO. Thus, lower levels of oxidized LDL following training may reduce platelet activation.

Coagulation

Coagulation potential appears to change little with endurance training in healthy adults. A number of investigators have reported no effect of training on thrombin time, prothombin time, and activated partial thromboplastin time, or on coagulation Factor VIII activity and antigen levels at rest (El-Sayed, Lin, and Rattu, 1995; Van den Burg et al., 1997) or in response to exercise (El-Sayed, Lin, and Rattu, 1995; Ferguson et al., 1987). The effect of chronic exercise on fibrinogen remains unclear, with data from cross-sectional studies supporting lower fibrinogen levels in trained compared to untrained subjects and data from training studies showing increased, decreased, or unchanged fibrinogen levels following training (Womack, Nagelkirk, and Coughlin, 2003). Endurance training, however, may play an important protective role in cardiac patients, where exercise training has been shown to lengthen activated partial thromboplastin time and decrease coagulation Factor VIII activity and fibrinogen levels (Suzuki et al., 1992)—all indices of reduced coagulability.

Fibrinolysis

Fibrinolytic activity is enhanced with regular exercise. Several researchers report greater activity of tissue plasminogen activator following acute exercise in trained compared to untrained subjects (Speiser et al., 1988; Szymanski, Pate, and Durstine, 1994). Such changes may be related to the greater release of tissue plasminogen activator or the reduced formation of plasminogen activator inhibitor complexes reported in trained subjects at rest (De Paz et al., 1992). Further, plasminogen activator inhibitor activity is lower at rest, and tends to decrease more following exercise, in trained

CAN EXERCISE PROTECT AGAINST EXERCISE-INDUCED SUDDEN CARDIAC DEATH?

We have all heard the story of the apparently healthy, middle-aged man who suffers a heart attack while shoveling snow in his driveway. Chapter 9 discussed the cascade of hemostatic events that could lead to such a situation. But could this sudden heart attack have been prevented? An aerobically trained individual is less likely to suffer a sudden exercise-induced cardiac event than someone who is sedentary.

A myocardial infarction, or heart attack, results when there is blockage of a coronary artery. Often, this occurs in a diseased artery when an atherosclerotic plaque ruptures, triggering the formation of a blood clot. As discussed in this chapter, regular aerobic exercise inhibits platelet adhesion and aggregation, both at rest and during an acute bout of exercise. Thus, an acute exercise bout results in a smaller prothrombotic stimulus in trained individuals compared to untrained individuals. Furthermore, in people with heart disease, aerobic exercise training reduces the tendency for coagulation. So, beginning a regular exercise program may provide an even greater protective effect for those with heart disease. Research also supports a greater potential for fibrinolysis in trained individuals. Together, this information suggests that regular aerobic exercise inhibits blood clot formation while enhancing the ability to break down clots. Thus, even though an acute bout of exercise, like shoveling snow, stimulates thrombotic processes, this effect is smaller in aerobically trained subjects.

individuals compared to their untrained counterparts (Speiser et al., 1988; Szymanski, Pate, and Durstine, 1994). Interestingly, Szymanski and associates (1994) reported that plasminogen activator inhibitor activity reached zero in 30% of their active subjects, suggesting a possible floor effect that may have blunted the effect of training on this variable. Together, these changes set the stage for greater plasmin formation and subsequent fibrin degradation. Consistent with this notion, De Paz and colleagues (1992) found higher fibrin and fibrinogen degradation products in trained runners compared to inactive controls. While the majority of evidence supports enhanced fibrinolysis following aerobic training, these findings are not universal. Baynard and colleagues (2007) reported no difference in either plasminogen activator or plasminogen activator inhibitor in trained versus untrained individuals.

Implications for Cardiovascular Disease

The changes in endothelial cell function and hemostasis described here have important implications for cardiovascular health. Healthy endothelial cells support an anticoagulatory, antithrombotic, and antiproliferative state that is instrumental in protecting against atherosclerosis and coronary artery disease. Nitric oxide and prostacyclin released from the endothelium prevent adhesion of platelets and monocytes to the arterial wall. Further, NO inhibits proliferation and migration of smooth muscle cells and also opposes the actions of endothelin, a potent vasoconstrictor and activator of smooth muscle cell proliferation (Maeda et al., 2001; Ruschitzka, Noll, and Luscher,

1997). Healthy endothelial cell function protects against the various stages of arterial plaque formation.

Atherosclerosis is a multifaceted process that has been linked to endothelial dysfunction. Damage to endothelial cells, which may result from factors such as hypertension, hyperlipidemia, or smoking, results in abnormal expression of adhesion molecules. This, in turn, increases the binding of various leukocytes, particularly monocytes and T-lymphocytes, to the endothelial cells (Libby, Ridker, and Masari, 2002; Ruschitzka, Noll, and Luscher, 1997). Inflammatory mediators cause monocytes and T-lymphocytes to migrate into the intima of the vessel (Libby, Ridker, and Masari, 2002). Meanwhile, LDL move into the vessel wall at the site of injury and become oxidized. Subsequently, activated macrophages accumulate oxidized LDL, forming foam cells that are the base of the atherosclerotic plaque. Another important component to the plaque's development is proliferation and migration of smooth muscle cells to the intima of the vessel. Figure 10.5 provides a schematic representation of these events. For a more detailed explanation, see Libby and Theroux, 2005.

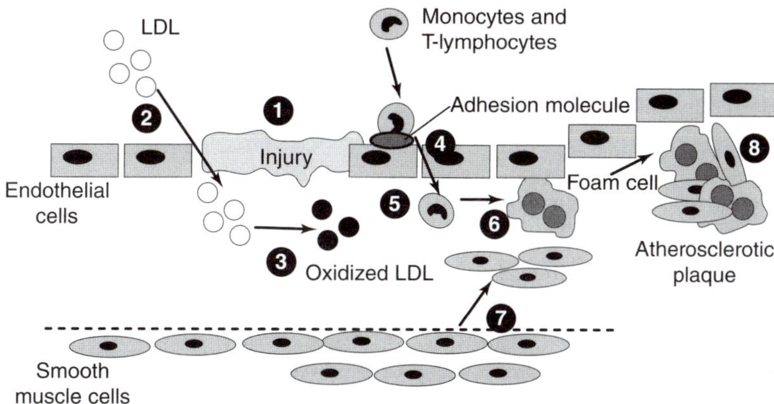

Figure 10.5 Schematic representation of the formation of an atherosclerotic plaque. Atherosclerosis begins with damage to the endothelium (1), which allows movement of low-density lipoprotein (LDL) beneath the endothelial cell layer (2) and subsequent oxidation of LDL (3). Concurrently, endothelial cells express adhesion molecules (4), resulting in monocyte and T-lymphocyte adhesion and movement into the arterial wall (5). Macrophages are activated and form foam cells containing oxidized LDL (6). Smooth muscle cell proliferation and migration is activated (7), and smooth muscles cells become incorporated into the foam cells of the growing plaque (8).

In the early stages, remodeling of the vessel wall will accommodate the growing plaque with little change in the vessel lumen diameter (Glagov et al., 1987). Thus, blood flow may not be significantly altered at this time. As the plaque continues to grow, the vessel will narrow and stiffen progressively, increasing vascular resistance and potentially limiting coronary blood flow. Additionally, the reduced availability of NO and increased release of endothelin associated with atherosclerosis will offset the normal balance of vascular tone, causing greater vasoconstriction and a further increase in vascular resistance.

Even in the early stages of atherosclerosis, when vessel narrowing may be minimal, the plaque is vulnerable to rupture. Rupture of an atherosclerotic plaque, and subsequent thrombus formation, is one of the most common causes of acute myocardial infarction. When the plaque ruptures, collagen and various procoagulant factors released from the plaque are exposed to flowing blood, with several important consequences. First, platelets become activated, adhere to the vessel wall, and aggregate to begin the formation of a thrombus. Second, the release of procoagulant factors stimulates coagulation (see chapter 8 for a complete discussion of the process of coagulation). Platelets also transport fibrinogen to the injury site, where it can then be converted to fibrin. The fibrin matrix reinforces and strengthens the thrombus. The end result is the formation of a thrombus that may occlude flow through a coronary artery and result in a myocardial infarction.

Aerobic training can protect against atherosclerosis by maintaining normal endothelial cell function and NO release. Additionally, moderate aerobic exercise has been shown to improve endothelial cell function in patients with coronary artery disease (Edwards et al., 2004). The progression of atherosclerosis may be slowed or even reversed through lifestyle modification, including regular exercise and management of other risk factors, such as hypertension, hyperlipidemia, diabetes, smoking, and obesity. Further, positive adaptations in platelet activity, coagulation, and fibrinolysis associated with endurance training may reduce the chance of an acute cardiac event in patients with atherosclerotic disease.

SUMMARY

Aerobic exercise training results in a number of anatomical and physiological adaptations that lead to enhanced cardiovascular function and improved aerobic exercise capacity. A larger, stronger heart contributes to a larger peak SV and $\dot{Q}$, while lower submaximal HRs subject the heart to less stress at any given submaximal exercise intensity. The expanded blood volume enhances venous return, thus augmenting SV with training. Substantial vascular remodeling and improved endothelial cell function contribute to the reduction in total peripheral resistance and higher blood flow to muscle during high-intensity exercise. Inhibition of platelet function along with enhanced fibrinolytic potential is cardioprotective. Overall, the cardiovascular adaptations to aerobic exercise training enable a higher level of exercise performance and foster a healthier cardiovascular system.

CHAPTER 11

Cardiovascular Responses to Acute Resistance Exercise

By Dan Drury, Bo Fernhall, and Denise Smith

During intense resistance training, the anaerobic system provides a large portion of the energy needed for muscle contraction. The aerobic energy system, however, contributes to energy production during resistance exercise; and the cardiovascular system responds to resistance exercise in a coordinated and integrated way to ensure that blood is delivered to the working muscle. Central and peripheral cardiovascular factors and hemostatic balance are affected by resistance exercise, demonstrating the connection between local muscle activity and the overall function of the cardiovascular system. This chapter describes the acute responses of the cardiovascular system to resistance exercise.

Cardiovascular responses to dynamic resistance exercise are largely determined by the exercise protocol used, specifically the number of exercises, the specific exercises employed, the number of repetitions, the number of sets, and the total work (volume) performed. The intensity of resistance exercise is described as percentage of the maximal amount of weight that can be lifted by an individual in one maximal repetition (1RM).

Cardiovascular responses to resistance exercise are not as well studied as cardiovascular responses to aerobic exercise. The reason in part is that resistance exercise has not been traditionally employed as a training regimen designed to improve cardiovascular health, and in part due to methodological difficulties in obtaining cardiovascular measures during resistance training exercises. In general, cardiovascular responses to resistance exercise are similar to the responses that occur during static exercise—that is, resistance exercise is characterized by a relatively large increase in blood pressure and a modest change in cardiac output. This is in contrast to aerobic exercise, which is characterized by a large increase in cardiac output and a modest increase in blood pressure (chapter 9).

CARDIAC RESPONSES

Resistance exercise is seldom used with the expressed purpose of enhancing cardiovascular function, but it does result in acute and chronic cardiovascular changes. This section details the acute cardiac responses to resistance exercise.

Cardiac Output and Its Components

A recent study showed that cardiac output increased during mild dynamic exercise involving lifting and extending the leg (Elstad et al., 2009). Researchers had participants perform 2 min of dynamic leg exercise that involved alternating contracting and relaxing the quadriceps for 2 s. As the quadriceps were contracted, the knee was extended and the heel lifted 3 to 5 cm. During the exercise, one leg was contracted while the other leg was relaxed. Bilateral weights of 2 to 5.5 kg (equal to 25% of maximal voluntary contraction force) were added to the ankles to increase muscular work. Throughout the leg exercise, heart rate was recorded and stroke volume was measured using Doppler ultrasound. In this study, heart rate increased by approximately 40% (from 55.3 to 78.0 beats/min), and stroke volume decreased by about 5% (from 86.5 to 82.2 ml). Thus, cardiac output increased by about 35% (from 4.59 to 6.18 L/min). Therefore, in this study, mild dynamic exercise with a light resistance caused a small increase in cardiac output resulting from a small decrease in stroke volume that was more than offset by the increase in heart rate.

Miles and colleagues (1987) reported that stroke volume decreased significantly (~20%) during leg extension exercises that involved 12 repetitions to fatigue (lasting about 90 s). In this study, the leg extension included a 3 s lifting motion, a 1 s pause, and a 3 s lowering motion, and stroke volume was assessed using impedance cardiography. Heart rate increased approximately 50 beats/min (from 70 to 120 beats/min), and cardiac output increased from approximately 5.4 to 6.3 L/min (17%), although this increase did not achieve statistical significance.

Cardiac output responses to more intense resistance training have been reported by Lentini and colleagues (1993), who had healthy male subjects perform a double leg press to failure at 95% of their maximum dynamic strength. Stroke volume was determined using echocardiography and was reported preexercise, at the end of the lift phase, during the "lockout," and during the lowering phase of the lift. Cardiac output increased significantly during the lifting phase and increased further during the lockout phase (figure 11.1*a*). The increase in cardiac output, however, is modest compared to that with aerobic exercise—and is due almost entirely to an increase in heart rate, which reached approximately 140 beats/min, as stroke volume was relatively unchanged or decreased slightly during the exercise (figure 11.1*b*).

Heart rate responses to resistance exercise have been more widely reported than changes in stroke volume and cardiac output. Most studies indicate that heart rate increases modestly during resistance exercise to volitional fatigue. In the study just cited, double leg press performed to failure with 95% of maximal strength resulted in a peak average heart rate of 143 beats/min. When taken to volitional failure, low-intensity resistance exercise results in a larger volume of work and produces heart

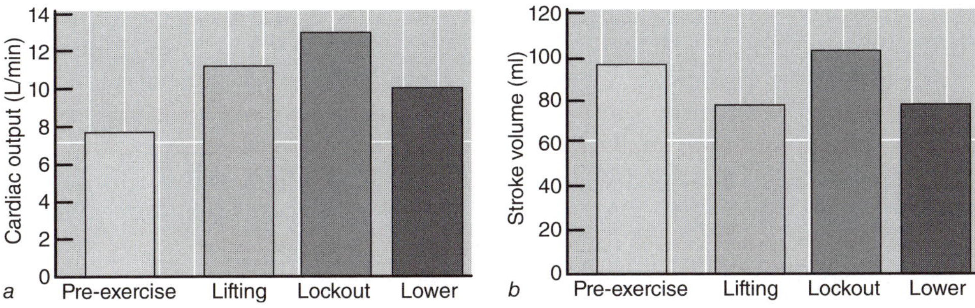

Figure 11.1 *(a)* Cardiac output and *(b)* stroke volume response to resistance exercise

Data from A. Lentini et al., 1993, "Left ventricular response in healthy young men during heavy-intensity weight-lifting exercise," *Journal of Applied Physiology* 75(6): 2703-2710..

rates that are higher than for a single 1RM (Falkel, Fleck, and Murray, 1992; Fleck and Dean, 1987). Some authors, however, have reported greatly elevated heart rates during high-intensity resistance exercise. Peak heart rates as high as 170 beats/min have been reported during performance of bilateral and unilateral lifts of the upper and lower body using weights equivalent to 80%, 90%, 95%, and 100% of maximum, with the highest heart rates occurring just before muscle fatigue prevented further contractions (MacDougall et al., 1985). Heart rate increases during acute resistance exercise are due to vagal withdrawal and stimulation of the sympathetic nervous system. It is likely that the sympathetic nervous system is stimulated by central command and from muscle chemo- and mechanoreceptors.

The relatively unchanged or decreased stroke volume that has been reported during resistance exercise is due to a combination of decreased preload, increased afterload, and enhanced contractility. Preload may be lower than baseline because of decreased filling time (due to increased heart rate) and a decrease in venous return. Venous return is likely decreased due to mechanical occlusion to the muscle during contraction and the performance of the Valsalva maneuver. High intramuscular pressure generated during contraction can temporarily occlude flow through the active muscles, thus decreasing stroke volume. Supporting the hypothesis that high intramuscular pressure may occlude venous return and thus blunt stroke volume during resistance exercise, Miles and coworkers (1987a) reported that stroke volume and cardiac output were significantly lower during the concentric phase of the exercise than during the eccentric phase. Lentini and colleagues (1993) found that end-diastolic ventricular volume decreased during both the lifting and lowering phase of the exercise. High intrathoracic pressure associated with performing the Valsalva maneuver can also impede venous return and thus lead to a decrease in stroke volume during resistance exercise. In the studies mentioned here, participants were told to avoid the Valsalva maneuver and were watched to confirm that they did so. Weightlifters, however, commonly perform the Valsalva maneuver during heavy lifting.

The **Valsalva maneuver,** defined as forcefully exhaling against a closed glottis, is performed as a natural tendency to stabilize the torso during resistance exercise (Gaffney, Sjøgaard, and Saltin, 1990; Sale et al., 1993). As intrathoracic pressure is dramatically increased to stabilize the spine, force is more efficiently transferred through the flexible spinal column. Because the lungs remain inflated against a closed glottis, the pressure within the thorax increases dramatically, providing additional rigidity for the spine. Concurrently, the pressure in and around the heart increases and venous return is decreased. Upon release of the air that has been temporarily trapped in the lungs, participants may feel light-headed because of a sudden decrease in blood pressure. The decrease in blood pressure may also be partly explained by a baroreceptor-mediated drop in heart rate and vasodilation. Although there are few data on post-Valsalva decreases in blood pressure, the sudden drop in blood pressure may be the result of rapid redistribution of blood to the periphery after the pressure within the thorax has been released. Any rapid redistribution of blood could result in syncope or light-headedness until the systemic pressure has been equalized.

The large increase in blood pressure that is associated with resistance exercise (discussed in the following section) results in an increase in afterload, which has a lowering effect on stroke volume. Activation of the sympathetic nervous system during resistance exercise would be expected to increase heart contractility. Evidence for increased contractility comes from the study by Lentini and colleagues (1993) that showed a decrease in end-systolic volume during the lifting and lowering phase of the exercise, and an increase in ejection fraction. The stroke volume response is a result of the balance of changes in preload, afterload, and contractility. The small-to-modest increase in cardiac output during resistance exercise is the result of a modest increase in heart rate and an unchanged or decreased stroke volume (figure 11.2).

The acute cardiovascular responses to resistance exercise just described are in stark contrast to those seen during aerobic exercise. Cardiac output increases dramatically during heavy aerobic exercise (five- to sevenfold) but modestly during resistance exercise (20-100%). More specifically, during aerobic exercise both heart rate and stroke volume increase to achieve a greater cardiac output. During resistance exercise, heart rate increases modestly but stroke volume decreases; thus cardiac output is only modestly increased.

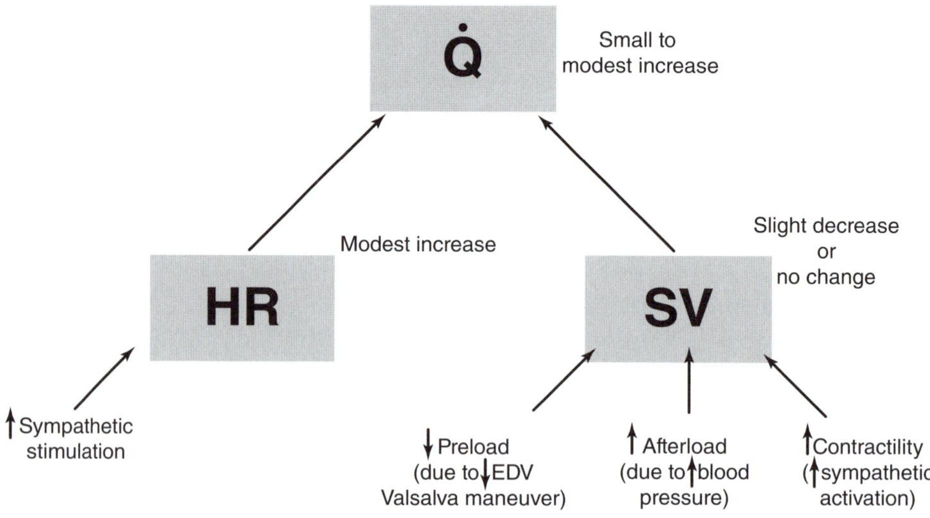

Figure 11.2 Factors affecting cardiac output during resistance exercise.

Myocardial Oxygen Consumption

The rate–pressure product (RPP) or double product ([HR × SBP]/100) has been used as an indirect measure of myocardial oxygen consumption. The RPP has been shown to correlate well with myocardial oxygen consumption under both static and dynamic exercise conditions (Nelson et al., 1974). Because of increases in both heart rate and systolic blood pressure, the RPP can rise to high levels during intense resistance exercise (MacDougall et al., 1985). However, many authors have found that RPP does not reach extremely high levels because heart rate increases are generally modest. Fleck and Dean (1987) assessed heart rate and blood pressure responses to one-knee extension exercises performed to volitional fatigue in trained bodybuilders, novice bodybuilders, and sedentary controls. In this study, the subjects achieved a RPP less than 250×10^2. Furthermore, the results indicated that the trained bodybuilders had a lower RPP than novice lifters or sedentary controls.

Historically, the high RPP was a primary reason that intense resistance exercise was considered contraindicated for persons with known cardiovascular disease (McCartney, 1999). However, revised guidelines from the American Heart Association suggest that resistance training may indeed be beneficial for those with known cardiovascular disease if contemporary prescriptive guidelines are employed with close supervision (Thompson et al., 2007). In a statement published by the American Heart Association (Braith and Stewart, 2006) regarding the use of resistance exercise in those with and without cardiovascular disease, the authors detail what is currently known about the safety of resistance exercise. They acknowledge that "excessive" blood pressure elevations have been documented with high-intensity resistance exercise (80-100% of 1RM performed to exhaustion), but note that such elevations are generally not a concern with low- to moderate-intensity resistance training performed with correct breathing technique and avoidance of the Valsalva maneuver. Furthermore, there

MUSCLE-STRENGTHENING ACTIVITY FOR OLDER ADULTS: RECOMMENDATIONS FROM THE AHA AND THE ACSM

To promote and maintain health and physical independence, older adults will benefit from performing activities, that maintain or increase muscular strength and endurance for a minimum of two days each week. The recommendation is to perform 8 to 10 exercises using the major muscle groups on two or more nonconsecutive days per week. To maximize strength development, a resistance (weight) should be used that allows 10 to 15 repetitions for each exercise. The level of effort for muscle-strengthening activities should be moderate to high. On a 10-point scale, where no movement is 0 and maximal effort of a muscle group is 10, moderate-intensity effort is a 5 or 6, and high-intensity effort is a 7 or 8. Muscle-strengthening activities include a progressive weight training program, weight-bearing calisthenics, and similar resistance exercises that use the major muscle groups.

From P.D. Thompson et al., 2007, "Exercise and acute cardiovascular events placing the risks into perspective: A Scientific statement from the American heart association council on nutrition, physical activity, and metabolism and the council on clinical cardiology," *Circulation* 114(17): 2358-68.

is indirect evidence that resistance exercise results in a more favorable balance in myocardial oxygen supply and demand than aerobic exercise because of the lower heart rate and higher myocardial (diastolic) perfusion pressure (Braith and Stewart, 2006).

In a study comparing the physiological responses to weightlifting and aerobic exercise, Featherstone and coworkers (Featherstone, Holly, and Amsterdam, 1993) tested 12 men with known cardiovascular disease. Participants performed both a maximal treadmill exercise and maximal resistance exercise at intensities of 40%, 60%, 80%, and 100% of maximal voluntary contraction. During the treadmill test, over 40% of the subjects experienced ST depression, whereas no such ischemia was observed during any of the resistance exercises. The RPP was higher during the treadmill test as compared to any of the lifting conditions. Although systolic pressures were similar between the conditions, the heart rates achieved during weightlifting were significantly lower than during the treadmill exercise.

Public health guidelines also suggest that resistance training is appropriate for older adults and should be part of an overall exercise program (Williams et al., 2007).

VASCULAR RESPONSES

There are several vascular responses to resistance exercise, including changes in blood pressure, vascular resistance, and arterial stiffness.

Blood Pressure Response to Resistance Exercise

The blood pressure (BP) response to resistance exercise is markedly higher than the BP response to aerobic exercise. During aerobic exercise, there are large increases in cardiac output (increased flow) concomitant with large decreases in vascular resistance. This leads to a modest increase in mean arterial BP, as SBP (systolic) increases but DBP (diastolic) remains stable. Conversely, cardiac output increases more modestly in resistance compared to aerobic exercise; but there is no change, or potentially an increase, in vascular resistance, producing a large increase in mean arterial pressure. Furthermore, the increase in mean arterial pressure is due to substantial increases in both SBP and DBP. Dynamic resistance exercise also includes an "isometric component" at the beginning of each lift, which contributes further to the increase in BP through the pressor response.

Peak BP during resistance exercise can reach exceedingly high levels, as depicted in figure 11.3. MacDougall and colleagues (1985b) measured intra-arterial BP during arm and leg exercise. The highest pressures were found during double leg press at 95% of 1RM, with the average pressure in excess of 300/240 mmHg and one subject reaching a BP of 480/350 mmHg! Subsequent research has confirmed large increases in BP, although not to the same level (McCartney, 1999; Narloch and Brandstater, 1995; de Vos et al., 2008).

When evaluating the BP response to exercise, it is important to ensure that the measurements are performed during the exercise and not after the exercise has been completed. The BP falls very quickly after exercise cessation, and may actually fall below resting levels during recovery (MacDougall et al., 1985a). It is difficult to measure BP using standard sphygmomanometry during a resistance exercise bout because of the time it takes to complete a measurement. Therefore, most studies have used intra-arterial measurements or continuous BP measurements using a finger cuff.

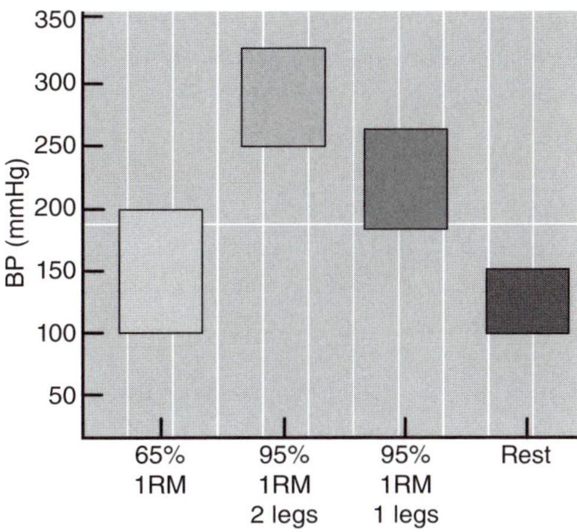

Figure 11.3 The blood pressure response to acute resistance exercise. The top of each box shows systolic blood pressure, and the bottom of each box shows diastolic blood pressure. There is a large increase in blood pressure during acute resistance exercise that is primarily related to the relative intensity of the exercise.

The traditional view has been that the BP response to resistance exercise is primarily a function of muscle size and the relative effort produced. More recent research has shown that the relative effort is probably the most important contributor. The higher the relative effort, the greater the BP response, and this is independent of muscle size (MacDougall et al., 1992). Although the BP response is greater during leg exercise than arm exercise at the same relative intensity in the same person, providing some evidence for the importance of muscle size, this relationship is not linear. For instance, the BP responses during one-leg compared to two-leg exercise are only slightly different, even though the muscle mass is double during two-legged exercise (McCartney, 1999; MacDougall et al., 1992). In addition, there was no difference in the BP response between individuals performing leg press exercise at the same relative intensity, even though muscle size varied considerably and was double in some subjects compared to others (MacDougall et al., 1992). There was also no effect of absolute muscle strength on the BP response. This shows that the influences of muscle size and absolute muscle strength on the BP response are minimal compared to the influence of relative contraction force.

During a set of weightlifting exercises, the BP varies in a cyclic manner with the different phases of a lift (figure 11.4). Blood pressure increases during the concentric portion of the lift, then decreases as the limbs are in a lockout position, and increases again during the eccentric phase. The decrease in BP during the lockout phase is due to a decrease in muscle force, since the weight is now supported by the skeletal structure as well as muscle. The BP may be slightly lower during the eccentric phase; however, this is due to the lower relative contraction intensity produced during the eccentric portion, since eccentric 1RM is much greater than concentric 1RM. At the same relative intensity of contraction, there is no difference in the BP response between eccentric and concentric contractions (McCartney, 1999). Also, as demonstrated in figure 11.4, BP increases from the beginning to end of a set, with the highest pressures observed at the conclusion of the set. This is likely due to increasing levels of fatigue, resulting in higher relative force production at the end of a set (MacDougall et al., 1985a, 1992).

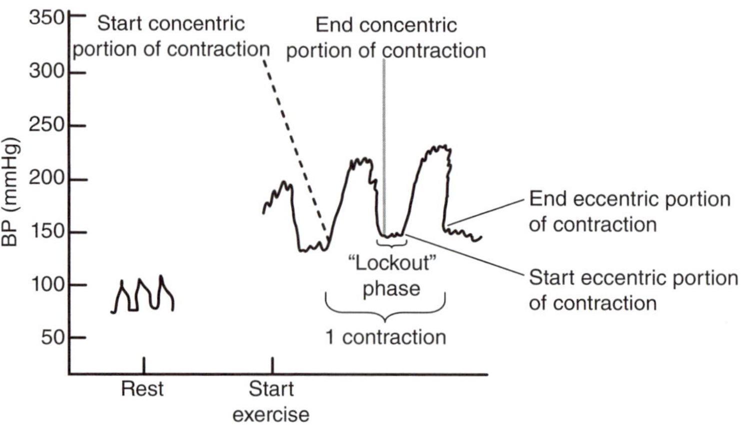

Figure 11.4 Beat-to-beat blood pressure response to acute resistance exercise. Blood pressure increases during the concentric portion of the lift, decreases during the "lockout" phase, and increases again during the eccentric portion of the lift.

Data from A.C. Lentini et al., 1993, "Left ventricular response in healthy young men during heavy-intensity weight-lifting exercise," *Journal of Applied Physiology* 75(6): 2703-2710.

The Valsalva maneuver is commonly used during resistance exercise, as it helps to stabilize the upper body during the lift. It is almost impossible to avoid the Valsalva maneuver at intensities above 85% of 1RM. When the Valsalva is invoked, the increase in BP is substantially greater than during resistance exercise alone (McCartney, 1999; Narloch and Brandstater, 1995; Haykowsky et al., 2003). This additional increase is due to the increase in intrathoracic pressure caused by the Valsalva maneuver. This has traditionally been interpreted as potentially dangerous, resulting in recommendations to avoid the Valsalva maneuver during lifting (Williams et al., 2007). Instead, exhaling slowly is recommended, as this can decrease the BP response by as much as 40% to 50% (Narloch and Brandstater, 1995). However, recent research suggests that the increase in intrathoracic pressure and subsequent increases BP may be protective of the cerebral circulation. The cerebrovascular transmural pressure was much lower during performance of resistance exercise with the use of the Valsalva maneuver compared to without, suggesting that invoking the Valsalva maneuver during a lift may decrease the stress on the cerebral arterial walls (Haykowsky et al., 2002). Similarly, performing a lift using the Valsalva maneuver decreased left ventricular end-systolic wall stress, suggesting that the maneuver may actually decrease the load on the heart (Haykowsky et al., 2001). Consequently, the use of the Valsalva maneuver during an acute resistance exercise bout may not be as detrimental as previously thought.

Vascular Resistance

Vascular resistance during an acute resistance exercise bout is dependent on the exercise intensity used. At low exercise intensities, the response is similar to that observed during aerobic exercise, as total peripheral resistance (TPR) decreases. However, the decrease is smaller compared to that with aerobic exercise, decreasing about 30% to 40% compared to the resting value during performance of double leg contractions at 25% of maximal voluntary contraction (Lewis et al., 1985). Similar reductions have been observed during contractions at 60% of 1RM (Meyer et al., 1999). However, at higher exercise intensities (95% of 1RM), TPR increased by 26% (Lentini et al., 1993), which was similar to the response to a 12RM exercise bout (Miles et al., 1987b). The higher TPR is probably due to the increase in intramuscular pressure during the active portion of the lift, as TPR returns to near resting levels during the lockout phase (e.g., full extension during a leg press), when the force of muscle contraction is less (Lentini et al., 1993).

Endothelial Function

Acute aerobic exercise increases endothelial function in healthy individuals, probably as a result of an increase in nitric oxide (NO) metabolism This appears to be a universal effect, as it is present in sedentary as well as aerobically trained and resistance-trained individuals (Baynard, Miller, and Fernhall, 2003; Baynard et al., 2007). However, the effects of resistance exercise on endothelial function have not been explored as much. Considering that shear stress within the microvasculature has been identified as one of the key factors leading to increases in NO production (Davies, 1995; Davies, Spaan, and Krams, 2005; Noris et al., 1995), resistance training may also increase endothelial function because resistance exercise increases blood flow and shear stress (Haram, Kemi, and Wisloff, 2008). However, available data suggest a more complex relationship.

A recent report (Jurva et al., 2006) showed that only experienced resistance-trained subjects exhibited improved endothelial function following an acute resistance exercise bout. Conversely, sedentary individuals actually showed a decrease in endothelial function after the acute bout. Thus, acute resistance exercise may temporarily impair the function of the endothelium in untrained individuals, while chronic exposure to resistance training may have a protective role in preventing vascular dysfunction. However, these authors did not measure changes in NO as part of their determination of endothelial function. Since NO production has been shown to exhibit dynamic changes after resistance exercise in humans (Kalliokoski et al., 2006), it is possible that NO metabolism may be differentially affected in resistance-trained compared to sedentary individuals.

It is also possible that acute resistance exercise has differential effects on conduit arteries (brachial artery) compared to the microvasculature. Two recent studies showed increases in microvascular function following an acute bout of resistance exercise (Fahs, Heffernan, and Fernhall, 2009), independent of whether the subjects were resistance trained or sedentary. Interestingly, endothelially mediated increases in blood flow were shown for changes that were potentially mediated by NO processes, as well as endothelial processes mediated by multiple mechanisms including NO metabolism (Fahs, Heffernan, and Fernhall, 2009). These findings suggest that endothelial function following acute resistance exercise may be different depending on where in the arterial tree the measurements are made. It also appears that endothelial vasodilatory mediators other than NO make important contributions to vasodilation.

Arterial Stiffness

While it is well established that acute aerobic exercise decreases arterial stiffness, the effects of resistance exercise are less uniform. First, acute resistance exercise has differential effects on the central (primarily aorta and carotid artery) and peripheral arteries (arm and leg arteries). It appears that a high-intensity whole body resistance exercise bout increases central artery stiffness by approximately 10% (Fahs, Heffernan, and Fernhall, 2009; DeVan et al., 2005; Heffernan et al., 2006). The increase in stiffness lasts for 30 min or more and returns to baseline levels within 60 min following exercise. This increase in stiffness is consistent across sedentary and resistance-trained individuals; thus the pretraining state does not appear to be important.

Although the exact mechanism for the increase in central artery stiffness is unknown, it has been speculated that the high BP during a resistance exercise bout alters the load bearing from more compliant elastin fibers to the stiffer collagen fibers (Fahs, Heffernan, and Fernhall, 2009; O'Rourke and Safar, 2005). It has also been suggested that increased sympathetic tone produces a less compliant artery by increasing vasoconstrictor tone of the smooth muscle (Heffernan et al., 2007) or that there is a decrease in NO bioavailability, thus reducing endothelial function and indirectly inducing arterial stiffness through increased constrictor tone in the smooth muscle. Unfortunately, there is little experimental evidence for any of these mechanisms. The BP 15 min after cessation of exercise is similar to resting pressure, but the increase in arterial stiffness remains (DeVan et al., 2005; Heffernan et al., 2006). Central arteries have relatively little smooth muscle, and it has also been shown that aerobic and resistance exercise elevate peripheral sympathetic modulation in a similar manner (Heffernan et al., 2007). Considering that acute resistance exercise elevates endothelial function, it is unlikely that reduced NO bioavailability can explain the increase in stiffness. Instead, it may be that high-intensity resistance exercise invokes the Valsalva maneuver and that the Valsalva maneuver independently increases aortic stiffness (Heffernan et al., 2007). This is thought to be related to the increase in thoracic pressure, which produces both an increase in BP and an external (outside-in) force on the artery. But how the effect of the Valsalva maneuver during the exercise bout would be maintained for 30 min in recovery is unknown.

Only high-intensity whole body resistance exercise produces an increase in central artery stiffness. Resistance exercise restricted to the use of only one or two legs has no effect on aortic stiffness (Heffernan et al., 2007c, 2007) (figure 11.5). Interestingly, whole body resistance exercise has no effect or actually decreases peripheral artery stiffness, while leg-only resistance exercise decreases peripheral stiffness in the legs (Fahs, Heffernan, and Fernhall, 2009; Heffernan et al., 2006, 2007a). This is thought to be due to an increase in peripheral artery vasodilation in response to the exercise, since arterial stiffness decreases only in the exercised leg and not in the limbs that were not exercised (Heffernan et al., 2006). However, the mechanical effect of muscle contraction, producing rhythmic compressions of the peripheral arteries, may also play a role, as external mechanical compression alone (achieved by inflating and deflating BP cuffs around the thigh) decreases peripheral artery stiffness in the leg while having no effect on the noninvolved leg (Heffernan et al., 2007b). It is possible that mechanical compression of the arteries creates a compression-related hyperemia, leading to local vasodilation. Thus, the smooth muscle relaxation leading to vasodilation results in a less stiff artery (Heffernan et al., 2007b).

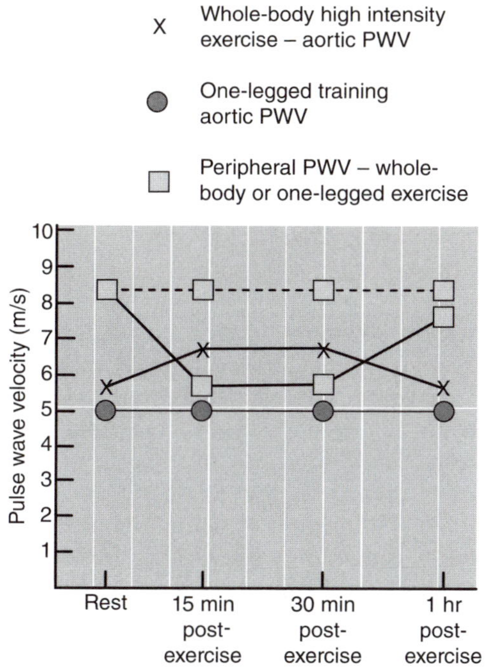

Figure 11.5 The effect of acute resistance exercise on arterial pulse wave velocity. Pulse wave velocity is a measure of arterial stiffness. Acute high-intensity, whole body resistance exercise increases aortic arterial stiffness, whereas one-legged resistance exercise has no effect. Peripheral arterial stiffness is either decreased, or does not change, in response to an acute resistance exercise bout.

HEMOSTATIC RESPONSES

Resistance exercise results in a large increase in mean arterial pressure and a small increase in cardiac output. The hemostatic system also responds acutely to resistance exercise. As with the other cardiovascular variables discussed here, less information is available on hemostatic responses to resistance exercise than to aerobic exercise. Additionally, the hemostatic responses are likely dependent on a number of factors, including factors related to the exercise stress (exercise protocol, intensity of contraction, number of lifts) and characteristics of the test participants (age, training profile, health status).

Blood Volume

Resistance exercise results in a decrease in plasma volume in the range of 13% to 22% (Collins et al., 1986; Craig, Byrnes, and Fleck, 2008; Kraemer, Kilgore, and Kraemer, 1993; Ploutz-Snyder, Convertino, and Dudley, 1995). Changes in plasma volume mirror changes in hematocrit and hemoglobin and are dependent on the exercise protocol. A recent study examined the effect of exercise protocol on changes in plasma volume and showed that a program using a 10RM resulted in a 22% decrease in plasma volume, whereas a protocol that used a 5RM caused a 13% loss of plasma volume (Craig,

Byrnes, and Fleck, 2008). It appears that the reduction in plasma volume caused by resistance exercise is responsible for selective increase in cross-sectional area (CSA) of exercising muscle. Ploutz-Snyder and colleagues (Ploutz-Snyder, Convertino, and Dudley, 1995) compared changes in active and inactive muscle CSA and plasma volume after barbell squat exercise. Muscle involvement in the exercise and muscle CSA were determined using magnetic resonance imaging (MRI). The authors reported a 22% decrease in plasma volume after six sets of squats. On the basis of the MRI scans, the authors estimated that 550 ml of fluid had moved into the active muscles and that the remaining fluid was delivered into other tissues not included in the scan. Furthermore, the absolute loss of plasma volume was highly correlated (.86) with absolute increase in muscle CSA immediately following exercise. These data suggest that increased muscle size after resistance exercise reflects a movement of fluid from the vascular space into active, but not inactive, muscle tissue. Plasma volume returns to preexercise levels during 30 min of recovery from resistance exercise.

Transient hypertrophy is a temporary increase in muscular size that is often evidenced after resistance exercise and can last anywhere from minutes to hours (Collins et al., 1986, 1989). The degree to which the muscles experience transient hypertrophy is affected by the volume of work performed over multiple contractions. Isometric contractions do not elicit the same response even if the durations of the muscular contractions are similar. It appears that the intermittent nature of multiple repetitions creates a pump-like transcapillary volume shift, forcing the fluids from the plasma into the interstitial spaces (Collins et al., 1989). Furthermore, the active tissues can potentially produce a significant amount of lactate, increasing the osmolarity of the muscle in comparison to that of the capillary fluid. Consequently, fluid would be drawn into the interstitial spaces as well as into the muscles, contributing to the plasma volume shift (Collins et al., 1986; Sjøgaard and Saltin, 1982). Together, the increases in hydrostatic pressure caused by elevations in BP, combined with the accumulation of osmotic solutes in the tissues, cause fluid to move from the vascular space into the interstitial space, resulting in a larger muscle CSA. Bodybuilders often seek to achieve a temporary increase in the size of their muscles, commonly referred to as a "pump," just prior to competition by performing low-intensity exercise for many repetitions. From a physiological perspective, this transient shift in plasma volume needs to be considered in studies that report the blood concentrations of specific variables.

Platelets

Resistance exercise increases platelet number and function. In a study of platelet activation and function after a single bout of resistance exercise, Ahmadizad and coworkers (Ahmadizad, El-Sayed, and Maclaren, 2006) reported an increase in both platelet count and function as a result of resistance exercise of varying intensities. Healthy young males performed three sets of resistance exercises at 40%, 60%, and 80% of 1RM. Platelet number increased following each trial, but the increase was not related to the intensity of exercise. In contrast to resistance exercise, isometric exercise does not appear to alter platelet number (Röcker et al., 2000).

Several mechanisms have been proposed to explain the increase in platelet number with exercise. First, the alterations in platelet count might be the result of the transient decrease in plasma volume seen during and shortly after exercise (Chamberlain, Tong, and Penington, 1990) leading to hemoconcentration (Ahmadizad and El-Sayed, 2005).

These dynamic shifts in plasma volume may also explain the findings that platelet count rapidly decreases within 30 min of exercise, mirroring the return of plasma volume (Ahmadizad, El-Sayed, and Maclaren, 2006; Ahmadizad and El-Sayed, 2005). Exercise-induced increases in circulating catecholamines can play a role in the sensitization of platelets as well, and may also explain the highly transient increases and decreases of platelets during and after exercise, respectively (Ahmadizad, El-Sayed, and Maclaren, 2006). The fresh supply of platelets during exercise is most likely a product of a catecholamine-derived platelet release from the lungs, spleen, and bone marrow (Chamberlain, Tong, and Penington, 1990; Gonzales et al., 1996). The release of catecholamines during resistance training is related to both the intensity of the exercise and the total amount of work performed. However, the relationship between resistance exercise and catecholamine release with platelet number and activation is not fully understood.

Platelet function has been reported to be increased following resistance exercise using various markers of platelet function. Platelet aggregation and increased levels of β-thromboglobulin (a platelet-specific protein released from the α-granules) have been reported to be increased following high-intensity (80% 1RM) resistance exercise (Ahmadizad, El-Sayed, and Maclaren, 2006; Ahmadizad and El-Sayed, 2003). Increased levels of circulating catecholamines have been suggested as the mechanism by which platelets are activated during high-intensity exercise, but studies measuring catecholamines or blocked α2-adrenoreceptor sites have demonstrated that platelet hyperaggregation cannot be entirely attributed to the adrenoreceptor pathway. Other potential mechanisms for hypercoagulability with exercise include an increase in H^+ ions, hemoconcentration, and the release of more metabolically active platelets from the reticular tissue (El-Sayed, El-Sayed Ali, and Ahmadizad, 2004). In contrast to resistance exercise, it does not appear that isometric exercise alters platelet function (Vind et al., 1993).

Coagulation

Coagulation is the process by which a platelet plug is reinforced with fibrin to form a clot. Researchers have reported that several coagulatory factors are altered following resistance exercise. Increased coagulatory potential during and following exercise is particularly important to understand, given that exercise-induced sudden death and myocardial infarctions are the result of a thrombus that forms and occludes a coronary artery in approximately 70% of the cases (Bartsch, 1999).

Factor VIII is the coagulatory variable thought to be responsible for the hypercoagulable state following exercise. Resistance exercise leads to an increase in Factor VIII that is positively correlated with the volume of weight lifted (El-Sayed, 1993). Although the mechanism responsible for the increase in Factor VIII with exercise is unclear, it appears to be related to the β-adrenergic receptor pathway because β-blockade blunts this response.

One study has reported an increase in fibrinogen immediately after resistance exercise with values returning to baseline within 30 min (Ahmadizad and El-Sayed, 2005). The researchers hypothesized that the increase in fibrinogen was most likely due to a hemoconcentration that was reversed in recovery when plasma volume returned to normal. However, given the wide variability in reports of fibrinogen response to aerobic exercise, it is fair to say that the fibrinogen response to resistance exercise is not firmly established.

Fibrinolysis

Fibrinolysis is the process by which fibrin is degraded into a soluble form, preventing it from becoming part of a clot (Smith, 2003). Fibrinolysis is stimulated during exercise by the release of tissue plasminogen activator (tPA) from the vascular endothelium. Tissue plasminogen activator is responsible for the formation of plasmin that degrades the thrombus and prevents excessive clotting (see chapter 8). Tissue plasminogen activator increases following many exercise protocols, including resistance exercise, and the response is intensity dependent. In addition to an increase in tPA, a concomitant decrease in plasminogen activator inhibitor (PAI) has been reported following resistance exercise (El-Sayed et al., 2000). Plasminogen activator inhibitor inhibits the formation of plasmin and hence causes a decrease in the breakdown of blood clots.

It is known that exercise in general, and resistance exercise in particular, alters many hemostatic variables, including platelets, coagulatory variables, and fibrinolytic variables. However, the effect of these changes on thrombus formation and breakdown is not well understood. Furthermore, it is clear that the responses are dependent upon the exercise protocol employed and the population studied. This is an important area of research that is likely to garner a great deal of attention in the future.

SUMMARY

Resistance exercise leads to acute changes in all components of the cardiovascular system. The magnitude of cardiovascular changes during exercise is dependent upon the intensity (load) and duration of the activity and individual characteristics of the exerciser (age, health and fitness status). During resistance exercise, there is a modest increase in cardiac output as a result of modest increases in heart rate and relatively little change (or slight decreases) in stroke volume. Blood pressure responses to resistance exercise are markedly higher than are normally seen during aerobic exercise, largely because vascular resistance does not decrease during resistance exercise as it does during aerobic exercise. The Valsalva maneuver also increases BP during resistance exercise. High-intensity resistance exercise appears to cause an increase in central artery stiffness that persists for approximately 30 min. Researchers have documented acute changes in coagulatory and fibrinolytic factors with a bout of resistance exercise; but in general, much less is known about how the hemostatic system responds to resistance exercise than is known in relation to aerobic exercise.

CHAPTER 12

Cardiovascular Adaptations to Resistance Training

By Dan Drury, Denise Smith, and Bo Fernhall

Until recently, the benefits of regular resistance training and other forms of anaerobic exercise were completely overshadowed by the profound cardiovascular benefits of aerobic exercise. Although aerobic exercise is—and should be—one of the cornerstones of a well-rounded fitness program, research evidence supporting the general health benefits of resistance training is accumulating (Braith and Stewart, 2006; Thompson et al., 2007). The focus of this chapter is on the chronic adaptations of the cardiovascular system to a regimented resistance training program.

CARDIAC ADAPTATIONS

While it is generally acknowledged that resistance training leads to improved general health and is a safe and effective way to attenuate age-related decline in muscle mass and functional capacity, the effect of resistance training on cardiac structure is equivocal (Haykowsky et al., 2002).

Cardiac Structure

Chronic exposure to resistance training can alter the morphology of the heart. In the past, some structural changes seen among those with consistent resistance training experience were misconstrued as maladaptations similar to those resulting from chronic hypertension. More recent data, however, have demonstrated that the morphological changes following resistance training are similar to those found with aerobic exercise and not those associated with pathological left ventricular hypertrophy (Pluim et al., 2000). **Athlete's heart** is a common term that refers to the increased size of the heart as a result of exercise training. This structural adaptation is associated with increased maximal stroke volume and cardiac output. In addition to the recognition that an athlete's heart increases in size as a result of training, a common belief is that there are divergent patterns of hypertrophy as a result of sport-specific training.

In the mid-1970s, Morganroth and colleagues hypothesized that an athlete's heart responds differently based on the hemodynamic stimulus imposed by the specific training regimen to which it is exposed. Aerobic training was purported to lead to an eccentric form of hypertrophy, characterized primarily by increased left ventricular cavity dimensions and thus ventricular mass. It was proposed that the increase in dimension and mass was a consequence of prolonged, repetitive "volume overload" due to increased venous return during rhythmical aerobic exercise. Conversely, Morganroth's hypothesis suggests that resistance training is associated with a form of **concentric hypertrophy** in which left ventricular mass increases due to increased ventricular wall thickness, but no changes occur in left ventricular cavity size. The increase in left ventricular mass resulting from resistance exercise is attributed to "pressure overload" caused by the need to overcome high pressures experienced during weight lifting (Naylor et al., 2008).

Several studies have shown that resistance-trained athletes have greater posterior wall and ventricular septal thickness, and greater left ventricular mass, than sedentary controls. These findings, however, are not universal. Pluim and colleagues (2000) conducted a meta-analysis of 31 studies involving aerobically trained athletes, 23 studies of combined aerobically trained + resistance-trained athletes, and 24 studies of resistance-trained athletes. In total, the analysis included 1,551 athletes and 813 control subjects. The analysis revealed that all the athletic groups had greater left ventricular mass than the sedentary controls, but there was no statistical significance among the athletic groups (figure 12.1*a*). Figure 12.1*b* presents a comparison of posterior wall and ventricular septal thickness among the athletes included in the meta-analysis.

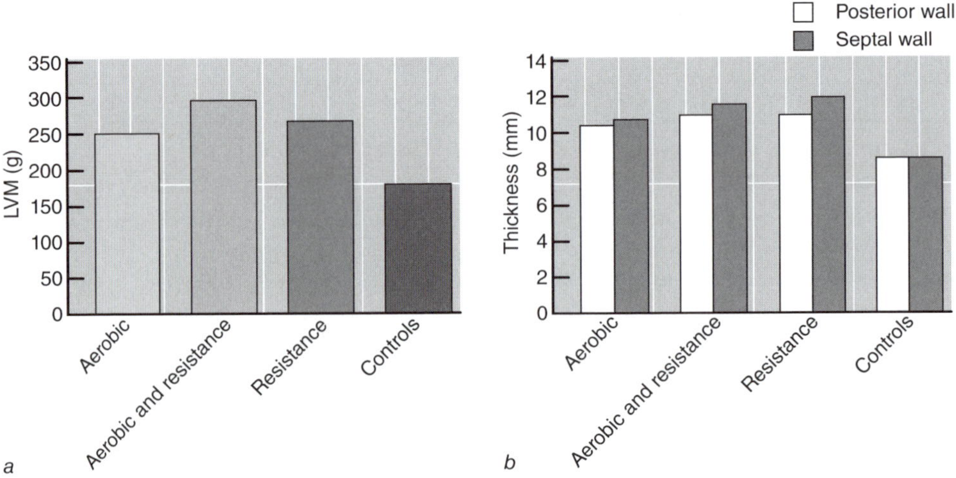

Figure 12.1 Comparison of *(a)* left ventricular mass and *(b)* wall thickness among aerobically trained, strength-trained, aerobically trained + strength-trained athletes, and control subjects. All athletes exhibit enhanced left ventricular mass (LVM) and wall thickness, but strength-trained athletes tend to have greater relative wall thickness than aerobically trained athletes and controls.

Ventricular septal wall thickness was significantly higher in the resistance-trained athletes versus the aerobically trained athletes. Some studies indicating that wall thickness is greater in resistance-trained athletes have shown no difference in wall thickness when it is reported relative to body size. In the meta-analysis conducted by Pluim and coworkers, the mean relative wall thickness (not shown) was significantly lower in the control subjects (0.36 mm) versus the aerobically trained (0.39 mm), the aerobically trained + resistance-trained (0.40 mm), or the resistance-trained (0.44 mm) subjects. Furthermore, the relative wall thickness of the resistance-trained subjects was significantly greater than for the aerobically trained athletes. In addition to greater wall thickness, the meta-analysis revealed that left ventricular internal diameter was significantly greater for the athletes than for the sedentary controls, with aerobically trained athletes tending to have a larger diameter than the resistance-trained athletes (53.7 mm vs. 52.1 mm, respectively).

The studies included in the meta-analysis were primarily cross-sectional studies. Inherently, the findings of these studies are limited based on the predilection of athletes to be drawn to certain training methods (aerobic or anaerobic) and also partially based on genetically determined factors. Further confounding this issue is the tendency for many athletes to train using both aerobic and resistance training components. Despite the challenges with cross-sectional studies, the majority of such studies comparing athletes to sedentary controls report that both aerobic training and resistance training lead to increased left ventricular mass.

Although resistance athletes have been shown to have increased left ventricular wall thickness as compared to controls, this is not a universal finding. Several authors have failed to find significant differences in normalized ventricular wall thickness with resistance exercise (Haykowsky et al., 2001; George et al., 1995; Pearson et al., 1986; Snoeckx et al., 1982). A review by Haykowsky and colleagues (2002) suggests that resistance training does not lead to an obligatory increase in left ventricular wall thickness or mass. These authors note that alteration in left ventricle geometry with resistance training may be dependent on the specific type of resistance training, such as Olympic-style weightlifting, powerlifting, or bodybuilding. In their review, they found that 37.5% of resistance-trained athletes exhibited normal left ventricular geometry; another 37.5% exhibited concentric hypertrophy (an increase in both left ventricular mass [LVM] and relative wall thickness); and 25% exhibited eccentric hypertrophy (increased LVM with preserved relative wall thickness). Interestingly, only bodybuilders and weightlifters (serious weightlifters, but not classified in one of the other categories) exhibited eccentric hypertrophy similar to what might be observed in aerobically trained athletes. Those athletes with normal geometry were either powerlifters or Olympic weightlifters, while those with concentric hypertrophy were 80% weightlifters and 20% Olympic weightlifters. These data might suggest that the type of resistance training may influence cardiac structure; however, no longitudinal studies exist to confirm this. Furthermore, the authors note that anabolic steroids are known to cause left ventricular hypertrophy and that underlying steroid use may influence the data.

Longitudinal studies assessing the effect of resistance training on cardiac structure are relatively sparse and have produced conflicting results (Naylor et al., 2008). However, there is some agreement regarding adaptations in previously sedentary individuals. In general, previously sedentary individuals who engage in resistance training respond by increasing cavity size **(eccentric hypertrophy)** or increasing wall thickness (Naylor, 2008; Haykowsky et al., 2005), but rarely both. Thus, additional training studies are necessary to accurately determine left ventricular changes as a result of resistance training.

Cardiac Function

Several longitudinal studies have reported small decreases or no change in resting heart rate after resistance training, and resting heart rates among serious resistance-trained athletes are comparable to those of sedentary individuals (Fleck and Dean, 1987; Haennel et al., 1992). It appears, therefore, that resistance training does not produce the same training-induced bradycardia (decrease in heart rate) often observed after aerobic training. At the same time, it is important to note that the resting heart rates of resistance-trained athletes do not appear to increase after resistance training either. This is noteworthy because many strength athletes carry a body mass that far exceeds the norm (Fleck and Dean, 1987; Longhurst and Stebbins, 1992). In the sedentary population, additional body mass has been associated with a higher resting heart rate and is often used as an indirect measure of cardiovascular fitness.

Stroke volume may be slightly higher in resistance-trained athletes than in sedentary controls, but these differences are likely to disappear when adjusted for body size and surface area (Adler et al., 2008). The meta-analysis by Pluim and colleagues (2000) showed no significant difference in ejection fraction (figure 12.2*a*) or fractional shortening (common measures of systolic function) between sedentary controls and athletes with different training backgrounds. Thus, it appears that there is no difference in systolic heart function at rest as a result of resistance training.

Pluim and colleagues (2000) also reported no difference in E/A ratio among sedentary controls and athletes (figure 12.2*b*). These data suggest that diastolic function at rest is not affected by exercise training in healthy adults.

Although systolic and diastolic function at rest do not seem to be altered as a result of a resistance training program, there is evidence that both systolic and diastolic function are enhanced in response to an isometric challenge in resistance-trained individuals. Adler and colleagues (2008) conducted a study of 96 men (48 experienced weightlifters, 48 sedentary) who performed a 2 min isometric exercise at 50% maximal

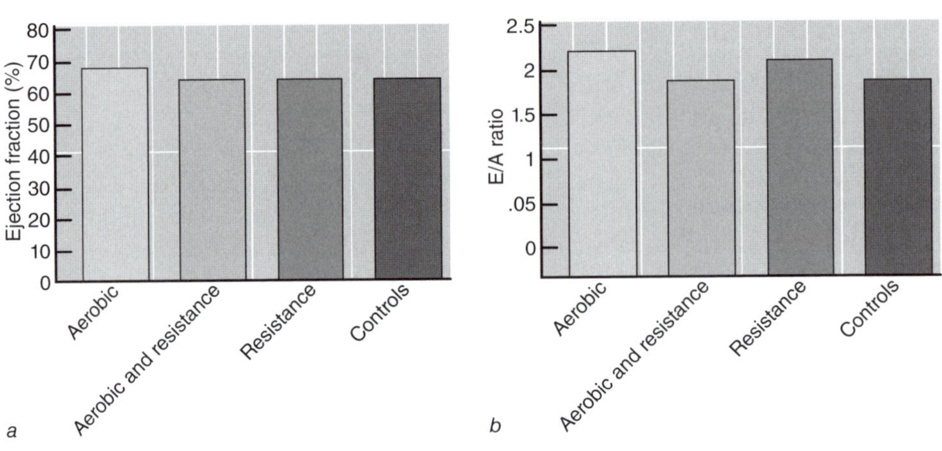

Figure 12.2 *(a)* Systolic (ejection fraction) and *(b)* diastolic (E/A ratio) function at rest in athletes and sedentary controls. There are no discernable differences in heart function between these groups.

voluntary contraction using a two-hand bar dynamometer. The resistance-trained athletes (all different types of resistance training) had a greater increase in stroke volume (figure 12.3*a*), indicating enhanced systolic function. The resistance-trained athletes also demonstrated enhanced diastolic function in a number of variables, including peak early filling (E wave), acceleration time and acceleration rate, deceleration time and deceleration rate, and E/A ratio (figure 12.3*b*). The authors concluded that despite marked left ventricular hypertrophy among the strength athletes, diastolic function was enhanced to the point of being "supernormal" with regard to the enhanced early diastolic filling. Isovolumic relaxation time was significantly less among the strength athletes, who also had significantly lower diastolic pressures. With a lower diastolic pressure, the left ventricle is able to begin filling earlier in the cardiac cycle via suction as the ventricles passively pulling blood from the atria early in the diastolic filling phase. Consequently, peak atrial velocity time was found to be less among the strength athletes largely because the ventricle had already received additional blood prior to the atrial kick.

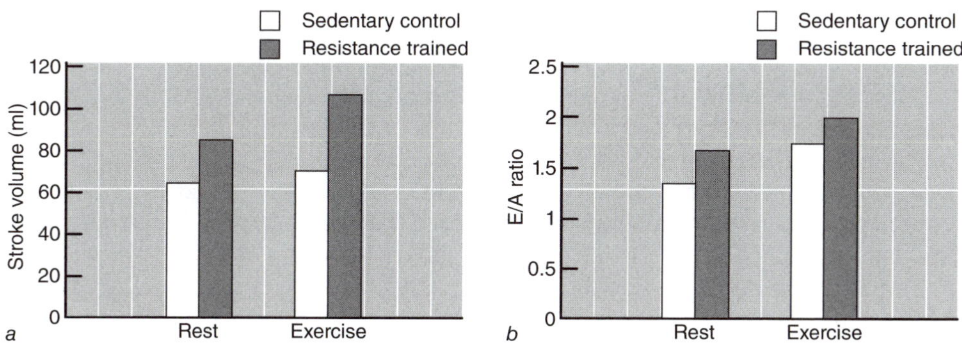

Figure 12.3 *(a)* Stroke volume at rest and during exercise in resistance-trained and sedentary subjects. *(b)* E/A ratio at rest and during exercise in resistance-trained and sedentary subjects.

In summary, cardiac function does not appear to be enhanced—or impaired—at rest following resistance training. There is evidence, however, that both systolic and diastolic function are enhanced during exercise (isometric) following resistance training. The finding of enhanced diastolic function among resistance-trained athletes is particularly important because in cardiac disease, hypertrophy is associated with diastolic dysfunction.

VASCULAR FUNCTION

Resistance training is associated with a decrease in blood pressure and a potential increase in arterial stiffness. These adaptations are discussed next.

Blood Pressure

Although it is clear that aerobic exercise training can reduce resting blood pressure (BP), traditionally it had been assumed that resistance training would not produce this response. However, recent meta-analyses of the effect of resistance exercise training on BP confirm this beneficial effect. All of these meta-analyses showed a reduction of 3 to 4 mmHg in both systolic (SBP) and diastolic (DBP) blood pressures following resistance training (Cornelissen and Fagard, 2005; Fagard, 2006; Kelley and Kelley, 2000). Although this reduction in resting BP is smaller than that seen after aerobic exercise, it is still physiologically and clinically significant. Interestingly, the type of resistance exercise program had no effect on the amount of BP reduction, as high-intensity standard programs elicited the same effect as lower-intensity circuit-based programs (Cornelissen and Fagard, 2005). Furthermore, short-duration programs (<15 weeks) elicited stronger effects on BP than long-term programs (>15 weeks) (Cornelissen and Fagard, 2005), with significant reductions in as little as four weeks of training (Collier et al., 2008), which is also similar to findings from aerobic training. Contrary to the situation with aerobic exercise training, the effects of resistance exercise training are similar between hypertensive and normotensive individuals (Collier et al., 2008; Cornelissen and Fagard, 2005; Fagard, 2006).

It is possible that measuring only traditional brachial BP may not elucidate the effect of resistance exercise training on BP. Several recent studies have shown no changes in brachial BP but significant reductions in central BP following resistance training (Heffernan et al., 2009; Rakobowchuck et al., 2005; Taaffe et al., 2007). This suggests that resistance exercise may have differential effects on central versus peripheral arteries; and the central BP changes are likely more important, since reduced central BP is associated with reduced afterload and myocardial work.

The mechanisms responsible for the BP reduction in response to resistance exercise training are not well understood. Reductions in sympathetic stimulation resulting in reduced peripheral resistance have been proposed (Fagard, 2006). Although peripheral resistance is reduced after resistance training, this is probably not caused by a decrease in sympathetic stimulation (Heffernan et al., 2009). Consequently, the reduced BP in response to resistance training is likely caused by reduced peripheral resistance, since resting cardiac output is unaltered by resistance training (Heffernan et al., 2009); but the mechanism behind this effect is still under investigation.

The BP response to an acute resistance exercise can also be modified with resistance exercise training, but only a few studies have investigated this. In older individuals, when the same absolute weight (80% of pretraining 1-repetition maximum, 1RM) was lifted after training as before training, both SBP and DBP were significantly reduced (figure 12.4). Since muscle strength also increased, the decrease in BP was likely a result of a decrease in the relative weight lifted after training. When the same relative weight

RESISTANCE EXERCISE AND BLOOD PRESSURE

A 45-year-old sedentary male with slightly elevated blood pressure wanted to start a resistance exercise program to try lowering his blood pressure because his physician had suggested he consider going on blood pressure–lowering medication. After three months of exercise training, using a standard exercise program of three times per week, three sets of 12RM for eight major body parts, his blood pressure had normalized, from 138/88 at the start to 128/80 mmHg. These changes were also coupled with improved endothelial function. As a result, his physician decided to take a "wait and see" approach before initiation of any blood pressure medication.

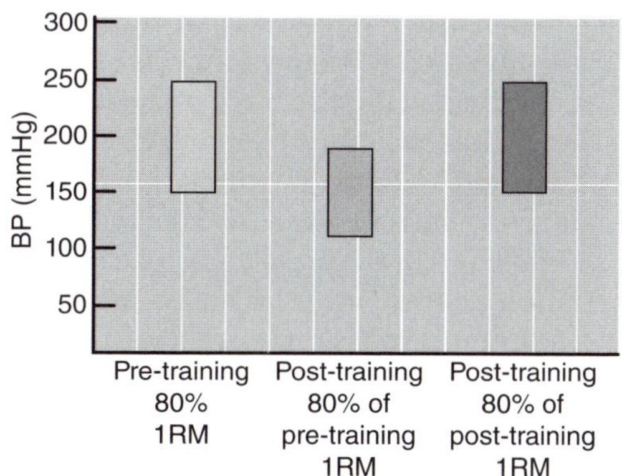

Figure 12.4 The effect of resistance exercise training on blood pressure during an acute resistance exercise session. Blood pressure is lower after training when the same absolute weight is lifted as during pretraining but not altered when the same relative weight is lifted.

was lifted before and after training (80% of 1RM) for 10 reps, there was no difference in BP (figure 12.4). This confirms that the lower BP during lifting of the same absolute weight after training was due to the increase in muscle strength, causing a decrease in the relative intensity of the weight lifted (McCartney et al., 1993). Similar results were shown in young individuals, with the exception that BP actually increased post-training versus pretraining when subjects lifted a weight at 85% of 1RM (Sale et al., 1994). Thus, when testing the effect of training on the BP response to acute resistance exercise, it is important to test the effects of both absolute and relative weight lifted.

Vascular Resistance

Peripheral vascular resistance is decreased following resistance exercise training in young healthy (Heffernan et al., 2009), older healthy (Anton et al., 2006), and middle-aged hypertensive individuals (Collier et al., 2008), suggesting that this is a common and universal effect. This decrease in peripheral resistance is also likely the cause of the decrease in BP already discussed. However, the mechanism for the decrease in peripheral resistance is unknown at this point. Although a decrease in arterial sympathetic constraint has been suggested, sympathetic modulation does not change, or it changes differentially in different groups with the same BP response. Thus, sympathetic constraint is an unlikely mechanism of changes in peripheral resistance (Heffernan et al., 2009, 2008). It is possible that changes in myogenic modulation of vascular tone may produce the decrease in peripheral resistance, but this needs further evaluation (Heffernan et al., 2009).

Endothelial Function

Resistance exercise increases shear rate of peripheral arteries (Heffernan et al., 2009), which should improve endothelial function. However, the effect of resistance training on endothelial function has been mixed, with several studies finding no effect (Casey, Beck, and Braith, 2007; Kawano et al., 2008; Rakobowchuk et al., 2005) and others showing increased endothelial function (Collier et al., 2008; Heffernan et al., 2009; Olson et al., 2006). It appears that conduit artery endothelial function is not improved in normal healthy individuals, whereas microvascular endothelial function is improved in most populations. Only one study has shown increased conduit artery endothelial function, and this study investigated the effect of resistance in obese women (Olson et al., 2006). This is consistent with the literature on the effect of endurance training, as brachial artery function is improved only in populations with significant cardiovascular risk factors and not in normal, healthy individuals (Fernhall and Angiovlasitis, 2008). Conversely, most of the research shows an increase in microvascular function regardless of the population tested, suggesting that the microvasculature is more sensitive to change as a result of exercise. It is generally believed that the increase in shear rate that occurs during each exercise session produces an increase in nitric oxide (NO) production, thus facilitating an increase in endothelial function. Since shear rate was also increased in the resting state following resistance training, it is possible that baseline NO availability is increased, also facilitating the increase in endothelial function in response to reactive hyperemia.

Arterial Stiffness

Interest in the effect of resistance exercise training on arterial stiffness was generated from observations in cross-sectional studies showing that resistance-trained men exhibited stiffer arteries than sedentary men (Bertovic et al., 1999; Miyachi et al., 2003). However, more recent studies suggest that arterial stiffness is not different between resistance-trained and sedentary men (Heffernan et al., 2007; Lane et al., 2006), and age is probably a greater modulator of arterial stiffness than resistance training per se.

The first longitudinal study on resistance training showed that two months of whole body resistance training increased arterial stiffness and decreased arterial compliance of the carotid artery in young, healthy Japanese men (Miyachi et al., 2004). These changes occurred without any changes in BP, and stiffness was increased further after four months of training but returned to baseline following two months of detraining. Subsequent studies have shown increases in arterial stiffness in women (Cortez-Cooper et al., 2005) participating in high-intensity resistance training, and following as little as four weeks of resistance training in hypertensives (Collier et al., 2008). Since moderate-intensity resistance training does not alter arterial stiffness (Casey et al., 2007; Taaffe et al., 2007; Yoshizawa et al., 2009), it was thought that only high-intensity training caused increases in arterial stiffness. However, several recent studies in young individuals used high-intensity whole body resistance exercise training without producing any changes in arterial stiffness (Heffernan et al., 2009; Rakobowchuk et al., 2005). Thus, it is possible that resistance exercise training may produce increases in arterial stiffness, but more evidence suggests that it does not change arterial stiffness. Certainly, moderate (60-80% of maximal)-intensity resistance training programs, or combining resistance and aerobic training (Kawano, 2006), does not appear to increase arterial stiffness.

The potential mechanisms for an increase in arterial stiffness in response to resistance training remain elusive. It is possible that the large pressure increases during the acute training bouts alter the arterial wall, or the properties of the elastin and collagen, causing a stiffer artery (Miyachi et al., 2004). Considering that concentric, but not eccentric, resistance training produces increased arterial stiffness (Okamoto et al., 2006), this may be a plausible explanation, since blood pressure elevations during eccentric contractions are lower than during concentric contractions. However, this does not explain why arterial stiffness increases in some but not other studies. Increased sympathetic constraint of the arteries has also been proposed, but there is little support for this, as discussed earlier.

HEMOSTATIC ADAPTATIONS WITH RESISTANCE TRAINING

Researchers have consistently found that acute exercise (static, resistance, and aerobic) leads to alterations in platelets, coagulation, and fibrinolytic factors; and studies investigating aerobic training programs have reported training adaptations in hemostatic variables (chapter 10). Thus, it seems reasonable to hypothesize that resistance training results in hemostatic adaptations. There is, however, a paucity of research on the chronic effects of resistance training on hemostatic variables (El-Sayed et al., 2005).

In a cross-sectional study, Baynard and colleagues (2003) compared fibrinolytic responses to a maximal treadmill test in aerobically trained athletes, resistance-trained athletes, and untrained controls. The researchers found no significant differences in tissue plasminogen activator (tPA) (the key enzyme that stimulates fibrinolysis) or plasminogen activator inhibitor (PAI-1) (the primary fibrinolytic inhibitor) activity at rest. Tissue plasminogen activator activity increased as a result of the maximal treadmill test, but there was no statistical difference among the groups. Plasminogen activator inhibitor activity decreased in all groups following maximal exercise, but it decreased slightly less in the resistance-trained group (70% decrease versus 86% decrease for aerobically trained and 82% decrease for the untrained). These data suggest that there is little difference in fibrinolytic potential between healthy young men who are resistance trained and those who are strength trained. Additional research is needed to characterize the effect of different training on hemostatic variables and to elucidate the role of age, fitness, and health status.

SUMMARY

Resistance exercise training may produce changes in cardiac morphology, usually measured as either an increase in left ventricular diameter or an increase in left ventricular wall thickness, but not both. Cross-sectional data on athletes suggest that different types of resistance exercise training may produce different types of adaptations; however, this has not yet been shown in longitudinal studies. While some studies have shown that resistance training may increase arterial stiffness, most studies show no change in arterial stiffness. There is general agreement that resistance exercise training decreases blood pressure and increases endothelial function. These positive changes in arterial function may partially explain why resistance training is associated with improved health outcomes. There is little information on the effects of chronic resistance exercise training on hemostatic adaptations, but initial data suggest little effect.

Glossary

afterload—The pressure that opposes the ejection of blood and is usually taken to be equal to systemic arterial pressure, as reflected by mean arterial pressure.

angiogenesis—The growth of new blood vessels.

athlete's heart—A common term used to refer to the increased size of the heart as a result of exercise training.

Bainbridge reflex—When receptors in the atria and the venous system detect an increase in blood volume, the right atrium is stretched, causing heart rate to accelerate.

basement membrane—A single-cell barrier; well suited for the function of exchange.

blood clot (thrombus)—The formation of a platelet plug that is reinforced with fibrin.

calcium-induced calcium release—Release of calcium stored in the SR through the calcium release channels; influx of extracellular calcium through the L-type calcium channel triggers this release of calcium.

calcium-induced calcium release channels or **ryanodine receptors**—Protein channels that extend from the junctional SR, release calcium when stimulated with calcium, and can bind to the drug ryanodine. Alternately called calcium release channels.

capacitance vessels—Another name for venules and veins due to their ability to accommodate large quantities of blood, which is directly related to the distensible characteristics of the vessel wall.

cardiac output—The amount of blood ejected from the ventricles in one minute (L/min); a measure of the heart's ability to pump blood to support the needs of the body on a per minute basis.

cardiovascular drift—Phenomenon in which central cardiovascular variables that had previously been at steady state levels begin to drift during prolonged aerobic exercise, especially at high intensities.

central command—A coordinated autonomic response initiated in the medulla in response to muscle contraction.

coagulation—The series of highly regulated steps resulting in the formation of fibrin that provides the framework for the loose platelet plug.

concentric hypertrophy—An increase in both LVM and relative wall thickness.

conducted vasodilation—Phenomenon in which vasodilation in distal vessels spreads proximally through the vasculature.

contractile myocytes—Cardiac muscle cells that shorten to produce the force that ejects blood from the heart.

contractility—The force of myocardial contraction independent of preload or afterload.

desmosome—A specialized junction to bind adjacent myocytes together.

diad—The area where the junctional SR and T-tubules nearly meet.

eccentric hypertrophy—An increase in LVM with unchanged relative wall thickness.

ectopic beat—A beat that does not originate from the SA node; the impulse to contract is initiated someplace else in the heart.

elastic (central) arteries—The largest, very distensible arteries (e.g., aorta and iliac arteries), whose distensibility is due to large amounts of elastin in the tunica media.

electrochemical equilibrium potential (E_m)—The balance between the electrical and chemical forces, where an electrical force is created by a separation of ion pairs.

endothelial dysfunction—Alterations in endothelial function; directly associated with early manifestation of atherosclerosis and with the increased incidence of cardiac events.

endothelium—A single layer of epithelial cells that line all blood vessels.

exchange vessels—Composed primarily of capillaries, but also including the smallest vessels on either side of the capillaries, these vessels perform the ultimate function of the cardiovascular system—gas and metabolite exchange.

excitation–contraction coupling (ECC)—The series of electrical, chemical, and mechanical events in the sarcolemma that leads to force generation in the myofilaments.

exercise hyperemia—The rise in muscle blood flow observed during exercise.

fibrinolysis—Degradation of fibrin, allowing a clot to dissolve.

flow-mediated dilation—Non-invasive test to asses endothelial function by temporarily occluding arterial inflow, thus inducing reactive hyperemia, and measuring the resulting increase in blood vessel diameter.

Frank-Starling law of the heart—Law stating that increased stretch of the myocardium (reflected in increased end-diastolic volume or pressure) enhances the contractile force of the myocardium, and therefore causes more blood to be ejected.

functional sympatholysis—The ability of exercise to overcome sympathetic vasoconstriction in active muscle beds.

heart rate variability (HRV)—The beat-to-beat variability of the R-R interval.

heat exhaustion—A condition whose symptoms include fatigue, weakness, disorientation, and nausea; results from severe dehydration.

heatstroke—A life-threatening condition in which body temperature rises to a level that leads to multiple system failure.

hemostasis—The stoppage of blood.

hemostatic balance—The dynamic balance between blood clot formation (coagulation) and blood clot dissolution (fibrinolysis).

isovolumetric contraction period—A brief period during which ventricles contract, causing pressure in the ventricles to increase dramatically while ventricular volume remains unchanged; one portion of systole.

isovolumetric relaxation period—A brief period during which the ventricles relax but there is no change in ventricular volume because the AV valves are still closed.

latch state—Phenomenon in which contracted smooth muscle is locked into a contracted state in order to maintain vascular tone while consuming very little ATP.

left ventricular hypertrophy—Increased left ventricular end-diastolic diameter and increased ventricular wall thickness.

L-type receptor—A channel protein located in the sarcolemma and T-tubules that allows calcium to enter the cell.

muscular, or **conduit, arteries**—Arteries that contain large amounts of smooth muscle; located in the tunica media layer (i.e., radial, ulnar, and popliteal arteries and the cerebral and coronary arteries).

myocardial relaxation—An active, energy-dependent process that causes pressure in the left ventricle to decrease rapidly after the contraction and during early diastole.

myocytes—Cardiac muscle cells.

myogenic autoregulation—A process in which perfusion pressure in the arterioles regulates blood flow to the tissues in order to control local blood flow.

pacemaker-conduction cells—Modified muscle cells that initiate the electrical signal and conduct it through the heart.

phospholamban—A protein that stimulates the reuptake of calcium into the sarcoplasmic reticulum and thus enhances myocardial relaxation.

platelet activation—Release of platelet contents; occurs in response to platelet adherence to the damaged vascular wall and humoral factors (epinephrine, ADP, thrombin).

platelet aggregation—The process by which activated platelets bind to one another; the final step in platelet plug formation.

Poiseuille's law—The mathematical relationship between pressure, flow, and resistance; states that resistance depends primarily on the size of the blood vessel.

preload—The amount of blood returned to the heart during diastole.

pressure–volume loop—The relationship between ventricular pressure and volume throughout the cardiac cycle.

primary hemostasis—Stoppage of blood flow due to formation of a platelet plug that occludes the vessel and stops blood flow.

rate–pressure product (RPP)—The product of SBP and HR; reflects myocardial oxygen consumption.

reactive hyperemia—Increased blood flow following release of blood flow inhibition (occlusion) to a particular tissue.

resistance vessels—Sites of greatest resistance within the vascular tree, primarily arterioles, which are responsible for determining local blood flow; vasodilate and vasoconstrict in order to ensure that local blood flow matches local metabolic demand.

sarcomere—The basic contractile unit of the myocyte; composed primarily of overlapping thick and thin filaments.

sarcoplasmic reticulum (SR)—An extensive series of tubular structures within the myocyte that store and release calcium.

sarcoplasmic reticulum calcium-ATPase (SERCA$_2$)—Located in the network SR within the muscle cell; actively pumps calcium back into the SR.

secondary hemostasis—Formation of fibrin strands that reinforce the platelet plug.

shear stress—The force exerted on the vessel wall by the sliding action of blood flow.

skeletal muscle pump—Pumping effect resulting from mechanical forces imparted to the vasculature during rhythmic contractions.

sports anemia—A condition in which, despite an increase in the total number of red blood cells, trained individuals often have a lower hematocrit or hemoglobin concentration.

stroke volume—The amount of blood ejected from each ventricle; can be calculated by subtracting the amount of blood that is in the ventricles after contraction (end-systolic volume) from the volume of blood that filled the ventricles at the end of ventricular filling (end-diastolic volume).

syncytium—A connected network of cells that function as a unit; ensures the effective pumping action of the heart.

systolic function—The general term given to the ability of the heart to adequately produce the force needed to eject blood from the ventricle.

tachycardic rhythm—A rhythm in which all of the beats are from the ectopic focus.

transverse tubules (T-tubules)—Deep invaginations in the sarcolemma at each Z disc that provide the mechanism for transmitting an electrical impulse (action potential) from the sarcolemma into the myocyte.

Valsalva maneuver—A forceful exhalation against a closed glottis.

vascular endothelium—A single layer of epithelial cells that line all blood vessels.

vascular spasm—Powerful constriction of the injured vessel.

ventricular ejection period—The period during which blood is ejected from the heart; one portion of systole.

ventricular fibrillation—A lethal dysrhythmia in which the muscle fibers of the ventricles are fibrillating.

ventricular filling period—The portion of diastole during which the ventricles fill with blood.

ventricular suction—The suction-like movement of blood from the left atrium into the left ventricle.

ventricular tachycardia—A potentially life-threatening dysrhythmia that originates in one of the ventricles of the heart, which may lead to ventricular fibrillation.

viscosity—The friction between fluid layers as they slide past each other in conditions of laminar flow.

Recommended Readings

Chapter 2

Allen, D.G., and J.C. Kentish. 1985. The cellular basis of the length-tension relation in cardiac muscle. *Journal of Molecular and Cellular Cardiology* 17:821-840.

Bonow, R.O. 1994. Left ventricular response to exercise. In: G.F. Fletcher (ed.), *Cardiovascular response to exercise.* American Heart Association Monograph. New York: Futura Press, 31-47.

Casadei, B. 2001. Vagal control of myocardial contractility in humans. *Experimental Physiology* 86:817-823.

Chemla, D., C. Coirault, J.L. Hebert, and Y. Lecarpentier. 2000. Mechanics of relaxation of the human heart. *News in Physiological Science* 15:78-89.

Fuchs, F., and S.S. Smith. 2001. Calcium, crossbridges, and the Frank-Starling relationship. *News in Physiological Science* 16:5-10.

Koch, W.J., R.J. Lefkowitz, and H.A. Rockman. 2000. Functional consequences of altering myocardial adrenergic receptor signaling. *Annual Review of Physiology* 62:237-260.

Sagawa, K. 1981. The end-systolic pressure-volume relationship of the ventricles: Definition, modification and clinical use. *Circulation* 63:1223-1227.

Sagawa, K., L. Maughan, H. Suga, and K. Sunagawa. 1988. *Cardiac contraction and the pressure-volume relationship.* New York: Oxford University Press.

Shepherd, J.T., and P.M. Vanhoutte. 1980. *The human cardiovascular system: Facts and concepts.* New York: Raven Press.

Starnes, J.W. 1994. Myocardial metabolism during exercise. In: G.F. Fletcher (ed.), *Cardiovascular response to exercise.* American Heart Association Monograph. New York: Futura Press, 3-13.

Chapter 4

Berne, R.M., and M.N. Levy. 2001. *Cardiovascular physiology,* 8th ed. St. Louis: Mosby.

Fozzard, H.A., and W.R. Gibbons. 1973. Action potential and contraction of heart muscle. *American Journal of Cardiology* 31:182.

Guyton, A.C., and J.E. Hall. 2000. *Textbook of medical physiology,* 10th ed. Philadelphia: Saunders.

Irisawa, H., H.F. Brown, and W. Giles. 1993. Cardiac pacemaking in the sinoatrial node. *Physiology Reviews* 73:197.

Katz, A.M. 1995. Regulation of cardiac contraction and relaxation. In: J.T. Willerson and J.T. Cohn (eds.), *Cardiovascular medicine.* New York: Churchill Livingstone.

Little, R.C. 1985. *Physiology of the heart and circulation.* Chicago: Year Book Medical.

Raven, P.B., J.T. Potts, X. Shi, and J. Pawelzyk. 2000. Baroreceptor-mediated reflex regulation of blood pressure during exercise. In: B. Saltin, R. Boushel, N. Secher, and J. Mitchell, *Exercise and circulation in health and disease.* Champaign, IL: Human Kinetics.

Task Force of the European Society of Cardiology and the North American Society of Pacing and Electrophysiology. 1996. Heart rate variability. Standards of measurement, physiological interpretation, and clinical use. *European Heart Journal* 17(3):354-381.

Trautwein, W. 1973. Membrane currents in cardiac muscle fibers. *Physiology Reviews* 53:793.

Chapter 9

Baynard, T., H.M. Jacobs, C.M. Kessler, J.A. Kanaley, and B. Fernhall. 2007. Fibrinolytic markers and vasodilatory capacity following acute exercise among men of differing training status. *European Journal of Applied Physiology* 101:595-602.

Gallo Jr., L., B.C. Maciel, J.A. Marin-Neto, and L.E. Martins. 1989. Sympathetic and parasympathetic changes in heart rate control during dynamic exercise induced by endurance training in man. *Brazilian Journal of Medical and Biological Research* 22:631-643.

Kalliokoski, K.K., V. Oikonen, T.O. Takala, H. Sipila, J. Knuuti, and P. Nuutila. 2001. Enhanced oxygen extraction and reduced flow heterogeneity in exercising muscle in endurance-trained men. *American Journal of Physiology* 280:E1015-E1021.

Chapter 11

Collier, S.R., J.A. Kanaley, R. Carhart Jr., V. Frechette, M.M. Tobin, A.K. Hall, et al. 2008. Effect of 4 weeks of aerobic or resistance exercise training on arterial stiffness, blood flow and blood pressure in pre- and stage-1 hypertensives. *Journal of Human Hypertension* 22(10):678-686.

Chapter 12

Alomari, M.A., and M.A. Welsch. 2007. Regional changes in reactive hyperemic blood flow during exercise training: Time-course adaptations. *Dynamic Medicine* 6(1):1-6.

Brett, S.E., J.M. Ritter, and P.J. Chowienczyk. 2000. Diastolic blood pressure changes during exercise positively correlate with serum cholesterol and insulin resistance. *Circulation* 101(6):611-615.

Camargo, M.D., R. Stein, J.P. Ribeiro, P.R. Schvartzman, M.O. Rizzatti, and B.D. Schaan. 2008. Circuit weight training and cardiac morphology: A trial with magnetic resonance imaging. *British Journal of Sports Medicine* 42(2):141.

Fagard, R. 2003. Athlete's heart. *Heart* (British Cardiac Society) 89(12):1455-1461.

References

Chapter 1

Fye, B.W. 2006. Profiles in cardiology: Ernest Henry Starling. *Clinical Cardiology* 29:181-182.

Chapter 2

Lester, S.J., A.J. Tajik, R.A. Nishimura, J.K. Oh, B.K. Khandheria, and J.B. Seward. 2008. Unlocking the mysteries of diastolic function. *Journal of the American College of Cardiology* 51:679-689.

Oh, J.K., J.B. Seward, and A.J. Tajik. 2007. *The echo manual,* 3rd ed. Philadelphia: Lippincott, Williams & Wilkins.

Shepherd, J.T., and P.M. Vanhoutte. 1980. *The human cardiovascular system: Facts and concepts.* New York: Raven Press.

Chapter 3

Gertz, E.W., J.A. Wisneski, W.C. Stanley, and R.A. Neese. 1988. Myocardial substrate utilization during exercise in humans. Dual carbon-labeled carbohydrate isotope experiments. *Journal of Clinical Investigation* 82:2017-2025.

Hasenfuss, G. 1998. Calcium pump overexpression and myocardial function: Implications for gene therapy of myocardial failure. *Circulation Research* 83:966-968.

Massie, B.M., G.G. Schwartz, J. Garcia, J.A. Wisneski, M.W. Weiner, and T. Owens. 1994. Myocardial metabolism during increased work states in the porcine left ventricle in vivo. *Circulation Research* 74:64-73.

Chapter 4

Akselrod, S., D. Gordon, F.A. Ubel, D.C. Shannon, A.C. Barger, and R.J. Cohen. 1981. Power spectrum analysis of heart rate fluctuation: A quantitative probe of beat to beat cardiovascular control. *Science* 213:220-222.

Jalife, J., and D.C. Michaels. 1994. Neural control of sinoatrial pacemaker activity. In: M.N. Levy and P.J. Schwartz (eds.), *Vagal control of the heart: Experimental basis and clinical implications.* New York: Futura.

Malliani, A., F. Lombardi, and M. Pagani. 1994. Power spectral analysis of heart rate variability: A tool to explore neural regulatory mechanisms. *British Heart Journal* 71:1-2.

Task Force of the European Society of Cardiology and the North American Society of Pacing and Electrophysiology. 1996. Heart rate variability. Standards of measurement, physiological interpretation, and clinical use. *European Heart Journal* 17(3):354-381.

Chapter 5

American College of Sports Medicine. 2000. *ACSM's guidelines for exercise testing and prescription,* 6th ed. Baltimore, MD: Lippincott, Williams & Wilkins.

American College of Sports Medicine. 2001. *ACSM's resource manual for guidelines for exercise testing and prescription,* 4th ed. Baltimore, MD: Lippincott, Williams & Wilkins.

Corrado, D., A. Biffi, C. Basso, A. Pelliccia, and G. Thiene. 2009. 12-lead ECG in the athlete: Physiological versus pathological abnormalities. *British Journal of Sports Medicine* 43:669-676.

Corrado, D., A. Pelliccia, H. Heidbuchel, et al. 2010. Recommendation for interpretation of the 12-lead electrocardiogram in the athlete. *European Heart Journal* 31:243-259.

Guyton, A.C., and J.E. Hall. 2000. *Textbook of medical physiology,* 10th ed. Philadelphia: Saunders.

Stein, E. 1992a. *Rapid analysis of arrhythmias,* 2nd ed. Philadelphia: Lea & Febiger.

Stein, E. 1992b. *Rapid analysis of electrocardiograms,* 2nd ed. Philadelphia: Lea & Febiger.

Thaler, M. 1988. *The only ECG book you will ever need.* Philadelphia: Lippincott, Williams & Wilkins.

Chapter 6

Badeer, H.S., and J.W. Hicks. 1992. Hemodynamics of vascular "waterfall": Is the analogy justified? *Respiration Physiology* 87:205-217.

Berne, R.M., and M.N. Levy. 2001. *Cardiovascular physiology,* 8th ed. St. Louis: Mosby.

Chobanian, A.V., G.L. Bakris, H.R. Black, W.C. Cushman, L.A. Green, J.L. Izzo, D.W. Jones, B.J. Masterson, S. Oparil, J.T. Wright, E.J. Rocella, and National High Blood Pressure Education Program Coordinating Committee. 2003. The seventh report of the Joint National Committee on Prevention, Detection, Evaluation, and Treatment of High Blood Pressure. *Journal of the American Medical Association* 289:2560-2572.

Germann, W.J., and C.L. Stanfield. 2002. *Principles of human physiology.* San Francisco: Benjamin Cummings.

Guyton, A.C., and J.E. Hall. 2000. *Textbook of medical physiology,* 10th ed. Philadelphia: Saunders.

Hainsworth, R. 1995. Cardiovascular reflexes from ventricular and coronary receptors. *Advances in Experimental Medicine and Biology* 381:157-74.

Koller, A., D. Sun, A. Huang, and G. Kaley. 1994. Corelease of nitric oxide and prostaglandins mediates flow-dependent dilation of rat gracilis muscle arterioles. *American Journal of Physiology* 267:H326-H332.

Little, R.C. 1985. *Physiology of the heart and circulation.* Chicago: Year Book Medical.

Marieb, E.N. 2004. *Human anatomy and physiology,* 6th ed. San Francisco: Pearson/Benjamin Cummings.

McGill, J.B. 2009. Improving microvascular outcomes in patients with diabetes through management of hypertension. *Postgraduate Medicine* 121:89-101.

Mulvany, M.J., and C. Aalkjaer. 1990. Structure and function of small arteries. *Physiology Reviews* 70:921-961.

O'Rourke, M.F., and W.W. Nichols. 2007. Timing and amplitude of wave reflection. *Hypertension* 49:E3.

Perloff, D., C. Grim, J. Flack, E.D. Frolich, M. Hill, M. McDonald, and B.Z. Morgenstern. 1993. Human blood pressure determination by sphygmomanometry. *Circulation* 88:2460-2470.

Pickering, T.G., J.E. Hall, L.J. Appel, B.E. Falkner, J. Graves, M.N. Hill, D.W. Jones, T. Kurtz, S.G. Sheps, and E.J. Roccella. 2005. Recommendations for blood pressure measurement in humans and animals, part 1: Blood pressure measurement in humans. *Hypertension* 45:142-161.

Raven, P.B., J.T. Potts, X. Shi, and J. Pawelzyk. 2000. Baroreceptor-mediated reflex regulation of blood pressure during exercise. In: B. Saltin, R. Boushel, N. Secher, and J. Mitchell, *Exercise and circulation in health and disease.* Champaign, IL: Human Kinetics.

Ray, C.A., and M. Sito. 2000. The cardiopulmonary baroreflex. In: B. Saltin, R. Boushel, N. Secher, and J. Mitchell, *Exercise and circulation in health and disease.* Champaign, IL: Human Kinetics.

Remington, J.W., and L.J. O'Brien. 1970. Construction of aortic flow pulse from pressure pulse. *American Journal of Physiology* 218:437-447.

Rowell, L.B. 1986. *Human circulation: Regulation during physical stress.* New York: Oxford University Press.

Williams, C.A., and A.R. Lind. 1979. Measurement of forearm blood flow by venous occlusion plethysmography: Influence of hand blood flow during sustained and intermittent isometric exercise. *European Journal of Applied Physiology and Occupational Physiology* 42:141-149.

Chapter 7

Akosah KO, Schaper A, Cogbill C,Schoenfeld. 2003. Preventing myocardial infarction in the young adult in the first place: How do the national cholesterol education panel iii guidelines perform? *Journal of the American College of Cardiology* 4:1475-1479.

Celermajer, D.S., K.E. Sorenson, C. Bull, J. Robinson, and J.E. Deanfield. 1994. Endothelium-dependent dilation in the systemic arteries of asymptomatic subjects relates to coronary risk factors and their interaction. *Journal of the American College of Cardiology* 24:1468-1474.

DiCorleto, P.E., and P.L. Fox. 2005. Vascular endothelium. In: V. Fuster, E.J. Topol, and E.G. Nabel (eds.), *Atherothrombosis and coronary artery disease.* Philadelphia: Lippincott, Williams & Wilkins, 389-400.

Krogh, A. 1929. *The anatomy and physiology of capillaries.* New Haven, CT: Yale University Press.

Levick, J.R. 2003. *Introduction to cardiovascular physiology,* 4th ed. London, England: Arnold.

Libby, P. 2005. The vascular biology of atherosclerosis. In: D.P. Zipes, P. Libby, R.O. Bonow, and E. Braunwald (eds.), *Braunwald's heart disease: A textbook of cardiovascular medicine,* 7th ed. Philadelphia: Elsevier Saunders, 2067-2092.

Maiorana, A., G. O'Driscoll, R. Taylor, and D. Green. 2003. Exercise and the nitric oxide vasodilator system. *Sports Medicine* 33(14):1013-1035.

Neunteufl, T., S. Heher, R. Katzenschalager, G. Wolft, and G. Maurer. 2000. Long-term prognostic value of flow-mediated dilation in the brachial artery of patients with chest pain. *Journal of the American College of Cardiology* 86:207-210.

Zipes, D.P. and H.J.J. Wellens. 1998. Sudden cardiac death. *Circulation* 98:2334-2351.

Chapter 8

Colman, R.W., A.W. Clowes, J.N. George, J. Hirsh, and V.J. Marder. 2001. Overview of hemostasis. In: R.W. Coleman, J. Hirsh, V.J. Marder, A.W. Clowes, and J.N. George (eds.), *Hemostasis and thrombo-*

sis: Basic principles and clinical practice, 4th ed. Philadelphia: Lippincott, Williams & Wilkins, 3-16.

Konkle, B.A., and A.I. Schafer. 2005. Hemostasis, thrombosis, fibrinolysis and cardiovascular disease. In: D.P. Zipes, P. Libby, R.O. Bonow, and E. Braunwald (eds.), *Braunwald's heart disease: A textbook of cardiovascular medicine,* 7th ed. Philadelphia: Elsevier Saunders, 2067-2092.

Rao, G.H.R. 1999. Platelet physiology and pharmacology: An overview. In: G.H.R. Rao (ed.), *Handbook of platelet physiology and pharmacology.* Boston: Kluwer Academic, 1-20.

Ryningen, A., and H. Holmes. 1999. Biochemistry of platelet activation. In: G.H.R. Rao (ed.), *Handbook of platelet physiology and pharmacology.* Boston: Kluwer Academic, 188-237.

Chapter 9

Acil, T., E. Atalar, L. Sahiner, B. Kaya, I.C. Haznedaroglu, L. Tokgozoglu, K. Ovunc, K. Aytemir, N. Ozer, A. Oto, F. Ozmen, N. Nazli, S. Kes, and S. Aksoyek. 2007. Effects of acute exercise on fibrinolysis and coagulation in patients with coronary artery disease. *International Heart Journal* 48:277-285.

Alam, M., and F.H. Smirk. 1937. Observations in man upon a blood pressure raising reflex arising from the voluntary muscles. *Journal of Physiology* 89:372-383.

Andrew, M., C. Carter, H. O'Brodovich, and G. Heigenhauser. 1986. Increases in factor VIII complex and fibrinolytic activity are dependent on exercise intensity. *Journal of Applied Physiology* 60:1917-1922.

Arai, M., H. Yorifuji, S. Ikematsu, H. Nagasawa, M. Fujimaki, K. Fukutake, T. Katsumura, T. Ishii, and H. Iwane. 1990. Influences of strenuous exercise (triathlon) on blood coagulation and fibrinolytic system. *Thrombosis Research* 57:465-471.

Bartlett, I.S., and S.S. Segal. 2000. Resolution of smooth muscle and endothelial pathways for conduction along hamster cheek pouch. *American Journal of Physiology* 278:H604.

Bartsch, P., B. Welsch, M. Albert, B. Friedmann, M. Levi, and E.K. Kruithof. 1995. Balanced activation of coagulation and fibrinolysis after a 2-h triathlon. *Medicine and Science in Sports and Exercise* 27:1465-1470.

Cadroy, Y., F. Pillard, K.S. Sakariassen, C. Thalamas, B. Boneu, and D. Riviere. 2002. Strenuous but not moderate exercise increases the thrombotic tendency in healthy sedentary male volunteers. *Journal of Applied Physiology* 93:829-833.

Clifford, P.S., and Y. Hellsten. 2004. Vasodilatory mechanisms in contracting skeletal muscle. *Journal of Applied Physiology* 97:393-403.

Cohen, R.J., S.E. Epstein, and L.S. Cohen. 1968. Alterations in blood fibrinolysis and blood coagulation induced by exercise and the role of beta-adrenergic receptor stimulation. *Lancet* 2:1264.

Coyle, E.F., and J. Gonzalez-Alonso. 2001. Cardiovascular drift during prolonged exercise: New perspectives. *Exercise and Sport Sciences Reviews* 29:88-92.

Davies, P.F. 1995. Flow-mediated endothelial mechanotransduction. *Physiological Reviews* 75:519-560.

Degens, H., S. Salmons, and J.C. Jarvis. 1998. Intramuscular pressure, force and blood flow in rabbit tibialis anterior muscles during single and repetitive contractions. *European Journal of Applied Physiology and Occupational Physiology* 78:13-19.

Delp, M.D. 1999. Control of skeletal muscle perfusion at the onset of dynamic exercise. *Medicine and Science in Sports and Exercise* 31:1011-1018.

El-Sayed, M.S., Z. El-Sayed Ali, and S. Ahmadizad. 2004. Exercise and training effects on blood haemostasis in health and disease: An update. *Sports Medicine* 34:181-200.

Ely, M.R., S.N. Cheuvront, W.O. Roberts, and S.J. Montain. 2007. Impact of weather on marathon-running performance. *Medicine and Science in Sports and Exercise* 39:487-493.

Emerson, G.G., and S.S. Segal. 2000a. Electrical coupling between endothelial cells and smooth muscle cells in hamster feed arteries: Role in vasomotor control. *Circulation Research* 87:474-479.

———. 2000b. Endothelial cell pathway for conduction of hyperpolarization and vasodilation along hamster feed artery. *Circulation Research* 86:94-100.

Ersoz, G., A.M. Zergeroglu, H. Ficicilar, H. Ozcan, P. Oztekin, S. Aytac, and S. Yavuzer. 2002. Effect of submaximal and incremental upper extremity exercise on platelet function and the role of blood shear stress. *Thrombosis Research* 108:297-301.

Ferguson, S., N. Gledhill, V.K. Jamnik, C. Wiebe, and N. Payne. 2001. Cardiac performance in endurance-trained and moderately active young women. *Medicine and Science in Sports and Exercise* 33:1114-1119.

Figures, W.R., L.M. Scearce, Y. Wachtfogel, J. Chen, R.F. Colman, and R.W. Colman. 1986. Platelet ADP receptor and alpha 2-adrenoreceptor interaction. Evidence for an ADP requirement for epinephrine-induced platelet activation and an influence of epinephrine on ADP binding. *Journal of Biological Chemistry* 261:5981-5986.

Fritzsche, R.G., T.W. Switzer, B.J. Hodgkinson, and E.F. Coyle. 1999. Stroke volume decline during prolonged exercise is influenced by the increase in heart rate. *Journal of Applied Physiology* 86:799-805.

Fuglevand, A.J., and S.S. Segal. 1997. Simulation of motor unit recruitment and microvascular unit perfusion: Spatial considerations. *Journal of Applied Physiology* 83:1223-1234.

Gledhill, N., D. Cox, and R. Jamnik. 1994. Endurance athletes' stroke volume does not plateau: Major advantage is diastolic function. *Medicine and Science in Sports and Exercise* 26:1116-1121.

Gonzalez-Alonso, J. 2008. Point: Stroke volume does/does not decline during exercise at maximal effort in healthy individuals. *Journal of Applied Physiology* 104:275-276; discussion 279-280.

Gorczynski, R.J., B. Klitzman, and B.R. Duling. 1978. Interrelations between contracting striated-muscle and precapillary microvessels. *American Journal of Physiology* 235:H494-H504.

Hamann, J.J., J.B. Buckwalter, and P.S. Clifford. 2004. Vasodilatation is obligatory for contraction-induced hyperaemia in canine skeletal muscle. *Journal of Physiology* 557:1013-1020.

Hamann, J.J., Z. Valic, J.B. Buckwalter, and P.S. Clifford. 2003. Muscle pump does not enhance blood flow in exercising skeletal muscle. *Journal of Applied Physiology* 94:6-10.

Harms, C.A., M.A. Babcock, S.R. McClaran, D.F. Pegelow, G.A. Nickele, W.B. Nelson, and J.A. Dempsey. 1997. Respiratory muscle work compromises leg blood flow during maximal exercise. *Journal of Applied Physiology* 82:1573-1583.

Hegde, S.S., A.H. Goldfarb, and S. Hegde. 2001. Clotting and fibrinolytic activity change during the 1 h after a submaximal run. *Medicine and Science in Sports and Exercise* 33:887-892.

Hester, R.L., A.C. Guyton, and B.J. Barber. 1982. Reactive and exercise hyperemia during high levels of adenosine infusion. *American Journal of Physiology* 243:H181-H186.

Higginbotham, M.B., K.G. Morris, R.S. Williams, P.A. McHale, R.E. Coleman, and F.R. Cobb. 1986. Regulation of stroke volume during submaximal and maximal upright exercise in normal man. *Circulation Research* 58:281-291.

Hjemdahl, P., P.T. Larsson, and N.H. Wallen. 1991. Effects of stress and beta-blockade on platelet function. *Circulation* 84:VI44-VI61.

Honig, C.R., C.L. Odoroff, and J.L. Frierson. 1980. Capillary recruitment in exercise: Rate, extent, uniformity, and relation to blood flow. *American Journal of Physiology* 238:H31-H42.

Jilma, B., E. Dirnberger, H.G. Eichler, B. Matulla, L. Schmetterer, S. Kapiotis, W. Speiser, and O.F. Wagner. 1997. Partial blockade of nitric oxide synthase blunts the exercise-induced increase of von willebrand factor antigen and of factor VIII in man. *Thrombosis and Haemostasis* 78:1268-1271.

Johnson, J.M., and M.K. Park. 1981. Effect of upright exercise on threshold for cutaneous vasodilation and sweating. *Journal of Applied Physiology* 50:814-818.

Jones, C.J., D.V. DeFily, J.L. Patterson, and W.M. Chilian. 1993. Endothelium-dependent relaxation competes with alpha 1- and alpha 2-adrenergic constriction in the canine epicardial coronary microcirculation. *Circulation* 87:1264-1274.

Kanstrup, I., and B. Ekblom. 1978. Influence of age and physical activity on central hemodynamics and lung function in active adults. *Journal of Applied Physiology* 45:709.

Kellogg, D.L. Jr., J.M. Johnson, and W.A. Kosiba. 1991a. Competition between cutaneous active vasoconstriction and active vasodilation during exercise in humans. *American Journal of Physiology* 261:H1184-H1189.

———. 1991b. Control of internal temperature threshold for active cutaneous vasodilation by dynamic exercise. *Journal of Applied Physiology* 71:2476-2482.

Kenney, W.L., C.G. Tankersley, D.L. Newswanger, and S.M. Puhl. 1991. Alpha 1-adrenergic blockade does not alter control of skin blood flow during exercise. *American Journal of Physiology* 260:H855-H861.

Kingwell, B.A., K.L. Berry, J.D. Cameron, G.L. Jennings, and A.M. Dart. 1997. Arterial compliance increases after moderate-intensity cycling. *American Journal of Physiology* 273:H2186-H2191.

Krip, B., N. Gledhill, V. Jamnik, and D. Warburton. 1997. Effect of alterations in blood volume on cardiac function during maximal exercise. *Medicine and Science in Sports and Exercise* 29:1469-1476.

Lee, K.W., A.D. Blann, J. Ingram, K. Jolly, G.Y. Lip, and BRUM Investigators. 2005. Incremental shuttle walking is associated with activation of haemostatic and haemorheological markers in patients with coronary artery disease: The birmingham rehabilitation uptake maximization study (BRUM). *Heart* 91:1413-1417.

Lin, H. and D.B. Young, 1995. Opposing effects of plasma epinephrine and norepinephrine on coronary Thrombosis in vivo. *Circulation* 91:1135-1142.

Little, T.L., E.C. Beyer, and B.R. Duling. 1995. Connexin 43 and connexin 40 gap junctional proteins are present in arteriolar smooth muscle and endothelium in vivo. *American Journal of Physiology* 268:H729-H739.

Maciel, B.C., L. Gallo Jr., J.A. Marin Neto, E.C. Lima Filho, and L.E. Martins. 1986. Autonomic nervous control of the heart rate during dynamic exercise in normal man. *Clinical Science* (Lond) 71:457-460.

Marshall, J.M., and H.C. Tandon. 1984. Direct observations of muscle arterioles and venules following contraction of skeletal muscle fibres in the rat. *Journal of Physiology* 350:447-459.

Martin, C.M., A. Beltran-Del-Rio, A. Albrecht, R.R. Lorenz, and M.J. Joyner. 1996. Local cholinergic mechanisms mediate nitric oxide-dependent flow-induced vasorelaxation in vitro. *American Journal of Physiology* 270:H442-H446.

McCloskey, D.I., and J.H. Mitchell. 1972. Reflex cardiovascular and respiratory responses originating in exercising muscle. *Journal of Physiology* 224:173-186.

Mo, M., S.G. Eskin, and W.P. Schilling. 1991. Flow-mediated changes in Ca+2 signaling of vascular endothelial cells: Effect of shear stress on ATP. *American Journal of Physiology* 260:H1698-H1707.

Montain, S.J., and E.F. Coyle. 1992. Fluid ingestion during exercise increases skin blood flow independent of increases in blood volume. *Journal of Applied Physiology* 73:903-910.

Montain, S.J., M.R. Ely, and S.N. Cheuvront. 2007. Marathon performance in thermally stressing conditions. *Sports Medicine* 37:320-323.

Mortensen, S.P., E.A. Dawson, C.C. Yoshiga, M.K. Dalsgaard, R. Damsgaard, N.H. Secher, and J. Gonzalez-Alonso. 2005. Limitations to systemic and locomotor limb muscle oxygen delivery and uptake during maximal exercise in humans. *Journal of Physiology* 566:273-285.

Nadel, E.R., E. Cafarelli, M.F. Roberts, and C.B. Wenger. 1979. Circulatory regulation during exercise in different ambient temperatures. *Journal of Applied Physiology* 46:430-437.

Naka, K.K., A.C. Tweddel, D. Parthimos, A. Henderson, J. Goodfellow, and M.P. Frenneaux. 2003. Arterial distensibility: Acute changes following dynamic exercise in normal subjects. *American Journal of Physiology* 284:H970-H978.

Niebauer, J., and J.P. Cooke. 1996. Cardiovascular effects of exercise: Role of endothelial shear stress. *Journal of the American College of Cardiology* 28:1652-1660.

Noris, M., M. Morigi, R. Donadelli, S. Aiello, M. Foppollo, M. Todeschini, S. Orisio, G. Remuzzi, and A. Remuzzi. 1995. Nitric oxide synthesis by cultured endothelial cells is modulated by flow conditions. *Circulation* 76:536-543.

Petidis, K., S. Douma, M. Doumas, I. Basagiannis, K. Vogiatzis, and C. Zamboulis. 2008. The interaction of vasoactive substances during exercise modulates platelet aggregation in hypertension and coronary artery disease. *BMC Cardiovascular Disorders* 8:11.

Phair, R.D., and H.V. Sparks. 1979. Adenosine content of skeletal muscle during active hyperemia and ischemic contraction. *American Journal of Physiology* 237:H1-H9.

Plotnick, G.D., L.C. Becker, M.L. Fisher, G. Gerstenblith, D.G. Renlund, J.L. Fleg, M.L. Weisfeldt, and E.G. Lakatta. 1986. Use of the frank-starling mechanism during submaximal versus maximal upright exercise. *American Journal of Physiology* 251:H1101-H1105.

Plow, E.F., and G.A. Marguerie. 1980. Induction of the fibrinogen receptor on human platelets by epinephrine and the combination of epinephrine and ADP. *Journal of Biological Chemistry* 255:10971-10977.

Ralevic, V., P. Milner, O. Hudlicka, F. Kristek, and G. Burnstock. 1990. Substance P is released from the endothelium of normal and capsaicin-treated rat hind-limb vasculature, in vivo, by increased flow. *Circulation Research* 66:1178-1183.

Rosenmeier, J.B., F.A. Dinenno, S.J. Fritzlar, and M.J. Joyner. 2003. Alpha1- and alpha2-adrenergic vasoconstriction is blunted in contracting human muscle. *Journal of Physiology* 547:971-976.

Rotto, D.M., and M.P. Kaufman. 1988. Effect of metabolic products of muscular contraction on discharge of group III and IV afferents. *Journal of Applied Physiology* 64:2306-2313.

Rowell, L.B. 1993. *Human cardiovascular control.* New York: Oxford University Press.

Rowland, T., K. Heffernan, S.Y. Jae, G. Echols, and B.O. Fernhall. 2006. Tissue doppler assessment of ventricular function during cycling in 7- to 12-yr-old boys. *Medicine and Science in Sports and Exercise* 38:1216-1222.

Rowland, T.W., and M.W. Roti. 2004. Cardiac responses to progressive upright exercise in adult male cyclists. *Journal of Sports Medicine and Physical Fitness* 44:178-185.

Ruschitzka, F.T., G. Noll, and T.F. Luscher. 1997. The endothelium in coronary artery disease. *Cardiology* 88(Suppl 3):3-19.

Sakita, S., Y. Kishi, and F. Numano. 1997. Acute vigorous exercise attenuates sensitivity of platelets to nitric oxide. *Thrombosis Research* 87:461-471.

Segal, S.S., and T.L. Jacobs. 2001. Role for endothelial cell conduction in ascending vasodilatation and exercise hyperaemia in hamster skeletal muscle. *Journal of Physiology* 536:937-946.

Shattil, S.J., A. Budzynski, and M.C. Scrutton. 1989. Epinephrine induces platelet fibrinogen receptor expression, fibrinogen binding, and aggregation in whole blood in the absence of other excitatory agonists. *Blood* 73:150-158.

Sheriff, D.D., and A.L. Hakeman. 2001. Role of speed vs. grade in relation to muscle pump function at locomotion onset. *Journal of Applied Physiology* 91:269-276.

Sheriff, D.D., L.B. Rowell, and A.M. Scher. 1993. Is rapid rise in vascular conductance at onset of dynamic exercise due to muscle pump? *American Journal of Physiology* 265:H1227-H1234.

Shoemaker, J.K., M.E. Tschakovsky, and R.L. Hughson. 1998. Vasodilation contributes to the rapid hyperemia with rhythmic contractions in humans. *Canadian Journal of Physiology and Pharmacology* 76:418-427.

Sjøgaard, G., B. Kiens, K. Jorgensen, and B. Saltin. 1986. Intramuscular pressure, EMG and blood flow during low-level prolonged static contraction in man. *Acta Physiologica Scandinavica* 128:475-484.

Spina, R.J., T. Ogawa, W.H. Martin 3rd, A.R. Coggan, J.O. Holloszy, and A.A. Ehsani. 1992. Exercise training prevents decline in stroke volume during exercise in young healthy subjects. *Journal of Applied Physiology* 72:2458-2462.

Szymanski, L.M., and R.R. Pate. 1994. Effects of exercise intensity, duration, and time of day on fibrinolytic activity in physically active men. *Medicine and Science in Sports and Exercise* 26:1102-1108.

Szymanski, L.M., R.R. Pate, and J.L. Durstine. 1994. Effects of maximal exercise and venous occlusion on fibrinolytic activity in physically active and inactive men. *Journal of Applied Physiology* 77:2305-2310.

Taylor, W.F., J.M. Johnson, and W.A. Kosiba. 1990. Roles of absolute and relative load in skin vasoconstrictor responses to exercise. *Journal of Applied Physiology* 69:1131-1136.

Tschakovsky, M.E., A.M. Rogers, K.E. Pyke, N.R. Saunders, N. Glenn, S.J. Lee, T. Weissgerber, and E.M. Dwyer. 2004. Immediate exercise hyperemia in humans is contraction intensity dependent: Evidence for rapid vasodilation. *Journal of Applied Physiology* 96:639-644.

Tschakovsky, M.E., J.K. Shoemaker, and R.L. Hughson. 1996. Vasodilation and muscle pump contribution to immediate exercise hyperemia. *American Journal of Physiology* 271:H1697-H1701.

Tune, J.D., M.W. Gorman, and E.O. Feigl. 2004. Matching coronary blood flow to myocardial oxygen consumption. *Journal of Applied Physiology* 97:404-415.

Uematsu, M., Y. Ohara, J.P. Navas, K. Nishida, T.J. Murphy, R.W. Alexander, R.M. Nerem, and D.G. Harrison. 1995. Regulation of endothelial cell nitric oxide synthase mRNA expression by shear stress. *American Journal of Physiology* 269:C1371-C1378.

VanTeeffelen, J.W., and S.S. Segal. 2000. Effect of motor unit recruitment on functional vasodilatation in hamster retractor muscle. *Journal of Physiology* 524 Pt 1:267-278.

Walker, K.L., N.R. Saunders, D. Jensen, J.L. Kuk, S.L. Wong, K.E. Pyke, E.M. Dwyer, and M.E. Tschakovsky. 2007. Do vasoregulatory mechanisms in exercising human muscle compensate for changes in arterial perfusion pressure? *American Journal of Physiology* 293:H2928-H2936.

Wang, J.S., and L.J. Cheng. 1999. Effect of strenuous, acute exercise on alpha2-adrenergic agonist-potentiated platelet activation. *Arteriosclerosis, Thrombosis, and Vascular Biology* 19:1559-1565.

Weiss, C., G. Seitel, and P. Bartsch. 1998. Coagulation and fibrinolysis after moderate and very heavy exercise in healthy male subjects. *Medicine and Science in Sports and Exercise* 30:246-251.

Welsh, D.G., and S.S. Segal. 1997. Coactivation of resistance vessels and muscle fibers with acetylcholine release from motor nerves. *American Journal of Physiology* 273:H156-H163.

Wigmore, D.M., K. Propert, and J.A. Kent-Braun. 2005. Blood flow does not limit skeletal muscle force production during incremental isometric contractions. *European Journal of Applied Physiology* 96:370-378.

Wolfe, L.A., and D.A. Cunningham. 1982. Effects of chronic exercise on cardiac output and its determinants. *Canadian Journal of Physiology and Pharmacology* 60:1089-1097.

Zhou, B., R.K. Conlee, R. Jensen, G.W. Fellingham, J.D. George, and A.G. Fisher. 2001. Stroke volume does not plateau during graded exercise in elite male distance runners. *Medicine and Science in Sports and Exercise* 33:1849-1854.

Chapter 10

Armstrong, R.B., and M.H. Laughlin. 1984. Exercise blood flow patterns within and among rat muscles after training. *American Journal of Physiology* 246:H59-H68.

Bartsch, P. 1999. Platelet activation with exercise and risk of cardiac events. *Lancet* 354:1747-1748.

Baynard, T., H.M. Jacobs, C.M. Kessler, J.A. Kanaley, and B. Fernhall. 2007. Fibrinolytic markers and vasodilatory capacity following acute exercise among men of differing training status. *European Journal of Applied Physiology* 101:595-602.

Bloor, C.M. 2005. Angiogenesis during exercise and training. *Angiogenesis* 8:263-271.

Breisch, E.A., F.C. White, L.E. Nimmo, M.D. McKirnan, and C.M. Bloor. 1986. Exercise-induced cardiac hypertrophy: A correlation of blood flow and microvasculature. *Journal of Applied Physiology* 60:1259-1267.

Brown, M.D. 2003. Exercise and coronary vascular remodelling in the healthy heart. *Experimental Physiology* 88:645-658.

Charifi, N., F. Kadi, L. Feasson, F. Costes, A. Geyssant, and C. Denis. 2004. Enhancement of microvessel tortuosity in the vastus lateralis muscle of old men in response to endurance training. *Journal of Physiology* 554:559-569.

Chesley, A., G.J. Heigenhauser, and L.L. Spriet. 1996. Regulation of muscle glycogen phosphorylase activity following short-term endurance training. *American Journal of Physiology* 270:E328-E335.

Clarkson, P., H.E. Montgomery, M.J. Mullen, A.E. Donald, A.J. Powe, T. Bull, M. Jubb, M. World, and J.E. Deanfield. 1999. Exercise training enhances endothelial function in young men. *Journal of the American College of Cardiology* 33:1379-1385.

Clausen, J.P., K. Klausen, B. Rasmussen, and J. Trap-Jensen. 1973. Central and peripheral circulatory changes after training of the arms or legs. *American Journal of Physiology* 225:675-682.

Cohen, J.L., and K.R. Segal. 1985. Left ventricular hypertrophy in athletes: An exercise-echocardiographic study. *Medicine and Science in Sports and Exercise* 17:695-700.

Convertino, V.A. 1991. Blood volume: Its adaptation to endurance training. *Medicine and Science in Sports and Exercise* 23:1338-1348.

Convertino, V.A., P.J. Brock, L.C. Keil, E.M. Bernauer, and J.E. Greenleaf. 1980. Exercise training-induced hypervolemia: Role of plasma albumin, renin, and vasopressin. *Journal of Applied Physiology* 48:665-669.

Convertino, V.A., L.C. Keil, and J.E. Greenleaf. 1983. Plasma volume, rennin, and vasopressin responses to graded exercise after training. *Journal of Applied Physiology* 54:508-514.

Currens, J.H., and P.D. White. 1961. Half century of running: Clinical, physiological and autopsy findings in the case of Clarence de Mar, "Mr. Marathoner." *New England Journal of Medicine* 265:988-993.

Delp, M.D. 1998. Differential effects of training on the control of skeletal muscle perfusion. *Medicine and Science in Sports and Exercise* 30:361-374.

De Paz, J.A., J. Lasierra, J.G. Villa, E. Vilades, M.A. Martin-Nuno, and J. Gonzalez-Gallego. 1992. Changes in the fibrinolytic system associated with physical conditioning. *European Journal of Applied Physiology and Occupational Physiology* 65:388-393.

Dinenno, F.A., H. Tanaka, K.D. Monahan, C.M. Clevenger, I. Eskurza, C.A. DeSouza, and D.R. Seals. 2001. Regular endurance exercise induces expansive arterial remodelling in the trained limbs of healthy men. *Journal of Physiology* 534:287-295.

Edwards, D.G., R.S. Schofield, S.L. Lennon, G.L. Pierce, W.W. Nichols, and R.W. Braith. 2004. Effect of exercise training on endothelial function in men with coronary artery disease. *American Journal of Cardiology* 93:617-620.

El-Sayed, M.S., X. Lin, and A.J. Rattu. 1995. Blood coagulation and fibrinolysis at rest and in response to maximal exercise before and after a physical conditioning programme. *Blood Coagulation and Fibrinolysis* 6:747-752.

Ferguson, E.W., L.L. Bernier, G.R. Banta, J. Yu-Yahiro, and E.B. Schoomaker. 1987. Effects of exercise and conditioning on clotting and fibrinolytic activity in men. *Journal of Applied Physiology* 62:1416-1421.

Ferguson, S., N. Gledhill, V.K. Jamnik, C. Wiebe, and N. Payne. 2001. Cardiac performance in endurance-trained and moderately active young women. *Medicine and Science in Sports and Exercise* 33:1114-1119.

Fisslthaler, B., S. Dimmeler, C. Hermann, R. Busse, and I. Fleming. 2000. Phosphorylation and activation of the endothelial nitric oxide synthase by fluid shear stress. *Acta Physiologica Scandinavica* 168:81-88.

Gallo Jr., L., B.C. Maciel, J.A. Marin-Neto, and L.E. Martins. 1989. Sympathetic and parasympathetic changes in heart rate control during dynamic exercise induced by endurance training in man. *Brazilian Journal of Medical and Biological Research* 22:631-643.

Glagov, S., E. Weisenberg, C.K. Zarins, R. Stankunavicius, and G.J. Kolettis. 1987. Compensatory enlargement of human atherosclerotic coronary arteries. *New England Journal of Medicine* 316:1371-1375.

Gledhill, N., D. Cox, and R. Jamnik. 1994. Endurance athletes' stroke volume does not plateau: Major advantage is diastolic function. *Medicine and Science in Sports and Exercise* 26:1116-1121.

Goto, C., Y. Higashi, M. Kimura, K. Noma, K. Hara, K. Nakagawa, M. Kawamura, K. Chayama, M. Yoshizumi, and I. Nara. 2003. Effect of different intensities of exercise on endothelium-dependent vasodilation in humans: Role of endothelium-dependent nitric oxide and oxidative stress. *Circulation* 108:530-535.

Hambrecht, R., V. Adams, S. Erbs, A. Linke, N. Krankel, Y. Shu, Y. Baither, S. Gielen, H. Thiele, J.F. Gummert, F.W. Mohr, and G. Schuler. 2003. Regular physical activity improves endothelial function in patients with coronary artery disease by increasing phosphorylation of endothelial nitric oxide synthase. *Circulation* 107:3152-3158.

Hammond, H.K., and V.F. Froelicher. 1985. The physiologic sequelae of chronic dynamic exercise. *Medical Clinics of North America* 69:21-39.

Heinicke, K., B. Wolfarth, P. Winchenbach, B. Biermann, A. Schmid, G. Huber, B. Friedmann, and W. Schmidt. 2001. Blood volume and hemoglobin mass in elite athletes of different disciplines. *International Journal of Sports Medicine* 22:504-512.

Hepple, R.T., S.L. Mackinnon, J.M. Goodman, S.G. Thomas, and M.J. Plyley. 1997. Resistance and aerobic training in older men: Effects on VO2peak and the capillary supply to skeletal muscle. *Journal of Applied Physiology* 82:1305-1310.

Higashi, Y., S. Sasaki, S. Kurisu, A. Yoshimizu, N. Sasaki, H. Matsuura, G. Kajiyama, and T. Oshima. 1999. Regular aerobic exercise augments endothelium-dependent vascular relaxation in normotensive as well as hypertensive subjects: Role of endothelium-derived nitric oxide. *Circulation* 100:1194-1202.

Hoppeler, H., H. Howald, K. Conley, S.L. Lindstedt, H. Claassen, P. Vock, and E.R. Weibel. 1985. Endurance training in humans: Aerobic capacity and structure of skeletal muscle. *Journal of Applied*

Physiology 59:320-327.

Huonker, M., A. Schmid, A. Schmidt-Trucksass, D. Grathwohl, and J. Keul. 2003. Size and blood flow of central and peripheral arteries in highly trained able-bodied and disabled athletes. *Journal of Applied Physiology* 95:685-691.

Imai, K., H. Sato, M. Hori, H. Kusuoka, H. Ozaki, H. Yokoyama, H. Takeda, M. Inoue, and T. Kamada. 1994. Vagally mediated heart rate recovery after exercise is accelerated in athletes but blunted in patients with chronic heart failure. *Journal of the American College of Cardiology* 24:1529-1535.

Kalliokoski, K.K., V. Oikonen, T.O. Takala, H. Sipila, J. Knuuti, and P. Nuutila. 2001. Enhanced oxygen extraction and reduced flow heterogeneity in exercising muscle in endurance-trained men. *American Journal of Physiology* 280:E1015-E1021.

Katona, P.G., M. McLean, D.H. Dighton, and A. Guz. 1982. Sympathetic and parasympathetic cardiac control in athletes and nonathletes at rest. *Journal of Applied Physiology* 52:1652-1657.

Kelley, G., and Z.V. Tran. 1995. Aerobic exercise and normotensive adults: A meta-analysis. *Medicine and Science in Sports and Exercise* 27:1371-1377.

Kenney, W.L. 1985. Parasympathetic control of resting heart rate: Relationship to aerobic power. *Medicine and Science in Sports and Exercise* 17:451-455.

Kozakova, M., F. Galetta, L. Gregorini, G. Bigalli, F. Franzoni, C. Giusti, and C. Palombo. 2000. Coronary vasodilator capacity and epicardial vessel remodeling in physiological and hypertensive hypertrophy. *Hypertension* 36:343-349.

Krip, B., N. Gledhill, V. Jamnik, and D. Warburton. 1997. Effect of alterations in blood volume on cardiac function during maximal exercise. *Medicine and Science in Sports and Exercise* 29:1469-1476.

Laughlin, M.H., J.S. Pollock, J.F. Amann, M.L. Hollis, C.R. Woodman, and E.M. Price. 2001. Training induces nonuniform increases in eNOS content along the coronary arterial tree. *Journal of Applied Physiology* 90:501-510.

Libby, P., P.M. Ridker, and A. Maseri. 2002. Inflammation and atherosclerosis. *Circulation* 105:1135-1143.

Libby, P., and P. Theroux. 2005. Pathophysiology of coronary artery disease. *Circulation* 111:3481-3488.

Lloyd, P.G., B.M. Prior, H. Li, H.T. Yang, and R.L. Terjung. 2005. VEGF receptor antagonism blocks arteriogenesis, but only partially inhibits angiogenesis, in skeletal muscle of exercise-trained rats. *American Journal of Physiology* 88:H759-H768.

Maciel, B.C., L. Gallo Jr., J.A. Marin Neto, E.C. Lima Filho, J.T. Filho, and J.C. Manco. 1985. Parasympathetic contribution to bradycardia induced by endurance training in man. *Cardiovascular Research* 19:642-648.

Maeda, S., T. Miyauchi, T. Kakiyama, J. Sugawara, M. Iemitsu, Y. Irukayama-Tomobe, H. Murakami, Y. Kumagai, S. Kuno, and M. Matsuda. 2001. Effects of exercise training of 8 weeks and detraining on plasma levels of endothelium-derived factors, endothelin-1 and nitric oxide, in healthy young humans. *Life Science* 69:1005-1016.

Mo, M., S.G. Eskin, and W.P. Schilling. 1991. Flow-mediated changes in Ca+2 signaling of vascular endothelial cells: Effect of shear stress on ATP. *American Journal of Physiology* 260:H1698-H1707.

Muller, J.M., P.R. Myers, and M.H. Laughlin. 1994. Vasodilator responses of coronary resistance arteries of exercise-trained pigs. *Circulation* 89:2308-2314.

Nakano, T., R. Tominaga, I. Nagano, H. Okabe, and H. Yasui. 2000. Pulsatile flow enhances endothelium-derived nitric oxide release in the peripheral vasculature. *American Journal of Physiology* 278:H1098-H1104.

Oltman, C.L., J.L. Parker, H.R. Adams, and M.H. Laughlin. 1992. Effects of exercise training on vasomotor reactivity of porcine coronary arteries. *Journal of Physiology* 263:H372-H382.

Oltman, C.L., J.L. Parker, and M.H. Laughlin. 1995. Endothelium-dependent vasodilation of proximal coronary arteries from exercise-trained pigs. *Journal of Applied Physiology* 79:33-40.

Oscai, L.B., B.T. Williams, and B.A. Hertig. 1968. Effect of exercise on blood volume. *Journal of Applied Physiology* 24:622-624.

Parker, J.L., C.L. Oltman, J.M. Muller, P.R. Myers, H.R. Adams, and M.H. Laughlin. 1994. Effects of exercise training on regulation of tone in coronary arteries and arterioles. *Medicine and Science in Sports and Exercise* 26:1252-1261.

Pescatello, L.S., B.A. Franklin, R. Fagard, W.B. Farquhar, G.A. Kelley, C.A. Ray, and American College of Sports Medicine. 2004. American college of sports medicine position stand. Exercise and hypertension. *Medicine and Science in Sports and Exercise* 36:533-553.

Plowman, S.A., and D.L. Smith. 2008. *Exercise physiology for health, fitness, and performance.* Baltimore: Lippincott, Williams, & Wilkins.

Pluim, B.M., A.H. Zwinderman, A. van der Laarse, and E.E. van der Wall. 2000. The athlete's heart. A meta-analysis of cardiac structure and function. *Circulation* 101:336-344.

Poole, D.C., O. Mathieu-Costello, and J.B. West. 1989. Capillary tortuosity in rat soleus muscle is not affected by endurance training. *American Journal of Physiology* 256:H1110-H1116.

Prior, B.M., P.G. Lloyd, H.T. Yang, and R.L. Terjung. 2003. Exercise-induced vascular remodeling. *Exercise and Sport Sciences Reviews* 31:26-33.

Proctor, D.N., J.D. Miller, N.M. Dietz, C.T. Minson, and M.J. Joyner. 2001. Reduced submaximal leg blood flow after high-intensity aerobic training. *Journal of Applied Physiology* 91:2619-2627.

Putman, C.T., N.L. Jones, E. Hultman, M.G. Hollidge-Horvat, A. Bonen, D.R. McConachie, and G.J. Heigenhauser. 1998. Effects of short-term submaximal training in humans on muscle metabolism in exercise. *American Journal of Physiology* 275:E132-E139.

Roca, J., A.G. Agusti, A. Alonso, D.C. Poole, C. Viegas, J.A. Barbera, R. Rodriguez-Roisin, A. Ferrer, and P.D. Wagner. 1992. Effects of training on muscle O2 transport at VO2max. *Journal of Applied Physiology* 73:1067-1076.

Rowland, T.W., and M.W. Roti. 2004. Cardiac responses to progressive upright exercise in adult male cyclists. *Journal of Sports Medicine and Physical Fitness* 44:178-185.

Ruschitzka, F.T., G. Noll, and T.F. Luscher. 1997. The endothelium in coronary artery disease. *Cardiology* 88(Suppl 3):3-19.

Sakuragi, S., and Y. Sugiyama. 2006. Effects of daily walking on subjective symptoms, mood and autonomic nervous function. *Journal of Physiological Anthropology* 25:281-289.

Schmidt, W., K. Heinicke, J. Rojas, J. Manuel Gomez, M. Serrato, M. Mora, B. Wolfarth, A. Schmid, and J. Keul. 2002. Blood volume and hemoglobin mass in endurance athletes from moderate altitude. *Medicine and Science in Sports and Exercise* 34:1934-1940.

Schmidt-Trucksass, A., A. Schmid, B. Dorr, and M. Huonker. 2003. The relationship of left ventricular to femoral artery structure in male athletes. *Medicine and Science in Sports and Exercise* 35:214-219; discussion 220.

Scott, A.S., A. Eberhard, D. Ofir, G. Benchetrit, T.P. Dinh, P. Calabrese, V. Lesiuk, and H. Perrault. 2004. Enhanced cardiac vagal efferent activity does not explain training-induced bradycardia. *Autonomic Neuroscience: Basic and Clinical* 112:60-68.

Sessa, W.C., K. Pritchard, N. Seyedi, J. Wang, and T.H. Hintze. 1994. Chronic exercise in dogs increases coronary vascular nitric oxide production and endothelial cell nitric oxide synthase gene expression. *Circulation Research* 74:349-353.

Shi, X., G.H. Stevens, B.H. Foresman, S.A. Stern, and P.B. Raven. 1995. Autonomic nervous system control of the heart: Endurance exercise training. *Medicine and Science in Sports and Exercise* 27:1406-1413.

Shin, K., H. Minamitani, S. Onishi, H. Yamazaki, and M. Lee. 1997. Autonomic differences between athletes and nonathletes: Spectral analysis approach. *Medicine and Science in Sports and Exercise* 29:1482-1490.

Shoemaker, J.K., H.J. Green, M. Ball-Burnett, and S. Grant. 1997. Relationships between fluid and electrolyte hormones and plasma volume during exercise with training and detraining. *Medicine and Science in Sports and Exercise* 30:497-505.

Smith, M.L., D.L. Hudson, H.M. Graitzer, and P.B. Raven. 1989. Exercise training bradycardia: The role of autonomic balance. *Medicine and Science in Sports and Exercise* 21:40-44.

Speiser, W., W. Langer, A. Pschaick, E. Selmayr, B. Ibe, P.E. Nowacki, and G. Muller-Berghaus. 1988. Increased blood fibrinolytic activity after physical exercise: Comparative study in individuals with different sporting activities and in patients after myocardial infarction taking part in a rehabilitation sports program. *Thrombosis Research* 51:543-555.

Spier, S.A., M.D. Delp, C.J. Meininger, A.J. Donato, M.W. Ramsey, and J.M. Muller-Delp. 2004. Effects of ageing and exercise training on endothelium-dependent vasodilatation and structure of rat skeletal muscle arterioles. *Journal of Physiology* 556:947-958.

Spina, R.J., M.M. Chi, M.G. Hopkins, P.M. Nemeth, O.H. Lowry, and J.O. Holloszy. 1996. Mitochondrial enzymes increase in muscle in response to 7-10 days of cycle exercise. *Journal of Applied Physiology* 80:2250-2254.

Starritt, E.C., D. Angus, and M. Hargreaves. 1999. Effect of short-term training on mitochondrial ATP production rate in human skeletal muscle. *Journal of Applied Physiology* 86:450-454.

Sun, D., A. Huang, A. Koller, and G. Kaley. 1994. Short-term daily exercise activity enhances endothelial NO synthesis in skeletal muscle arterioles of rats. *Journal of Applied Physiology* 76:2241-2247.

Suzuki, T., K. Yamauchi, Y. Yamada, T. Furumichi, H. Furui, J. Tsuzuki, H. Hayashi, I. Sotobata, and H. Saito. 1992. Blood coagulability and fibrinolytic activity before and after physical training during the recovery phase of acute myocardial infarction. *Clinical Cardiology* 15:358-364.

Szymanski, L.M., R.R. Pate, and J.L. Durstine. 1994. Effects of maximal exercise and venous occlusion on fibrinolytic activity in physically active and inactive men. *Journal of Applied Physiology* 77:2305-2310.

Uematsu, M., Y. Ohara, J.P. Navas, K. Nishida, T.J. Murphy, R.W. Alexander, R.M. Nerem, and D.G. Harrison. 1995. Regulation of endothelial cell nitric oxide synthase mRNA expression by shear stress. *American Journal of Physiology* 269:C1371-C1378.

Van den Burg, P.J., J.E. Hospers, M. van Vliet, W.L. Mosterd, B.N. Bouma, and I.A. Huisveld. 1997. Effect of endurance training and seasonal fluctuation on coagulation and fibrinolysis in young sedentary men. *Journal of Applied Physiology* 82:613-620.

Van Hoof, R., P. Hespel, R. Fagard, P. Lijnen, J. Staessen, and A. Amery. 1989. Effect of endurance training on blood pressure at rest, during exercise and during 24 hours in sedentary men. *American Journal of Cardiology* 63:945-949.

Vinereanu, D., N. Florescu, N. Sculthorpe, A.C. Tweddel, M.R. Stephens, and A.G. Fraser. 2001. Differentiation between pathologic and physiologic left ventricular hypertrophy by tissue doppler assessment of long-axis function in patients with hypertrophic cardiomyopathy or systemic hypertension and in athletes. *American Journal of Cardiology* 88:53.

Wang, J.S., C.J. Jen, and H.I. Chen. 1995. Effects of exercise training and deconditioning on platelet function in men. *Arteriosclerosis, Thrombosis, and Vascular Biology* 15:1668-1674.

———. 1997. Effects of chronic exercise and deconditioning on platelet function in women. *Journal of Applied Physiology* 83:2080-2085.

Wang, J., M.S. Wolin, and T.H. Hintze. 1993. Chronic exercise enhances endothelium-mediated dilation of epicardial coronary artery in conscious dogs. *Circulation Research* 73:829-838.

Warburton, D.E., M.J. Haykowsky, H.A. Quinney, D. Blackmore, K.K. Teo, D.A. Taylor, J. McGavock, and D.P. Humen. 2004. Blood volume expansion and cardiorespiratory function: Effects of training modality. *Medicine and Science in Sports and Exercise* 36:991-1000.

Weight, L.M., M. Klein, T.D. Noakes, and P. Jacobs. 1992. "Sports anemia"—a real or apparent phenomenon in endurance-trained athletes? *International Journal of Sports Medicine* 13:344-347.

White, F.C., C.M. Bloor, M.D. McKirnan, and S.M. Carroll. 1998. Exercise training in swine promotes growth of arteriolar bed and capillary angiogenesis in heart. *Journal of Applied Physiology* 85:1160-1168.

Wilmore, J.H., P.R. Stanforth, J. Gagnon, T. Rice, S. Mandel, A.S. Leon, D.C. Rao, J.S. Skinner, and C. Bouchard. 2001. Heart rate and blood pressure changes with endurance training: The HERITAGE family study. *Medicine and Science in Sports and Exercise* 33:107-116.

Wolfe, L.A., and D.A. Cunningham. 1982. Effects of chronic exercise on cardiac output and its determinants. *Canadian Journal of Physiology and Pharmacology* 60:1089-1097.

Womack, C.J., P.R. Nagelkirk, and A.M. Coughlin. 2003. Exercise-induced changes in coagulation and fibrinolysis in healthy populations and patients with cardiovascular disease. *Sports Medicine* 33:795-807.

Woodman, C.R., M.A. Thompson, J.R. Turk, and M.H. Laughlin. 2005. Endurance exercise training improves endothelium-dependent relaxation in brachial arteries from hypercholesterolemic male pigs. *Journal of Applied Physiology* 99:1412-1421.

Woodman, C.R., J.R. Turk, D.P. Williams, and M.H. Laughlin. 2003. Exercise training preserves endothelium-dependent relaxation in brachial arteries from hyperlipidemic pigs. *Journal of Applied Physiology* 94:2017-2026.

Zhou, B., R.K. Conlee, R. Jensen, G.W. Fellingham, J.D. George, and A.G. Fisher. 2001. Stroke volume does not plateau during graded exercise in elite male distance runners. *Medicine and Science in Sports and Exercise* 33:1849-1854.

Chapter 11

Ahmadizad, S., and M.S. El-Sayed. 2005. The acute effects of resistance exercise on the main determinants of blood rheology. *Journal of Sports Sciences* 23(3):243-249.

———. 2003. The effects of graded resistance exercise on platelet aggregation and activation. *Medicine and Science in Sports and Exercise* 35(6):1026-1032.

Ahmadizad, S., M.S. El-Sayed, and D.P.M. Maclaren. 2006. Responses of platelet activation and function to a single bout of resistance exercise and recovery. *Clinical Hemorheology and Microcirculation* 35(1-2):159-168.

Bartsch, P. 1999. Platelet activations with exercise and risk of cardiac events. *Lancet* 354:1747-1748.

Baynard, T., H.M. Jacobs, C.M. Kessler, J.A. Kanaley, and B. Fernhall. 2007. Fibrinolytic markers and vasodilatory capacity following acute exercise among men of differing training status. *European Journal of Applied Physiology* 101(5):595-602.

Baynard, T., W.C. Miller, and B. Fernhall. 2003. Effects of exercise on vasodilatory capacity in endurance- and resistance-trained men. *European Journal of Applied Physiology* 89(1):69-73.

Braith, R.W., and K.J. Stewart. 2006. Resistance exercise training: Its role in the prevention of cardiovascular disease. *Circulation* 113(22):2642-2650.

Chamberlain, K.G., M. Tong, and D.G. Penington. 1990. Properties of the exchangeable splenic platelets released into the circulation during exercise-induced thrombocytosis. *American Journal of Hematology* 34(3):161-168.

Collins, M.A., K.J. Cureton, D.W. Hill, and C.A. Ray. 1989. Relation of plasma volume change to intensity of weight lifting. *Medicine and Science in Sports and Exercise* 21(2):178-185.

Collins, M.A., D.W. Hill, K.J. Cureton, and J.J. DeMello. 1986. Plasma volume change during heavy-resistance weight lifting. *European Journal of Applied Physiology and Occupational Physiology* 55(1):44-48.

Craig, S.K., W.C. Byrnes, and S.J. Fleck. 2008. Plasma volume during weight lifting. *International Journal of Sports Medicine* 29(2):89-95.

Davies, P.F. 1995. Flow-mediated endothelial mechanotransduction. *Physiological Reviews* 75(3):519-560.

Davies, P.F., J.A. Spaan, and R. Krams. 2005. Shear stress biology of the endothelium. *Annals of Biomedical Engineering* 33(12):1714-1718.

DeVan, A.E., M.M. Anton, J.N. Cook, D.B. Neidre, M.Y. Cortez-Cooper, and H. Tanaka. 2005. Acute effects of resistance exercise on arterial compliance. *Journal of Applied Physiology* 98(6):2287-2291.

de Vos, N.J., N.A. Singh, D.A. Ross, T.M. Stavrinos, R. Orr, and M.A. Fiatarone Singh. 2008. Continuous hemodynamic response to maximal dynamic strength testing in older adults. *Archives of Physical Medicine and Rehabilitation* 89(2):343-350.

El-Sayed, M.S. 1993. Fibrinolytic and hemostatic parameter response after resistance exercise. *Medicine and Science in Sports and Exercise* 25(5):597-602.

El-Sayed, M.S., C. Sale, P.G. Jones, and M. Chester. 2000. Blood hemostasis in exercise and training. *Medicine and Science in Sports and Exercise* 32(5):918-925.

El-Sayed, M.S., Z. El-Sayed Ali, and S. Ahmadizad. 2004. Exercise and Training Effects of Blood Haemostatis in Health and Disease. *Sports Medicine* 34 (3): 181-200.

Elstad, M., I.H. Nådland, K. Toska, and L. Walløe. 2009. Stroke volume decreases during mild dynamic and static exercise in supine humans. *Acta Physiologica* 195(2):289-300.

Fahs, C.A., K.S. Heffernan, and B. Fernhall. 2009. Hemodynamic and vascular response to resistance exercise with L-arginine. *Medicine and Science in Sports and Exercise* 41(4):773-779.

Falkel, J.E., S.J. Fleck, and T.F. Murray. 1992. Comparison of central hemodynamics between powerlifters and bodybuilders during resistance exercise. *Journal of Applied Sport Science Research* 6(1):24-35.

Featherstone, J.F., R.G. Holly, and E.A. Amsterdam. 1993. Physiologic responses to weight lifting in coronary artery disease. *American Journal of Cardiology* 71(4):287-292.

Fleck, S.J., and L.S. Dean. 1987. Resistance-training experience and the pressor response during resistance exercise. *Journal of Applied Physiology* 63(1):116-120.

Gaffney, F.A., G. Sjøgaard, and B. Saltin. 1990. Cardiovascular and metabolic responses to static contraction in man. *Acta Physiologica Scandinavica* 138(3):249-258.

Gonzales, F., M. Mañas, I. Seiquer, J. Quiles, F.J. Mataix, J.R. Huertas, and E. Martinez-Victoria. 1996. Blood platelet function in healthy individuals of different ages. effects of exercise and exercise conditioning. *Journal of Sports Medicine and Physical Fitness* 36(2):112-116.

Haram, P.M., O.J. Kemi, and U. Wisloff. 2008. Adaptation of endothelium to exercise training: Insights from experimental studies. *Frontiers in Bioscience* 13:336-346.

Haykowsky, M.J., R. Dressendorfer, D. Taylor, S. Mandic, and D. Humen. 2002. Resistance training and cardiac hypertrophy: Unravelling the training effect. *Sports Medicine* 32(13):837-849.

Haykowsky, M., T. Dylan, T. Koon, A. Quinney, and H. Dennis. 2001. Left ventricular wall stress during leg-press exercise performed with a brief valsalva maneuver. *Chest* 119:150-154.

Haykowsky, M.J., N.D. Eves, D.E.R. Warburton, and M.J. Findlay. 2003. Resistance exercise, the valsalva maneuver, and cerebrovascular transmural pressure. *Medicine and Science in Sports and Exercise* 35(1):65-68.

Heffernan, K.S., S.R. Collier, E.E. Kelly, S.Y. Jae, and B. Fernhall. 2007a. Arterial stiffness and baroreflex sensitivity following bouts of aerobic and resistance exercise. *International Journal of Sports Medicine* 28(3):197-203.

Heffernan, K.S., D.G. Edwards, L. Rossow, S.Y. Jae, and B. Fernhall. 2007b. External mechanical compression reduces regional arterial stiffness. *European Journal of Applied Physiology* 101(6):735-741.

Heffernan, K.S., S.Y. Jae, D.G. Edwards, E.E. Kelly, and B. Fernhall. 2007c. Arterial stiffness following repeated valsalva maneuvers and resistance exercise in young men. *Applied Physiology, Nutrition, and Metabolism* 32(2):257-264.

Heffernan, K.S., L. Rossow, S.Y. Jae, H.G. Shokunbi, E.M. Gibson, and B. Fernhall. 2006. Effect of single-leg resistance exercise on regional arterial stiffness. *European Journal of Applied Physiology* 98(2):185-190.

Jurva, J.W., S.A. Phillips, A.Q. Syed, A.Y. Syed, S. Pitt, A. Weaver, and D.D. Gutterman. 2006. The effect of exertional hypertension evoked by weight lifting on vascular endothelial function. *Journal of the American College of Cardiology* 48(3):588-589.

Kalliokoski, K.K., H. Langberg, A.K. Ryberg, C. Scheede-Bergdahl, S. Doessing, A. Kjaer, M. Kjaer, and R. Boushel. 2006. Nitric oxide and prostaglandins influence local skeletal muscle blood flow during exercise in humans: Coupling between local substrate uptake and blood flow. *American Journal of Physiology: Regulatory, Integrative and Comparative Physiology* 291(3):R803-R809.

Kraemer, R.R., J.L. Kilgore, and G.R. Kraemer. 1993. Plasma volume changes in response to resistive exercise. *Journal of Sports Medicine and Physical Fitness* 33(3):246-251.

Lentini, A.C., R.S. McKelvie, N. McCartney, C.W. Tomlinson, and J.D. MacDougall. 1993. Left ventricular response in healthy young men during heavy-intensity weight-lifting exercise. *Journal of Applied Physiology* 75(6):2703-2710.

Lewis, S.F., P.G. Snell, W.F. Taylor, M. Hamra, R.M. Graham, W.A. Pettinger, and C.G. Blomqvist. 1985. Role of muscle mass and mode of contraction in circulatory responses to exercise. *Journal of Applied Physiology* 58(1):146-151.

MacDougall, J.D., R.S. McKelvie, D.E. Moroz, D.G. Sale, N. McCartney, and F. Buick. 1992. Factors affecting blood pressure during heavy weight lifting and static contractions. *Journal of Applied Physiology* 73(4):1590-1597.

MacDougall, J.D., D. Tuxen, D.G. Sale, J.R. Moroz, and J.R. Sutton. 1985. Arterial blood pressure response to heavy resistance exercise. *Journal of Applied Physiology* 58(3):785-790.

McCartney, N. 1999. Acute responses to resistance training and safety. *Medicine and Science in Sports and Exercise* 31(1):31-37.

Meyer, K., R. Hajric, S. Westbrook, S. Haag-Wildi, R. Holtkamp, D. Leyk, and K. Schnellbacher. 1999. Hemodynamic responses during leg press exercise in patients with chronic congestive heart failure. *American Journal of Cardiology* 83, (11):1537-1543.

Miles, D.S., J.J. Owens, J.C. Golden, and R.W. Gotshall. 1987a. Central and peripheral hemodynamics during maximal leg extension exercise. *European Journal of Applied Physiology and Occupational Physiology* 56(1):12-17.

Narloch, J.A., and M.E. Brandstater. 1995. Influence of breathing technique on arterial blood pressure during heavy weight lifting. *Archives of Physical Medicine and Rehabilitation* 76(5):457-462.

Nelson, R.R., F.L. Gobel, C.R. Jorgensen, K. Wang, Y. Wang, and H.L. Taylor. 1974. Hemodynamic predictors of myocardial oxygen consumption during static and dynamic exercise. *Circulation* 50(6):1179-1189.

Noris, M., M. Morigi, R. Donadelli, S. Aiello, M. Foppolo, M. Todeschini, S. Orisio, G. Remuzzi, and A. Remuzzi. 1995. Nitric oxide synthesis by cultured endothelial cells is modulated by flow conditions. *Circulation Research* 76(4):536-543.

O'Rourke, M.F., and M.E. Safar. 2005. Relationship between aortic stiffening and microvascular disease in brain and kidney: Cause and logic of therapy. *Hypertension* 46(1):200-204.

Ploutz-Snyder, L.L., V.A. Convertino, and G.A. Dudley. 1995. Resistance exercise-induced fluid shifts: Change in active muscle size and plasma volume. *American Journal of Physiology* 269(3):R536-R543.

Röcker, L., S. Günay, H.C. Gunga, W. Hopfenmüller, A. Ruf, H. Patscheke, and M. Möckel. 2000. Activation of blood platelets in response to maximal isometric exercise of the dominant arm. *International Journal of Sports Medicine* 21(3):191-194.

Sale, D.G., D.E. Moroz, R.S. McKelvie, J.D. MacDougall, and N. McCartney. 1993. Comparison of blood pressure response to isokinetic and weight-lifting exercise. *European Journal of Applied Physiology and Occupational Physiology* 67(2):115-120.

Sjøgaard, G., and B. Saltin. 1982. Extra- and intracellular water spaces in muscles of man at rest and with dynamic exercise. *American Journal of Physiology* 243(3):R271-R280.

Smith, J.E. 2003. Effects of strenuous exercise on haemostasis. *British Journal of Sports Medicine* 37(5):433-435.

Thompson, P.D., B.A. Franklin, G.J. Balady, S.N. Blair, D. Corrado, N.A. Mark Estes 3rd, J.E. Fulton, et al. 2007. Exercise and acute cardiovascular events placing the risks into perspective: A scientific statement from the American Heart Association Council on Nutrition, Physical Activity, and Metabolism and the Council on Clinical Cardiology. *Circulation* 115(17):2358-2368.

Vind, J., G. Gleerup, P.T. Nielsen, and K. Winther. 1993. The impact of static work on fibrinolysis and platelet function. *Thrombosis Research* 72(5):441-446.

Williams, M.A., W.L. Haskell, P.A. Ades, E.A. Amsterdam, V. Bittner, B.A. Franklin, M. Gulanick, S.T. Laing, and K.J. Stewart. 2007. Resistance exercise in individuals with and without cardiovascular disease: 2007 update: A scientific statement from the American Heart Association Council on Clinical Cardiology and Council on Nutrition, Physical Activity, and Metabolism. *Circulation* 116(5):572-584.

Chapter 12

Adler, Y., E.Z. Fisman, N. Koren-Morag, D. Tanne, J. Shemesh, E. Lasry, et al. 2008. Left ventricular diastolic function in trained male weight lifters at rest and during isometric exercise. *American Journal of Cardiology* 102:97-101.

Anton, M.M., M.Y. Cortez-Cooper, A.E. DeVan, D.B. Neidre, J.N. Cook, and H. Tanaka. 2006. Resistance training increases basal limb blood flow and vascular conductance in aging humans. *Journal of Applied Physiology* 101(5):1351-1355.

Baynard, T., W.C. Miller, and B. Fernhall. 2003. Effects of exercise on vasodilatory capacity in endurance- and resistance-trained men. *European Journal of Applied Physiology* 89(1):69-73.

Bertovic, D.A., T.K. Waddell, C.D. Gatzka, J.D. Cameron, A.M. Dart, and B.A. Kingwell. 1999. Muscular strength training is associated with low arterial compliance and high pulse pressure. *Hypertension* 33(6):1385-1391.

Braith, R.W., and K.J. Stewart. 2006. Resistance exercise training: Its role in the prevention of cardiovascular disease. *Circulation* 113(22):2642-2650.

Casey, D.P., D.T. Beck, and R.W. Braith. 2007. Progressive resistance training without volume increases does not alter arterial stiffness and aortic wave reflection. *Experimental Biology and Medicine* 232(9):1228-1235.

Casey, D.P., G.L. Pierce, K.S. Howe, M.C. Mering, and R.W. Braith. 2007. Effect of resistance training on arterial wave reflection and brachial artery reactivity in normotensive postmenopausal women. *European Journal of Applied Physiology* 100(4):403-408.

Collier, S.R., J.A. Kanaley, R. Carhart Jr., V. Frechette, M.M. Tobin, A.K. Hall, et al. 2008. Effect of 4 weeks of aerobic or resistance exercise training on arterial stiffness, blood flow and blood pressure in pre- and stage-1 hypertensives. *Journal of Human Hypertension* 22(10):678-686.

Cornelissen, V.A., and R.H. Fagard. 2005. Effect of resistance training on resting blood pressure: A meta-analysis of randomized controlled trials. *Journal of Hypertension* 23(2):251-259.

Cortez-Cooper, M.Y., A.E. DeVan, M.M. Anton, R.P. Farrar, K.A. Beckwith, J.S. Todd, and H. Tanaka. 2005. Effects of high intensity resistance training on arterial stiffness and wave reflection in women. *American Journal of Hypertension* 18(7):930-934.

El-Sayed, M.S., et al. 2005. Aggregation and activation of blood platelets in exercise and training. *Sports Medicine* 35:11-22.

Fagard, R.H. 2006. Exercise is good for your blood pressure: Effects of endurance training and resistance training. *Clinical and Experimental Pharmacology and Physiology* 33(9):853-856.

Fernhall, B., and S. Agiovlasitis. 2008. Arterial function in youth: Window into cardiovascular risk. *Journal of Applied Physiology* 105(1):325-333.

Fleck, S.J., and L.S. Dean. 1987. Resistance-training experience and the pressor response during resistance exercise. *Journal of Applied Physiology* 63(1):116-120.

Haennel, R.G., G.D. Snydmiller, K.K. Teo, P.V. Greenwood, H.A. Quinney, and C.T. Kappagoda. 1992. Changes in blood pressure and cardiac output during maximal isokinetic exercise. *Archives of Physical Medicine and Rehabilitation* 73(2):150-155.

Haykowsky, M.J., R. Dressendorfer, D. Taylor, S. Mandic, and D. Humen. 2002. Resistance training and cardiac hypertrophy: Unravelling the training effect. *Sports Medicine* 32(13):837-849.

Haykowsky, M., J. McGavock, I. Vonder Muhll, et al. 2005. Effect of exercise training on peak power, left ventricular morphology, and muscle strength in healthy older women. *Journals of Gerontology: Series A, Biological Sciences and Medical Sciences* 60:307-311.

Haykowsky, M., D. Taylor, K. Teo, A. Quinney, and D. Humen. 2001. Left ventricular wall stress during leg-press exercise performed with a brief valsalva maneuver. *Chest* 119:150-154.

Heffernan, K.S., C.A. Fahs, G.A. Iwamoto, S.Y. Jae, K.R. Wilund, J.A. Woods, B. Fernhall. 2009. Resistance exercise training reduces central blood pressure and improves microvascular function in African-American and white men. *Atherosclerosis* 207:220-226.

Heffernan, K.S., S. Y. Jae, G.H. Echols, N.R. Lepine, B. Fernhall. 2007. Aterial stiffness and wave reflection following exercise in resistance trained men. *Medicine and Science in Sports and Exercise* 39:842-849.

Heffernan, K.S., S.Y. Jae, M. Lee, J.A. Woods, and B. Fernhall. 2008. Arterial wave reflection and vascular autonomic modulation in young and older men. *Aging Clinical and Experimental Research* 20(1):1-7.

Kawano H, H. Tanaka, and M. Miyachi. 2006. Resistance training and arterial compliance: Keeping the benefits while minimizing the stiffening. *Journal of Hypertension* 24:1753-1759.

Kawano, H., M. Tanimoto, K. Yamamoto, K. Sanada, Y. Gando, I. Tabata, et al. 2008. Resistance training in men is associated with increased arterial stiffness and blood pressure but does not adversely affect endothelial function as measured by arterial reactivity to the cold pressor test. *Experimental Physiology* 93(2):296-302.

Kelley, G.A., and K.S. Kelley. 2000. Progressive resistance exercise and resting blood pressure. A meta-analysis of randomized controlled trials. *Hypertension* 35:838-843.

Lane, H.A., F. Grace, J.C. Smith, K. Morris, J. Cockcroft, M.F. Scanlon, and J.S. Davies. 2006. Impaired vasoreactivity in bodybuilders using androgenic anabolic steroids. *European Journal of Clinical Investigations* 36(7):483-488.

Longhurst, J.C., and C.L. Stebbins. 1992. The isometric athlete. *Cardiology Clinics* 10(2):281-294.

McCartney, N., R.S. McKelvie, J. Martin, D.G. Sale, and J.D. MacDougall. 1993. Weight training-induced attenuation of the circulatory response of older males to weight lifting. *Journal of Applied Physiology* 74(3):1056-1060.

McGuigan, M.R., R. Bronks, R.U. Newton, M.J. Sharman, J.C. Graham, D.V. Cody, et al. 2001. Resistance training in patients with peripheral arterial disease: Effects on myosin isoforms, fiber type distribution, and capillary supply to skeletal muscle. *Journals of Gerontology: Series A, Biological Sciences and Medical Science* 56(7):B302-B310.

Miyachi, M., A.J. Donato, K. Yamamoto, K. Takahashi, P.E. Gates, K.L. Moreau, et al. 2003. Greater age-related reductions in central arterial compliance in resistance-trained men. *Hypertension* 41(1):130-135.

Miyachi, M., H. Kawano, J. Sugawara, K. Takahashi, K. Hayashi, K. Yamazaki, et al. 2004. Unfavorable effects of resistance training on central arterial compliance: A randomized intervention study. *Circulation* 110(18):2858-2863.

Naylor, L.H., K. George, G. O'Driscoll, and D.J. Green. 2008. The athlete's heart: A contemporary appraisal of the "Morganroth hypothesis." *Sports Medicine* 38(1):69-90.

Okamoto, T., M. Masuhara, and K. Ikuta. 2006. Effects of eccentric and concentric resistance training on arterial stiffness. *Journal of Human Hypertension* 20(5):348-354.

Olson, T.P., D.R. Dengel, A.S. Leon, and K.H. Schmitz. 2006. Moderate resistance training and vascular health in overweight women. *Medicine and Science in Sports and Exercise* 38(9):1558-1564.

Pearson, A.C., M. Schiff, D. Mrosek, A.J. Leibovitz, and G.A. Williams. 1986. Left ventricular diastolic function in weight lifters. *American Journal of Cardiology* 58:1254-1259.

Pluim, B.M., A.H. Zwinderman, A. van der Laarse, and E.E. van der Wall. 2000. The athlete's heart: A meta-analysis of cardiac structure and function. *Circulation* 101:336-344.

Rakobowchuk, M., C.L. McGowan, P.C. de Groot, D. Bruinsma, J.W. Hartman, S.M. Phillips, and M.J. MacDonald. 2005. Effect of whole body resistance training on arterial compliance in young men. *Experimental Physiology* 90(4):645-651.

Rakobowchuk, M., C.L. McGowan, P.C. de Groot, J.W. Hartman, S.M. Phillips, and M.J. MacDonald. 2005. Endothelial function of young healthy males following whole body resistance training. *Journal of Applied Physiology* 98(6):2185-2190.

Sale, D.G., D.E. Moroz, R.S. McKelvie, J.D. MacDougall, and N. McCartney. 1994. Effect of training on the blood pressure response to weight lifting. *Canadian Journal of Applied Physiology* 19(1):60-74.

Snoeckx, L.H., H.F. Abeling, J.A. Lambregts, J.J. Schmitz, F.T. Verstappen, and R.S. Reneman. 1982. Echocardiographic dimensions in athletes in relation to their training programs. *Medicine and Science in Sports and Exercise* 14(6):428-434.

Taaffe, D.R., D.A. Galvao, J.E. Sharman, and J.S. Coombes. 2007. Reduced central blood pressure in older adults following progressive resistance training. *Journal of Human Hypertension* 21(1):96-98.

Thompson, P.D., B.A. Franklin, G.J. Balady, S.N. Blair, D. Corrado, N.A. Estes, et al. 2007. Exercise and acute cardiovascular events: Placing the risks into perspective: A scientific statement from the American Heart Association Council on Nutrition, Physical Activity, and Metabolism and the Council on Clinical Cardiology. *Circulation* 115(17):2358-2368.

Yoshizawa, M., S. Maeda, A. Miyaki, M. Misono, Y. Saito, K. Tanabe, S. Kuno, and R. Ajisaka. 2008. Effect of 12 weeks of moderate intensity resistance training on arterial stiffness: A randomized controlled trial in women aged 32-59. *British Journal of Sports Medicine* 43:615-618

Index

Note: The italicized *f* and *t* following pages numbers refer to figures and tables, respectively.

About the Authors

Denise L. Smith, PhD, is a professor in the department of health and exercise sciences and the class of 1961 term professor at Skidmore College. She also holds an appointment as a research scientist at the University of Illinois Fire Service Institute. She received her PhD from the University of Illinois in exercise physiology in 1990. For nearly two decades, Smith has conducted scientific research on cardiovascular responses to exercise. Her research is focused on the physiological strain associated with heat stress, with a specific emphasis on cardiovascular and thrombotic responses to firefighting. She has led several federally funded research projects dealing with the cardiovascular strain of firefighting.

Smith has published studies on heat stress, cardiovascular function, and the physiological aspects of firefighting in numerous peer-reviewed scientific journals, including the *American Journal of Cardiology*, *Medicine and Science in Sports and Exercise*, *Ergonomics*, *Journal of Thermal Biology*, and *Aviation, Space and Environmental Medicine*. Smith has collaborated extensively with fire service organizations, has served in leadership roles in the American College of Sports Medicine, and is a member of the American Physiological Society.

Bo Fernhall, PhD, is a professor in the department of kinesiology and community health at the University of Illinois at Urbana-Champaign. He received his PhD in exercise physiology from Arizona State University in 1984. Fernhall has nearly 30 years of experience in cardiovascular research, with a current focus on how exercise and diet affect heart, arterial, and autonomic function. He also directed cardiovascular rehabilitation programs for over 20 years, combining research and clinical experience.

Fernhall is a fellow of the American Heart Association, the American Association of Cardiopulmonary Rehabilitation, and the American College of Sports Medicine. He was elected to the American Academy of Kinesiology and Physical Education in 2005. He has won several national research awards, most recently the G. Lawrence Rarick National Research Award in 2006 for his research on the benefits of exercise in people with disabilities. Fernhall has published over 160 peer-reviewed manuscripts in scientific journals, including the *American Journal of Cardiology, Amercan Journal of Hypertension, American Journal of Physiology, Atherosclerosis, European Heart Journal,* and *Medicine and Science in Sports and Exercise.*